Helicobacter pylori

Hidekazu Suzuki • Robin Warren • Barry Marshall
Editors

Helicobacter pylori

Springer

Editors
Hidekazu Suzuki
Medical Education Center
Keio University School of Medicine
Tokyo, Japan

Robin Warren
The University of Western Australia and the Royal Perth Hospital
Western Australia, Australia

Barry Marshall
The University of Western Australia
Western Australia, Australia

ISBN 978-4-431-55704-3 ISBN 978-4-431-55705-0 (eBook)
DOI 10.1007/978-4-431-55705-0

Library of Congress Control Number: 2016930559

Printed on acid-free paper

This Springer imprint is published by Springer Nature
The registered company is Springer Japan KK

Preface

Before the 1970s, well-fixed specimens of the gastric mucosa were rare. Then the flexible endoscope was introduced. This enabled gastroenterologists to take numerous well-fixed small biopsies from the stomach. Gastric histology and pathology were clearly demonstrated. Whitehead accurately described it in 1972, including a feature he termed "active" gastritis. This involved only the superficial gastric epithelium, with polymorph infiltration and epithelial cell distortion.

In June 1979 I was examining a gastric biopsy showing chronic inflammation and the active change. A thin blue line on the surface showed numerous small curved bacilli. These were clearly visible with a Warthin-Starry silver stain. They appeared to grow on the surface of the foveolar epithelial cells.

Over the next 2 years I collected numerous similar cases. The changes were often much milder or more focal than the original biopsy, but the main features were usually similar, with chronic gastritis and usually some of the active change. These features could show considerable variation, from near normal to severe.

In 1981 I met Barry Marshal, and we completed a clinicopathological study of 100 outpatients referred for gastroscopy. There was little relation between the infection and the patients' symptoms. Peptic ulcers, particularly duodenal ulcers, were very closely related to the infection. We cultured *Helicobacter pylori*.

In 1986, with Marshall et al., I studied the effect of eradication of *H. pylori* on the recurrence of duodenal ulcer. I graded the gastritis (0–36) using the features seen with active gastritis. The range was 15–35 before treatment. After eradication of *H. pylori*, this changed to 5–20 within 2 weeks. This provides powerful evidence that *H. pylori* causes the active change.

Duodenal ulcer usually occurs in the duodenal cap. The gastric mucosa normally extends through the pylorus. In this study, the proximal border of all ulcers was either definite gastric mucosa or scarred and consistent with a gastric origin. This suggests duodenal ulcer is either actually a distal pyloric ulcer or gastroduodenal.

It may well arise in the damaged, inflamed, and infected mucosa in the position of maximum stress – the lip of the pyloric sphincter.

Professor Emeritus Robin Warren
The University of Western Australia
and the Royal Perth Hospital, Australia

Preface

This year marks the 33rd anniversary of the discovery of *Helicobacter pylori* by Dr. J. Robin Warren and Prof. Barry J. Marshall and the 10-year anniversary of their being awarded the Nobel Prize in Physiology and Medicine. This book is intended to be a comprehensive reference for *H. pylori* science, including its pathophysiology, diagnosis of infection, and treatment.

First, Prof. Barry Marshall introduces the history of *Helicobacter pylori* research, starting with its discovery in the early 1980s.

In the pathophysiology sections, Dr. Muhammad Miftahussurur and Prof. Yoshio Yamaoka describe the human migration of *H. pylori* and explain the correlation between dynamic human movement and *H. pylori* genotype. Dr. Yoshie Senda and Prof. Masanori Hatakeyama summarize the actions of CagA, a major oncoprotein derived from *H. pylori* through the type IV secretion system. Dr. Masayuki Nakano, Prof. Toshiya Hirayama, Prof. Joel Moss, and Prof. Kinnosuke Yahiro summarize the pleiotropic activity of *H. pylori* VacA in host cells. Dr. Hitoshi Tsugawa and I describe *H. pylori* infection-related autophagy and a specific autophagy escape mechanism in CD44v9-positive cancer stemlike cells. Dr. Yoon Jin Choi and Prof. Nayoung Kim describe ghrelin and gut hormone in *H. pylori* infection, suggesting these molecules as possible biomarkers of gastric cancer risk. Prof. Masahiko Nakamura, Dr. Anders Øverby, Prof. Somay Y. Murayama, Dr. Tetsufumi Takahashi, Prof. Shinichi Takahashi, Dr. Hidenori Matsui, and I describe gastric non-*Helicobacter pylori Helicobacter* species and emphasize their significance in human gastric diseases. Dr. Juntaro Matsuzaki and I describe symptoms generated by *H. pylori*-infected patients that indicate a novel disease, *H. pylori*-associated dyspepsia.

In the diagnosis section, Prof. Akiko Shiotani, Prof. Maria Pina Dore, and Prof. David Y. Graham summarize fundamentals underlying the use of two major urease-dependent diagnostic tools, the urea breath test and the rapid urease test. Prof. Mototsugu Kato displays characteristic endoscopic findings of *H. pylori* infection, and Prof. Kazuhiko Inoue describes a method for gastric cancer risk stratification following *H. pylori* infection known as the ABC gastric cancer risk medical check-up.

In the treatment section, Prof. Makoto Sasaki documents the trend in global eradication rates. Prof. Jyh-Ming Liou, Prof. Jaw-Town Lin, and Prof. Ming-Shiang Wu describe gastric cancer prevention through *H. pylori* eradication, an especially important issue in East Asia. Prof. Takahisa Furuta, Prof. Mitsushige Sugimoto, Dr. Mihoko Yamade, Dr. Takahiro Uotani, Dr. Shu Sahara, Dr. Hitomi Ichikawa, and Prof. Takuma Kagami describe the role of personalized therapy in *H. pylori* eradication, especially as stratified by polymorphisms in drug metabolic enzymes. Prof. Javier Molina-Infante explains the efficacy of quadruple regimens against *H. pylori* infection. Finally, Prof. Toshihiro Nishizawa and I describe recent advances in third-line eradication therapies against *H. pylori*.

I hope this book offers comprehensive information to all those who have an interest in *H. pylori*.

Professor Hidekazu Suzuki
Medical Education Center
Keio University School of Medicine
Tokyo, Japan

Contents

Part III Diagnosis

Part IV Treatment

Part I
History

Chapter 1
A Brief History of the Discovery of *Helicobacter pylori*

Barry Marshall

Abstract *Helicobacter pylori* (*H. pylori*) has infected humans in Africa since the early Stone Age. Prior to the twentieth century, the majority of the world's population was infected with the bacterium. Around the globe many researchers came close to discovering *H. pylori* and its role in gastric disease, and in Japan there were two strong areas of enquiry. The first was the study of "spirochaetes" in the stomach of mammals by Kasai and Kobayashi. The second was the work by Kimura and Takemoto on the histology of the gastric mucosa. Nearby in China, no specific pathogen was identified, but ulcer treatment with the antibiotic furazolidone was successfully trialled, and Dr. Yao Shi from Shanghai almost discovered *H. pylori* through electron microscopy studies. In Australia, in the 1980s Robin Warren and myself were able to make comparisons between healthy and unhealthy stomachs and show the striking correlation between gastritis and presence of the spiral bacterium. After drinking cultures derived from a patient, I was able to fulfil Koch's postulates for *H. pylori* and gastritis. The rationale for future treatments via antibiotic-based eradication therapy was in place.

Keywords *Helicobacter pylori* • Discovery • History • *Campylobacter* • Western Australia

1.1 Introduction

Medical scientists know that *H. pylori* was only cultured recently, in 1982 at Royal Perth Hospital in Western Australia. However, the interesting true story of that adventure and the pioneering work of many other research workers across the globe including Asia in the prior 100 years are less well known. This chapter will demonstrate the long-lived and ubiquitous association between *H. pylori* and

B. Marshall (✉)
The Marshall Centre for Infectious Diseases Research and Training, School of Pathology and Laboratory Medicine, The University of Western Australia, QEII Medical Centre, Hospital Avenue, Nedlands, Perth, WA 6009, Australia

UM Marshall Centre, High Impact Research Building, University of Malaya, Kuala Lumpur 50603, Malaysia
e-mail: Barry.Marshall@uwa.edu.au

H. Suzuki et al. (eds.), *Helicobacter pylori*, DOI 10.1007/978-4-431-55705-0_1

mankind and how despite, or perhaps because of this, the bacterium was overlooked for so many years. It will outline some of the early microbiology and clinical research that came close to the breakthrough. It will also show how an advance in one area of technology can unexpectedly solve a major problem in another area.

1.2 Prehistory

We now know that *H. pylori* is common, infecting around 50 % of the global human population. These epidemiological facts were unravelled in the 1980s and 1990s by serological surveys. It became evident that the bacterium was present in all human races, on all continents, although it was more common in developing countries and less common in western countries or affluent communities within countries.

By examining stored serum samples from the 1960s [1] and biopsy material stored for 20 years [2], it could be shown that the infection was decreasing in the latter half of the twentieth century as standard of living increased, water quality improved and families became smaller. In poor countries like Brazil and Africa, 80–90 % of the population was still infected. Extrapolating back therefore, it could be shown that at the beginning of the twentieth century, the infection rate with *H. pylori* must have been close to 100 % in every part of the globe. Since there did not appear to be an animal vector, *H. pylori* must have been solely a human pathogen, passed on from mother to child, for generations.

In the 1990s, as genomics and proteomics allowed us to differentiate between *H. pylori* strains from different geographic locations, it was discovered that people in South America were infected with "Spanish" strains of *H. pylori*, which by sequence analysis of the vacA toxin gene were designated as "Europe 1" strains [3]. So the questions then arose: "Did the *H. pylori* from Europe infect the people in South America? Were people in South America *H. pylori*-free before this infection?" The answers came from two sources. Firstly, dehydrated Andean mummies in Peru were found to have *H. pylori* antigen in their faeces [4]. This predated Columbus by hundreds of years showing that people in Peru were also colonised by *H. pylori* more than 1000 years ago. Additionally, since the colonisation of South America by humans occurred 15,000 BP, then *H. pylori* must have been quite ancient. To reinforce this story, later genomic studies showed that *H. pylori* in South America was of Asian origin, probably arriving there with the human migrations [5]. Finally, new genomic studies of hundreds of isolates reveal that all human strains of *H. pylori* had a common ancestor more than 60,000 years ago. Thus, *H. pylori* has colonised humans ever since they walked out of Africa [6, 7]. Evidence for colonisation long before this comes from the lion strain of *H. pylori*, which had a common ancestor to the human strain 200,000 years ago [8]. Since then it has mutated to a nonpathogenic form with "shredded" toxin genes.

Since apparently all human races became infected with *H. pylori* and remained colonised until now, we can speculate on possible reasons why *H. pylori* has been so tenacious. Is there a risk-benefit equation we should study? To start the discussion, I propose that *H. pylori*, by producing folic acid, benefits humans with borderline nutrition and poor access to fresh vegetables. This benefit might have been especially valuable during ice age migrations. Secondly, if recent observations hold true, *H. pylori* dampens immune system hyperactivity making allergic responses less troublesome [9]. As humans moved from a closeted African environment to occupy the whole planet, decreased reactivity to new environments and antigens might have been a second benefit conferred by *H. pylori*. Balancing this against the "risk" of having *H. pylori*, the risk appears rather small, especially for Stone Age humans who had short lives so they did not experience the gastric cancer risk, which takes 50 years or more to develop. Additionally, because of poor nutrition, gastric acidity was low, so peptic ulcer was rare.

In summary then, *H. pylori* has infected humans in Africa since the early Stone Age. Humans migrated throughout the world starting about 65,000 years ago and benefited from their cargo of *H. pylori*, which in those days, had low pathogenicity and on balance could have aided the global migration of man.

1.3 Japan

1.3.1 Spiral Bacteria in the Stomach of Mammals

In Japan, two lines of research came very close to discovering *H. pylori*. The first of these, reported by Kasai and Kobayashi in 1919, was the study of "spirochaetes" in the stomach of mammals [10]. The second close call came in the study of fibre-optic biopsies and gastritis in the Japanese population reported by Kimura, Takemoto and many others.

In brief, a strong microbiology tradition was founded in Japan by Kitasato, who had trained in microbiology under Robert Koch and, after returning to Japan, had started the Kitasato Institute. At around the same time, after also working in Germany, the Italian anatomist Giulio Bizzozero reported spirochaetes in the gastric mucosa of the dog [11]. Bizzozero's colour illustrations showed the corkscrew organisms living in the mucus layer, in gastric glands and within the canaliculi of the oxyntic cells (Fig. 1.1) [12]. This finding implied that the bacteria were acid tolerant or else might have turned off acid.

Extending this work, Kasai and Kobayashi (Fig. 1.2) observed the same spiral bacteria in the stomach of most cats. Although they could not culture these organisms in vitro, they were able to transmit them to mice and thence to other animals such as rabbits and guinea pigs. In the rabbits, they sometimes saw erosions and inflammation in the gastric mucosa [10]. At that time also, the first successes of Paul Ehrlich's arsenical treatments for syphilis had been reported, so it seemed logical to try these agents on the animal models of the gastric spirochaetes. These efforts were partially successful. Salvarsan could eliminate the bacteria from mice and protect rabbits from the pathogenic effects of the spiral organisms [10].

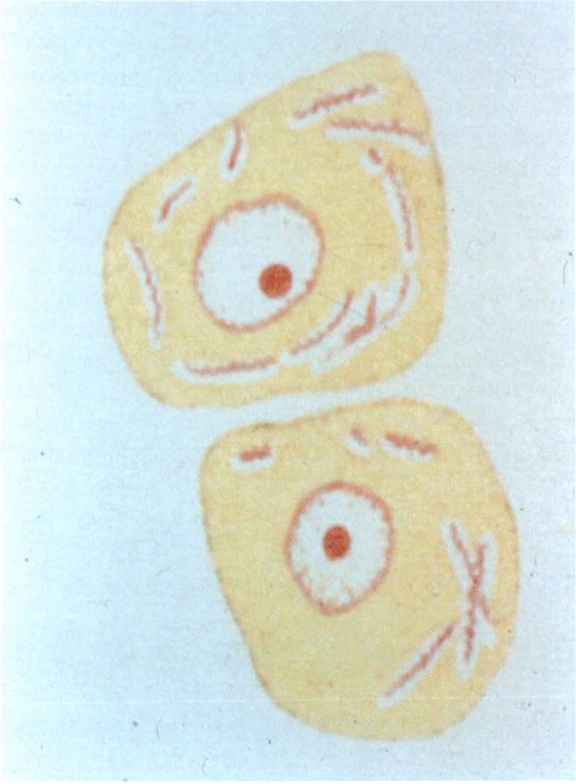

Fig. 1.1 Bizzozero's illustration showing numerous "spirilli" in the protoplasm of glandular neck cells (from [12])

With hindsight, we can see why *H. pylori* was not discovered by these investigators. Spiral bacteria appeared to be quite common in mammals, so the sighting of the bacteria in human specimens by several investigators in different countries was not surprising (Table 1.1). In fact, bacteria in the putrid gastric contents of gastric cancer patients were to be expected. Apart from gastrectomy specimens, which would have been mostly from patients with advanced gastric cancer, there was no easy way to study fresh stomach tissue from normal humans. After resection, autolysis of the gastric mucosa commences almost immediately making *H. pylori* hard to find. Additionally, by the time patients present with gastric cancer, the *H. pylori* have often disappeared from an atrophic stomach [14].

Although Kasai and Kobayashi's idea of arsenical treatment was never copied for peptic ulcer in humans, a very similar but less toxic metal, bismuth, had a long tradition as a human gastric medicine. In Europe, bismuth subnitrate had been a component of "gastritis" (dyspepsia) remedies and remains in use today. In the USA, bismuth subsalicylate (as ®Pepto-Bismol, Procter & Gamble) was used after 1900 for "infantile cholera" and "gastritis" and now remains a component of successful combination therapies for *H. pylori* [15].

Fig. 1.2 Kasai and Kobayashi (from [13])

In retrospect, Kasai's work was largely done in very turbulent times, i.e. 1910–1918, and during World War I. Like subsequent investigators, their funding agency may have had more important infectious disease priorities. Twenty years later, in the USA when World War II was imminent, Freedberg in Boston described spirochaetes in the human stomach but could not follow it up and ultimately became a cardiologist [16].

The original work of Kasai and Kobayashi is a fine example of how curiosity-driven research, in a well-supported institute, served to fulfil the vision of earlier pioneers such as Koch, Ehrlich and Kitasato. Their studies were incomplete only because they did not have the tools available to myself and Robin Warren, i.e. microaerophilic culture techniques [17] and the fibre-optic endoscope [18].

1.3.2 Gastritis and Gastric Cancer in Japan

For at least 50 years, gastric adenocarcinoma has been recognised as the major cause of cancer deaths in Japan. To combat this, the first mass-screening X-ray examination was implemented in 1953, and in 1958 the Japan Cancer Society was established and the mass screening was promoted. The blind gastrocamera, invented in 1950, was used for gastric screening until it was replaced by the glass-fibre camera in 1975. Nevertheless, gastric cancer remains very common in Japan, with 130,000 new cases and about 50,000 deaths each year. No doubt

Table 1.1 History of the discovery of *Helicobacter*

Date	Researcher	Discovery
1875	Bottcher/ Letulle	Bacteria in ulcer margin
1881	Klebs	Bacterial colonisation and inflammation
1888	Letulle	*Staphylococcus aureus* induces acute gastritis in guinea pigs
1889	Jaworski	*Vibrio rugula* in the stomach
1893	Bizzozero	Spirochaetes in dog stomach
1896	Salomon	Gastric spirochaetes in dogs, cats and rats and showed transmission to the mouse
1906	Krienitz	Spirochaetes in the stomach with gastric cancer
1908	Turck	*Escherichia coli* induces gastric ulcer in the dog
1909	Regaud	Spirochaetes in cat and dog stomachs
1916	Suda	Spirochaetes in dog gastric glands (Japan)
1916	Rosenow	Streptococcus induces gastric ulcer
1917	Dragstedt	Bacteria do not induce gastric ulcer
1917	Kasai, Kobayashi	Spirochaetes in wild rats and guinea pigs
1919	Kasai, Kobayashi	Spirochaetes in cat, dog, rat and monkey stomachs; gastric spirochaetes transmitted between species; Salvarsan eliminates spirochaetes in mice
1921	Edkins	Experiments with *Spirilla regaudi* (*H. felis*)
1924	Luck	Urease activity in the stomach
1925	Hofmann	"Hofmann's bacillus" induces ulceration
1930	Berg	Partial vagotomy inhibits secondary infections of ulcers
1938	Doenges	Spirochaetes induce gastritis in monkeys and humans
1940	Freedberg/ Barron	Gastric spirochaetes are not pathogenic
1940	Gorham	Acidophilic bacteria induce gastric ulcer
1954	Palmer	No spirochaetes detected using H&E in 1140 suction biopsies
1966	Aoyagi	Highest urease activity in the stomach
1975	Steer	*Pseudomonas aeruginosa* induces gastric inflammation in ulcer margin
1979	Warren	Spiral bacteria in the human stomach
1983	Warren	Gastric spiral bacteria associated with gastritis in humans
1983	Marshall	*H. pylori* associated with peptic ulcer isolated and cultured
1984	Inoue	First success in culturing *H. pylori* in Japan
1985–1987	Marshall/ Morris	Inoculation with *H. pylori* proved Koch's 3rd postulate
1989	Goodwin	New spiral bacteria named *H. pylori*

Adapted from [13] and [10]

thousands of papers were published looking at the histology of gastric cancer and gastric mucosa but the *H. pylori* were overlooked. I choose just to mention studies published by Kimura and Takemoto whose landmark studies of gastritis in the Japanese population were presented in detail at a session of the American Gastroenterological Association in 1995 in San Diego.

In their earlier paper, these investigators reported the natural history of atrophic gastritis in a large series of Japanese patients. Theirs was one of the first papers using fibre-optic endoscopy to obtain precisely located biopsy material from several locations within the stomach [19]. They reported then what we know to be true: (a) gastritis commences at a young age, (b) the lower half of the stomach is involved but the disease moves proximally with age, (c) inflammation gradually leads to intestinal metaplasia and gastric atrophy and (d) eventually "nearly all" Japanese develop metaplasia and atrophic gastritis. The logical conclusion was that the gastric mucosa aged so that Japanese people ultimately developed the atrophic mucosa, with low acid secretion, which had been recognised as a risk factor for patients with Type A "pernicious anaemia" autoimmune gastritis in western countries. Various causes for the Japanese findings were postulated and some of these were probably relevant, especially salty food, pickles containing nitrate and cigarette smoking; but the presence of *H. pylori* was overlooked. Why was this so? Why did Robin and I come to a different conclusion in a far smaller series of patients in Perth in 1982?

The clue to the presence of *H. pylori* was in the changing epidemiology of gastric cancer noticed in the USA during the twentieth century. In 1930, gastric cancer was the most common cancer in the USA. Between 1930 and 1970 however, the incidence declined from a rate similar to that in Japan (60 per 100,000 per annum) to the modern rate of about seven. One hypothesis is that the consumption of fresh fruit and vegetables increased in the USA with the advent of modern refrigerators after 1930. Vitamin C prevents the formation of nitrosamines in the stomach, so there is plausible scientific data to support this. Additionally, and perhaps less importantly, *H. pylori* declined in the USA during that time such that by 1966, the seroprevalence was 60 % in California [1].

So, like many investigators before and after, Kimura and Takemoto studied a population in which *H. pylori* was almost universally present. The association between the spiral bacteria and inflammation can only be seen when "normal controls" are included in the study. For Japan, "normal" was the *H. pylori*-positive state. When Robin and I collected our first 20 or so patients in Perth, Western Australia, however, we had the advantage of a declining prevalence of *H. pylori* in a modern affluent western county. The population had smaller families, clean food and water and free access to powerful antibiotics such as amoxycillin.

Thus, in our study of only 100 gastroscopy patients conducted in 1982, only 58 had *H. pylori*. The 42 patients without *H. pylori* were the control group. In these "controls" gastritis was almost completely absent and, in most cases, the endoscopic appearance of the stomach was also normal. Age, smoking, diet, alcohol and NSAIDs were unrelated to gastritis. *Helicobacter pylori* was the only associated factor. The association was so tight that the "p value" could not be calculated using the mainframe computers available in Western Australia. Robin Warren purchased a new Hewlett-Packard 11C calculator and found a one-tailed Fisher's test result of $p < 10^{-8}$ [20, 21].

Folding these findings back into the studies of Kimura and Takemoto, and with further studies since, we can say that the almost universal presence of *H. pylori* in the Japanese population prior to 1970 stimulated an ageing process which led to atrophic gastritis and gastric cancer. However, there is great optimism now because for the past 20 years, very few Japanese children have been infected with *H. pylori*. In addition, effective treatments have been available for decades, and the tools for *H. pylori* eradication, i.e. diagnostic tests and combination antibiotics, are freely available. By supplementing the endoscopic screening with *H. pylori* testing and treatment of the population, the gastric cancer epidemic may disappear from the Japanese population during the next generation [22, 23].

1.4 China

Numerically, China is still the country with the largest population of *H. pylori*-infected people. Even today it is estimated that 50 % of the population is infected, so half a billion people in China remain at risk of *H. pylori*-caused diseases, i.e. peptic ulcer and gastric cancer. It is not surprising therefore that pioneering researchers also came very close to discovering *H. pylori* in that country. Considering the various upheavals in China during the twentieth century, the investigators can be proud of the advances they made no doubt benefiting millions of patients.

In China, gastric cancer was always very common and remains so today, with about 35 % of all the world's cases occurring in that country. The contribution of environmental factors was always suspected in China because, in a relatively homogeneous population, cancer rates varied greatly in different provinces. Likely causes of this variation included diet and dietary carcinogens. Even today, we still think these factors modulate the effect of chronic *H. pylori* infection.

After 1970 however, several Chinese investigators became aware of the presence of gastritis in the stomach of Chinese people with peptic ulcer or gastric cancer. Similarly to the Japanese studies, intestinal metaplasia and atrophy were also quite common. In the 1970s several groups had considered that bacteria might be related in some way. A specific pathogen was not identified, but, since the healthy stomach was relatively sterile and people with gastric disease often had low acidity with a putrid gastric flora, broad-spectrum antibiotics seemed to be worth trying. There was a prior history to this. At the Mayo Clinic, oral neomycin had been used to decrease gastrointestinal ammonia production (from urease) in patients with hepatic encephalopathy [24]. Also, in Athens, Greece, Dr. John Lykoudis had used an antibiotic "brew" called "Elgaco" with great success in hundreds of duodenal ulcer patients [25, 26]. Finally, by the mid-1980s Spanish investigators had performed a study using metronidazole to treat duodenal ulcer in a small prospective controlled pilot study which showed improved radiologic healing in the treated group [27].

In China, furazolidone was used in peptic ulcer treatment from as far back as 1972, and thousands of patients all over the country were treated [28]. Professor Zhi-Tian Zheng in Beijing conducted a clinical trial to confirm its efficacy in late 1982 to early 1983. The ulcer healing rate was 73 % for the furazolidone group (exposed to a 2-week regimen of 200 mg t.i.d.) versus 24 % in the placebo group ($p < 0.001$) [29].

After seeing our results from Western Australia in 1983, Xiao and colleagues in Shanghai revisited the use of furazolidone and carried out studies of antibiotic use in gastric disease where *H. pylori* was also correlated with histologic and clinical findings.

This work transitioned into larger studies including the 5-year "Dutchigas Project" with Professor Guido Tytgat in Amsterdam, which began in 1995. The collaborators undertook a series of projects on the eradication of *H. pylori* using bismuth/furazolidone-based and PPI furazolidone-based triple or quadruple therapies.

Some of the early studies were difficult to interpret especially where metronidazole was used, and in vitro susceptibility studies of the offending *H. pylori* strains were not available. However, there is little doubt that these Chinese investigators were very close to the *H. pylori* discovery.

To reinforce this impression, Dr. Yao Shi, after performing electron microscopic studies of mucus structure in Shanghai, reported different appearances of the mucus, suggesting that its physical structure was somehow defective in ulcer patients. In Shi's illustrations however, the spiral shapes of *H. pylori* were definitely visible (Fig. 1.3). This work was similar to studies done contemporaneously by Steer and Colin Jones in the UK [30] and at Royal Perth Hospital by Fung and

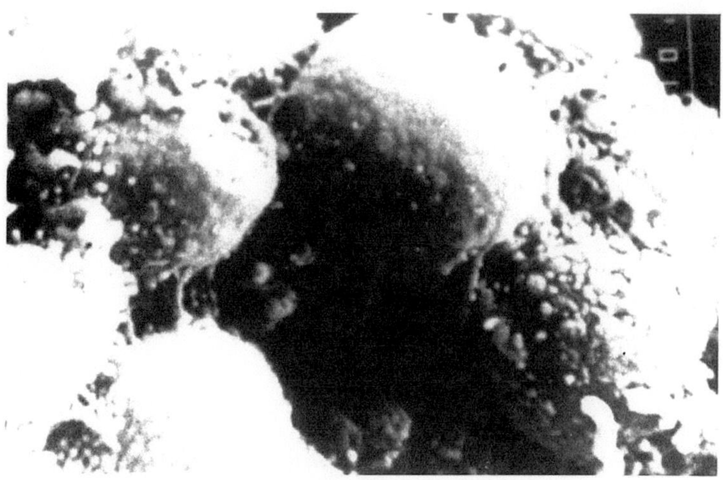

Fig. 1.3 Dr. Yao Shi's electron micrograph with visible *H. pylori*

Papadimitriou [31]. Certainly, *H. pylori* could have been discovered in China. Once again the story of the discovery reveals how curiosity-driven research, common sense clinical "pilot" studies and basic research could lead to a discovery benefiting many millions of people.

1.5 Perth, Australia, 1981–1984

The detailed history of the discovery of *H. pylori* by Robin and myself (and many others) has been well described in the book *Helicobacter Pioneers* [20, 32], so in this small chapter I can't repeat all of that. However, this is a good time to relate my personal recollection of the initial culture of the bacterium and then my later attempt to fulfil Koch's postulates for the bacterium by drinking it!

After some pilot studies where we confirmed Robin's initial impression of a strong association between the gastric *Campylobacter*-like organism (CLO) and active chronic gastritis (ACG), we carried out a prospective study of 100 patients coming to elective endoscopy. This is the study mentioned above and published in *The Lancet* in 1984. After informed consent (only one patient refused to participate), two biopsies were taken from the antrum, one for histological examination by Robin and a second one for Gram-stain and culture attempts in the microbiology department. I attended the endoscopy and prospectively coded the endoscopic findings as well as the clinical information from the patient. This information was sent to the statistician who was also blinded to all the other clinical and biopsy information. After 100 patients had taken part, the study was closed and the results were analysed by the statistician. The major outcome is shown in Table 1.2. Nevertheless, these findings failed to convince the sceptics that the association between spiral bacteria with gastritis was important. However, the study and the fact that the new organism could easily be cultured created an exponential increase in publications by microbiologists.

Table 1.2 Association of bacteria with endoscopic diagnoses

Endoscopic appearance[a]	Total	With bacteria	p
Gastric ulcer	22	18 (77 %)	0.0086
Duodenal ulcer	13	13 (100 %)	0.00044
All ulcers	31	27 (87 %)	0.00005
Oesophagus abnormal	34	14 (41 %)	0.996
Gastritis[b]	42	23 (55 %)	0.78
Duodenitis[b]	17	9 (53 %)	0.77
Bile in stomach	12	7 (58 %)	0.62
Normal	16	8 (50 %)	0.84
Total	100	(58 %)	

From Marshall and Warren [21]

[a]More than one description applies to several patients (e.g., four patients had both gastric and duodenal ulcers)

[b]Refers to endoscopic appearance, not histology

After failed attempts to infect piglets in 1984, I, after having a baseline endoscopy done, drank liquid broth containing the scrapings from two Petri dishes containing cultured *H. pylori*, expecting to develop, perhaps years later, an ulcer. I was surprised when, only three days later, I developed vague nausea and halitosis (due to the achlorhydria, there was no acid to kill the mouth flora in the stomach, and anaerobic waste products manifested as bad breath), noticed mainly by my wife and my mother. On days 5–8, I developed achlorhydric (no acid) vomiting. On day 8, I had a repeat endoscopy and biopsy, which showed massive inflammation (gastritis), and *H. pylori* was cultured. On the 14th day after ingestion, a third endoscopy was done, and I began to take antibiotics (tinidazole). However, by then the *H. pylori* had totally disappeared! This story was related in my Nobel acceptance lecture on 8 December 2005, available for viewing on the Nobel website [33]. Interestingly, I did not develop antibodies to *H. pylori*, suggesting that innate immunity can sometimes eradicate acute *H. pylori* infection. My illness and recovery, based on a culture of organisms extracted from a patient with gastritis, fulfilled Koch's postulates for *H. pylori* and gastritis but not for peptic ulcer. This experiment was published in 1985 in the *Medical Journal of Australia* [34] and is among the most cited articles from the journal.

References

1. Parsonnet J, Friedman GD, Vandersteen DP, Chang Y, Vogelman JH, Orentreich N, et al. *Helicobacter pylori* infection and the risk of gastric carcinoma. N Engl J Med. 1991;325(16):1127–31. doi:10.1056/NEJM199110173251603.
2. Sipponen P. *Helicobacter pylori*: a cohort phenomenon. Am J Surg Pathol. 1995;19 Suppl 1: S30–6.
3. Van Doorn LJ, Figueiredo C, Megraud F, Pena S, Midolo P, Queiroz DM, et al. Geographic distribution of vacA allelic types of *Helicobacter pylori*. Gastroenterology. 1999;116 (4):823–30.
4. Castillo-Rojas G, Cerbon MA, Lopez-Vidal Y. Presence of *Helicobacter pylori*in a Mexican pre-Columbian mummy. BMC Microbiol. 2008;8. doi:Artn 119 Doi 10.1186/1471-2180-8-119.
5. Ghose C, Perez-Perez GI, Dominguez-Bello MG, Pride DT, Bravi CM, Blaser MJ. East Asian genotypes of *Helicobacter pylori* strains in Amerindians provide evidence for its ancient human carriage. Proc Natl Acad Sci U S A. 2002;99(23):15107–11. doi:10.1073/pnas.242574599.
6. Falush D, Wirth T, Linz B, Pritchard JK, Stephens M, Kidd M et al. Traces of human migrations in *Helicobacter pylori* populations. Science. 2003;299(5612):1582–5. doi:10.1126/science.1080857.
7. Linz B, Balloux F, Moodley Y, Manica A, Liu H, Roumagnac P, et al. An African origin for the intimate association between humans and *Helicobacter pylori*. Nature. 2007;445 (7130):915–8. doi:10.1038/nature05562.
8. Eppinger M, Baar C, Linz B, Raddatz G, Lanz C, Keller H, et al. Who ate whom? Adaptive Helicobacter genomic changes that accompanied a host jump from early humans to large felines. PLOS Genet. 2006;2(7):1097–110. doi:ARTN e120 DOI 10.1371/journal.pgen.0020120.

9. Chen Y, Blaser MJ. *Helicobacter pylori* colonization is inversely associated with childhood asthma. J Infect Dis. 2008;198(4):553–60. doi:10.1086/590158.
10. Kasai K, Kobayashi R. Stomach spirochetes occurring in mammals. J Parasitol. 1919;6:1–11.
11. Bizzozero G. Uber die Schlauchformigen Drusen des Magendarmkanals und die Bezienhungen ihres Epithels zu dem Oberflachenepithel der Schleimhaut. Arch Mikr Anat. 1893;42:82–152.
12. Figura N, Bianciardi L. Chapter 1: Helicobacters were discovered in Italy in 1892: an episode in the scientific life of an eclectic pathologist, Giulio Bizzozero. In: Marshall BJ, editor. Helicobacter pioneers: firsthand accounts from the scientists who discovered helicobacters 1892–1982. Carlton South: Blackwell Publishing; 2002. p. 1–13.
13. Fukuda Y, Shimoyama T, Shimoyama T, Marshall BJ. Chapter 2. Kasai, Kobayashi and Koch's postulates in the history of Helicobacter pylori. In: Marshall BJ, editor. Helicobacter pioneers: Firsthand accounts from the scientists who discovered helicobacters 1892–1982. Carlton South: Blackwell Publishing; 2002. p. 15–24.
14. Asaka M, Kato M, Kudo M, Katagiri M, Nishikawa K, Yoshida J, et al. Relationship between *Helicobacter pylori* infection, atrophic gastritis and gastric carcinoma in a Japanese population. Eur J Gastroen Hepat. 1995;7 Suppl 1:S7–10.
15. Chey WD, Wong BC. American College of Gastroenterology guideline on the management of *Helicobacter pylori* infection. Am J Gastroenterol. 2007;102(8):1808–25. doi:10.1111/j.1572-0241.2007.01393.x.
16. Freedberg A, Barron L. The presence of spirochaetes in human gastric mucosa. Am J Dig Dis. 1940;28:639–46.
17. King EO. Human Infections with Vibrio Fetus and a Closely Related Vibrio. J Infect Dis. 1957;101(2):119–28. doi:10.1093/infdis/101.2.119.
18. Edmonson JM. History of the instruments for gastrointestinal endoscopy. Gastrointest Endosc. 1991;37(2 Suppl):S27–56.
19. Takemoto T. Endoscopic diagnosis of chronic gastritis. Diagnosis Treatment. 1966;54:1274–85.
20. Warren JR. Chapter 14: The discovery of *Helicobacter pylori* in Perth, Western Australia. In: Marshall BJ, editor. Helicobacter pioneers: Fisthand accounts from the scientists who discovered helicobacters 1892-1982. Carlton South: Blackwell Publishing; 2002. p. 151–63.
21. Marshall B, Warren JR. Unidentified curved bacilli in the stomach of patients with gastritis and peptic ulceration. Lancet. 1984;323(8390):1311–5. doi:10.1016/S0140-6736(84)91816-6.
22. Asaka M. A new approach for elimination of gastric cancer deaths in Japan. Int J Cancer. 2013;132(6):1272–6. doi:10.1002/ijc.27965.
23. Sugano K, Osawa H, Satoh K. Clinical management of *Helicobacter pylori* – the Japanese perspective. Digest Dis. 2014;32(3):281–9. doi:10.1159/000357859.
24. Wolpert E, Phillips SF, Summerskill WHJ. Ammonia production in the human colon. N Engl J Med. 1970;283(4):159–64. doi:10.1056/NEJM197007232830401.
25. Rigas B, Feretis C, Papavassiliou ED. John Lykoudis: an unappreciated discoverer of the cause and treatment of peptic ulcer disease. Lancet. 1999;354(9190):1634–5. doi:10.1016/S0140-6736(99)06034-1.
26. Rogas B, Papavassiliou ED. Chapter 7: John Lykoudis: The general practitioner in Greece who in 1958 discovered the etiology of, and a treatment for, peptic ulcer disease. In: Marshall BJ, editor. Helicobacter pioneers: Firsthand accounts from the scientists who discovered helicobacters 1892–1982. Carlton South: Blackwell Publishing; 2002. p. 75–87.
27. Diaz M, Escobar A. Metronidazole versus cimetidine in treatment of gastroduodenal ulcer. Lancet. 1986;327(8486):907.
28. Xiao S-D, Shi Y, Liu W-Z. Chapter 9: How we discovered in China in 1972 that antibiotics cure peptic ulcer. In: Marshall BJ, editor. Helicobacter pioneers: Firsthand accounts from the scientists who discovered helicobacters 1892-1982. Carlton South: Blackwell Publishing; 2002. p. 99–104.

29. Zheng ZT, Wang TY, Zhu YS. A double-blind short-term clinical trial of the effect of furazolidone on peptic ulcer. Chin J Intern Med. 1984;23:195–7.
30. Steer HW, Colin-Jones DG. Mucosal changes in gastric ulceration and their response to carbenoxolone sodium. Gut. 1975;16(8):590–7.
31. Fung WP, Papadimitriou JM, Matz LR. Endoscopic, histological and ultrastructural correlations in chronic gastritis. Am J Gastroenterol. 1979;71(3):269–79.
32. Marshall BJ. Chapter 15: The discovery that *Helicobacter pylori*, a spiral bacterium, caused peptic ulcer disease. In: Masrhall BJ, editor. Helicobacter pioneers: Firsthand accounts from the scientists who discovered helicobacters 1892-1982. Carlton South: Blackwell Publishing; 2002. p. 165–203.
33. Marshall BJ. Helicobacter Connections: Nobel Lecture December 8, 2005. http://www.nobelprize.org/nobel_prizes/medicine/laureates/2005/marshall-lecture.pdf#search='marshall. Accessed 27 Aug 2015.
34. Marshall BJ. The pathogenesis of non-ulcer dyspepsia. Med J Australia. 1985;143(7):319.

Part II
Pathogenesis

Chapter 2
Human Migration

Muhammad Miftahussurur and Yoshio Yamaoka

Abstract *Helicobacter pylori* strains from different geographic areas exhibit clear phylogeographical differentiation. The genotype of the virulence genes is useful as a tool to track human migration utilizing the high genetic diversity and frequent recombination between different *H. pylori* strains. Using combinations of the virulence genes, five major groups have been defined according to geographical associations. Multilocus sequence typing (MLST) analysis using seven housekeeping genes also are widely used markers for genomic diversity. It was revealed that seven modern population types of *H. pylori* which derived from six ancestral populations provide more detailed information on human migration than does the analysis of human genetics. Although approaches by MLST and virulence factors are effective, these methods focus on a small number of genes and may miss information conveyed by the rest of the genome. Genome-wide analyses using DNA microarray or whole-genome sequencing technology give a broad view on the genome of *H. pylori*. In particular, next-generation sequencers, which can read DNA sequences in less time and at lower costs than Sanger sequencing, enabled us to efficiently investigate not only the evolution of *H. pylori*, but also novel virulence factors and genomic changes related to drug resistance.

Keywords *Helicobacter pylori* • Asia • Virulence factors • Multilocus sequence typing • Next-generation sequencer • Migration

---

M. Miftahussurur
Institute of Tropical Disease, Airlangga University, Surabaya, Indonesia

Department of Environmental and Preventive Medicine, University Faculty of Medicine, 1-1 Idaigaoka, Hasama-machi, Yufu, Oita 879-5593, Japan

Y. Yamaoka (✉)
Department of Environmental and Preventive Medicine, University Faculty of Medicine, 1-1 Idaigaoka, Hasama-machi, Yufu, Oita 879-5593, Japan

Department of Medicine-Gastroenterology, Baylor College of Medicine, Houston, TX, USA
e-mail: yyamaoka@oita-u.ac.jp

H. Suzuki et al. (eds.), *Helicobacter pylori*, DOI 10.1007/978-4-431-55705-0_2

Abbreviations

cagA	Cytotoxin-associated gene
MLST	Multilocus sequence typing

2.1 Introduction

More than half of all humans are infected with *Helicobacter pylori*, a gram-negative spiral bacterium whose ecological niche is the human stomach which is also linked to severe gastritis-associated diseases, including peptic ulcer and gastric cancer [1]. *H. pylori* strains from different geographical areas show clear phylogeographic features; these features enabled us to assume the migration of human populations by phylogeographic analyses of *H. pylori*. In addition, the genetic diversity within *H. pylori* is greater than within most other bacteria [2] and about 50-fold greater than that of the human population [3]. Moreover, frequent recombination between different *H. pylori* strains [4] leads to only partial linkage disequilibrium between polymorphic loci, which provide additional information for population genetic analysis [5].

Several virulence factors of *H. pylori* have been demonstrated to be predictors of gastric atrophy, intestinal metaplasia, and severe clinical outcomes [6–11]. Currently, two most extensively studied virulence factors of *H. pylori*, *cagA* and *vacA*, are used as markers for genomic diversity within distinct populations [12]. In addition, multilocus sequence typing (MLST) analysis, which uses seven housekeeping genes, is also useful to predict the history of human migrations [2, 5, 13–15]. MLST was proposed in 1998 as a tool for the epidemiological study of bacteria [16]. Recently, the genomic diversity within *H. pylori* populations was examined by employing the MLST method using seven housekeeping genes (*atpA, efp, mutY, ppa, trpC, ureI,* and *yphC*) [5, 13, 14]. MLST analysis is reported to give more detailed information about human population structure than the method using human microsatellite or mitochondrial DNA [15]. Moreover recently the whole-genome sequencing technology is another powerful tool to study the evolution and pathogenicity of *H. pylori*. In this chapter we describe the current knowledge about the usefulness of virulence factors and housekeeping genes for elucidating the history of human migration and overview on the utilization of genome-wide information for advanced studies.

2.2 Migration out of Africa to the Pacific

It was believed that *H. pylori* was already established in human stomachs at least 100,000 years ago [17] from an unknown source. It is most likely transmitted from large felines which contained *H. acinonychis* to San peoples (hpAfrica2; very

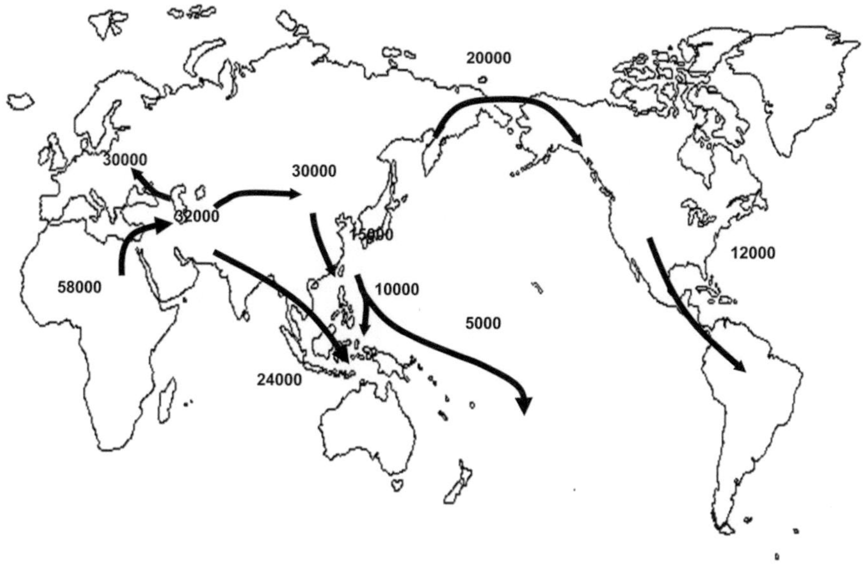

Fig. 2.1 Modern human migration out of Africa. *Black arrows* and *numbers* represent predicted paths and times of migration

distinct and has only been isolated in South Africa) and then widespread throughout Africa (hpAfrica1 and hpNEAfrica) [18]. hpAfrica1 divided into two subpopulations, hspWAfrica (West Africans, South Africans, and Afro-Americans) and hspSAfrica (South Africans). On the other hand, hpNEAfrica is predominant in isolates from Northeast Africa [19]. *H. pylori* is predicted to have spread from East Africa over the same time period as anatomically modern humans (~58,000 years ago) and mirrors the human pattern of increased genetic distance and decreased diversity with distance from Africa (Fig. 2.1) [13, 14]. Using MLST the modern populations derived from six ancestral populations (Table 2.1) which were designated ancestral European 1 (AE1), ancestral European 2 (AE2), ancestral East Asia, ancestral Africa 1, ancestral Africa 2 [5], and ancestral Sahul [14]. These ancestrals recently derived to seven population types based on geographical associations: hpEurope, hpEastAsia, hpAfrica1, hpAfrica2, hpAsia2, hpNEAfrica, and hpSahul (Fig. 2.2) [5, 13, 14].

By a southern coastal route, the ancestors of modern humans passed from India to the Southeast and Australasia [20] during their first "out of Africa" migration, which subsequently resulted in the Asian lineages (hpAsia2). Recently hpAsia2 strains have been isolated in South, Southeast, and Central Asia [19]. Most strains in India initially belonged to hpAsia2 [13], whereas some strains belonged to hpEurope [21]. However *H. pylori* in the Indian population is more heterogeneous in origin, reflecting perhaps both earlier common ancestry and recent imports. It is notable that hpAsia2 strains from Ladakh Indians and Malaysian Indians can be divided into two subpopulations, hspLadakh and hspIndia [22]. From mainland

Table 2.1 Multilocus sequence types of *Helicobacter pylori* according to geographical area

Geographical area/ethnic group	East Asia	Polynesians, Melanesians, Taiwan (aboriginals)	Amerindians	South/Central Asia	Europe	Northeast Africa	West/South Africa	South Africa	Australia, New Guinea (aboriginals)
Modern population type	hpEastAsia	hpEastAsia	hpEastAsia	hpAsia2	hpEurope	hpNEAfrica	hpAfrica1	hpAfrica2	hpSahul
Modern sub-population type	hspEAsia	hspMaori	hspAmerind				hspWAfrica	hspSAfrica	
Ancestral population type	Ancestral East Asia	Ancestral East Asia	Ancestral East Asia	Ancestral Europe 1 (AE1)	AE1/AE2	Ancestral Europe 2 (AE2)	Ancestral Africa 1	Ancestral Africa 2	

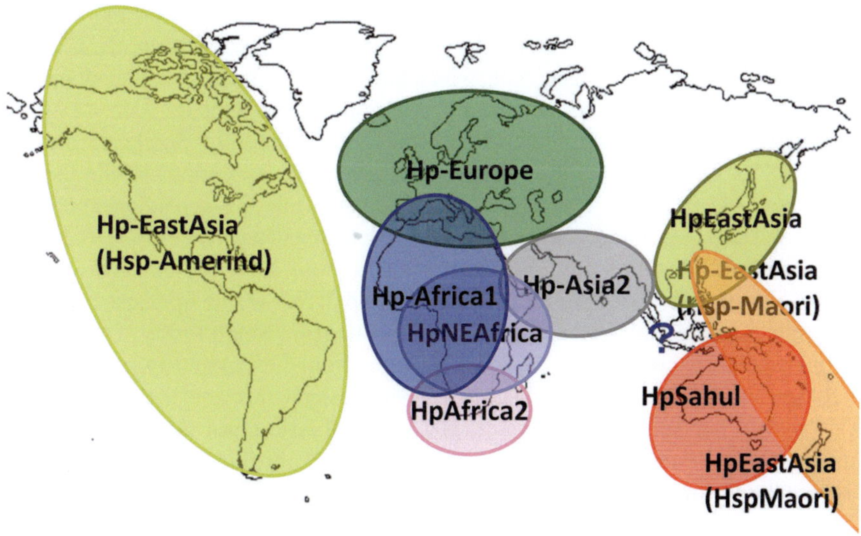

Fig. 2.2 Seven population types based on geographical associations. *Colored circles* illustrate the putative distribution of *H. pylori* before "the age of exploration." Six ancestries derived to seven populations: hpEurope, hpEastAsia, hpAfrica1, hpAfrica2, hpAsia2, hpNEAfrica, and hpSahul

Asia the route extended along the Pleistocene landmass, known as Sundaland (i.e., the Malay Peninsula, Sumatra, Java, Borneo, and Bali), that was joined to the Asian mainland as a result of low sea levels during the last ice age (12,000–43,000 years ago). Low sea levels also meant that Australia, New Guinea, and Tasmania were connected in a continent called Sahul, separated from Sundaland by a few narrow deep-sea channels [14]. Recently hpSahul strains are isolated from aborigines of Australia and highlanders of New Guinea [19].

Subsequent migrations of ancestors of the African hpNEAfrica and/or the Asian hpAsia2 populations resulted in the admixed hpEurope population which then became the predominant population of extant *H. pylori* in Europe, the Middle East, and Western Asia. The modern humans settled in Europe about 30,000–40,000 years ago, probably entering via two routes: from Turkey along the Danube corridor into Eastern Europe and along the Mediterranean coast [20]. hpEurope includes almost all *H. pylori* strains isolated from ethnic Europeans, including people from countries colonized by Europeans. The hpEurope can be divided into AE1 and AE2. AE1 originated in Central Asia, because it shares phylogenetic signals with isolates from Estonia, Finland, and Ladakh in India. It is not clear which population arrived first, but AE1 has a higher frequency in Northern Europe, while AE2 is more common in southern Europe. MLST analyses from Iran also provided evidence that *H. pylori* strains from Iran are similar to other isolates from Western Eurasia and can be placed in the previously described hpEurope population [23].

Human migrations in Southeast Asia have also been clarified on the basis of MLST analyses from Cambodia [24]. Cambodian strains have been classified in two groups, hpEurope and hspEAsia, which have resulted from three ancient human migrations: (1) from India, introducing hpEurope into Southeast Asia; (2) from China, carrying hspEAsia; and (3) from Southern China into Thailand carrying hpAsia2 [20, 24]. Their findings also support two recent migrations within the last 200 years: (1) from the Chinese to Thailand and Malaysia spreading hspEAsia strains and (2) from Indians to Malaysia carrying hpAsia2 and hpEurope [20, 24]. In concordance with this study, *H. pylori* isolates from Malaysia are classified as hpEastAsia, hpAsia2, or hpEurope. A new subpopulation within hpAsia2, hspIndia, may reflect as the Malaysian Indians mainly came from South India.

hpEastAsia is common in *H. pylori* isolates from East Asia. hpEastAsia also includes subpopulations, i.e., hspMaori (Polynesians, Melanesians, and native Taiwanese), hspAmerind (Amerindians), and hspEAsia (East Asians). Approximately 12,000 years ago, *H. pylori* (hspAmerind) accompanied humans when they crossed the Bering Strait from Asia to the Americas [12]. Our previous data showed that four strains isolated from the Ainu ethnic group, living in Hokkaido, a northern island of Japan, belong to the hspAmerind population [25]. Japanese aboriginal people, known as Jomon people, are thought to have migrated to the northern or southern area such as Hokkaido and Okinawa because of the immigration of the Yayoi people from the Korean Peninsula [26]. Finally around 5000 years ago, *H. pyllori* (hspMaori) accompanied several subgroup of the Austronesia language family spread from Taiwan through the Pacific [14] included several islands in east Indonesia [27] into Melanesia and Polynesia.

2.3 Virulence Factors for Tracking Human Migration

The relationships between MLST and virulence factors were reported [28, 29]. The phylogeny of most *cag* pathogenicity island (PAI) genes, an approximately 40-kilobase pair region that is thought to have been incorporated into the *H. pylori* genome by horizontal transfer from an unknown source [30], was similar to that of MLST, indicating that *cag* PAI was probably acquired only once by *H. pylori*, and its genetic diversity reflects the isolation by distance which has shaped this bacterial species since modern humans migrated out of Africa [29]. The *cagA* gene which encodes a highly immunogenic protein (CagA) is located at one end of the *cag* PAI. The *cag* PAI encodes a type IV secretion system, through which CagA is delivered into host cells [31–33]. After delivery into gastric epithelial cells, CagA is tyrosine phosphorylated at Glu-Pro-Ile-Tyr-Ala (EPIYA) motifs located in the 3′ region of the *cagA* gene [34]. Supporting that *H. pylori* mirrors the human pattern of increased genetic distance and decreased diversity, our group has reported that the structure of the 3′ region of the *cagA* gene varies between strains from East Asian and Western countries [9, 12, 35, 36]. In East Asian strains, two types of repeats are found: 57 bp repeats followed by 162 bp repeats (East Asian-type *cagA*). Western strains have similar 57 bp repeats; however, they are followed by a repeat region consisting of 102 bp repeats, which are completely different from those of East Asian strains (Western-type *cagA*). Previous reports also show that the structure of the 5′ region of the *cagA* gene varies between strains from East Asian and Western countries [12, 37]. East Asian-type *cagA* is only observed in *H. pylori* isolates from the East Asian population, whereas Western-type *cagA* is widely distributed among isolates from European, South and Central Asian, North and South American, and African populations [12]. Almost all *H. pylori* isolates from East Asia possess the *cagA* gene, whereas approximately 20–40 % of isolates from Europe and Africa are *cagA* negative. Thailand is at the cultural crossroads between East and South Asia and, indeed, approximately half of the strains in Thailand possess East Asian-type *cagA*, whereas others possess Western-type *cagA* [38]. Interestingly Western-type *cagA* detected in strains from Okinawa (J-Western-type *cagA*) formed a different cluster compared to the original Western-type *cagA* [39]. The pre-EPIYA region of *cagA* also shows geographic divergence [40].

Most strains isolated from East Asia have a 39-bp deletion, but this deletion was absent in most strains from Western countries. On the other hand, an 18-bp deletion was common in Vietnamese strains. In addition, we found that the frequencies of the EPIYT and ESIYT motifs are relatively high among the sequences of the Okinawa strains [41]. Amerindian-type *cagA* from part of Machiguenga-speaking residents of the Shimaa village in the remote Peruvian Amazon (AM-I) also contained ESIYT motifs, which supports the possibility that these populations share the same origin [42]. A recent study revealed the recombination processes of *cagA* [43]. Interestingly, the left half of the EPIYA-D segment of East Asian-type *cagA* was derived from the Western-type EPIYA, with the Amerindian-type EPIYA as intermediate, through rearrangement of specific sequences within the gene. J-Western type EPIYA is phylogenetically located between the Western-type EPIYA and Amerindian-type EPIYA. This finding suggests that the original *H. pylori* strain had a Western-type *cagA* sequence. Subsequently, they evolved to the J-Western-type *cagA*, to the Amerindian-type *cagA*, and then to the East Asian-type *cagA*.

The right end of the *cag* PAI has been divided into five subtypes according to deletion, insertion, and substitution motifs [44]. Type I is most common in isolates from ethnic European groups and from Africa, type II is predominant in those from East Asia, and type III is predominant in isolates from South Asia [12, 44]. Type IV is very rare and, therefore, has not been assigned to a specific geographical area. Type V is found in a few strains from Calcutta, India [12, 44]. Interestingly, our report showed that type V was present in 10 % of isolates from patients of Thailand, and the ratio was especially high in strains obtained from ethnic Thai (21 %) [38]. The presence of this genotype in Thailand suggests that it migrated to the east of Calcutta. Overall, these data might show that transmission of specific genotypes remains conserved within ethnic groups for at least several generations.

On the other hand, the overall topology of the *vacA* tree did not always match with that of MLST [28]. Furthermore, rooting the *vacA* tree with out-group sequences from the closely related *H. acinonychis* revealed that the ancestry of *vacA* is different from the African origin. VacA is a vacuolating cytotoxin that induces cytoplasmic vacuoles in various eukaryotic cells. Unlike the case of the *cagA* gene, all *H. pylori* strains carry a functional *vacA* gene. However, there is considerable variation in vacuolating activities among strains [45, 46], primarily as a result of differences in the *vacA* gene structure at the signal region (s1 and s2) and the middle region (m1 and m2) [47]. Interestingly, the *vacA* s1 genotype is closely correlated with the presence of the *cagA* gene [8, 48, 49]. The *vacA* gene may comprise any combination of signal and middle-region types, although the s2/m1 combination is rare [47, 50]. All East Asian *H. pylori* strains are of the *vacA* s1 type [8, 12]. Within East Asian countries, the m1 type is predominant in Japan and Korea, whereas the prevalence of m2 types gradually increases in southern parts of East Asia (Vietnam, Hong Kong) [12]. The *vacA* s1 type is subdivided into s1a, s1b, and s1c [37, 47], and the m1 type is subdivided into m1a, m1b, and m1c [51]. The *vacA* s1c and m1b types are typical of *H. pylori* from East Asia (i.e., more than 95 % of s1 and m1) and the s1a and m1c types are common in South Asia (i.e., approximately 85 % of s1 and nearly 100 % of m1) [12, 52]. The *vacA* m1c genotype is also found in strains from Central Asia (ethnic Kazakhs) [12]. The m1a type is typical of Africans and ethnic Europeans (i.e., nearly 100 % of m1) [12, 37]. Both the s1a and s1b types are common in ethnic European strains, and s1b types are especially common in strains from the Iberian Peninsula and Latin America (i.e., approximately 85 % of s1) [37, 53]. The s1b type is also predominant in Africa (i.e., approximately 90 % of s1) [50, 53]. The *H. pylori* genotypes circulating among ethnic groups (Blacks, White Hispanics, Whites, and Vietnamese) living in the same region (Houston, Texas, USA) [54] have been examined by our group. According to ethnicity genotypes, the most common were the following: Blacks, s1b-m1; Hispanics, s1b-m1; Whites, s1a-m2; and Vietnamese, s1c-m2, which completely overlap with the predominant genotypes of Africa, the Iberian Peninsula, Northern and Eastern Europe, and Vietnam, respectively. In Thailand, the predominant *vacA* genotypes among s1-m1 strains are s1a-m1c in ethnic Thai people and s1b-m1b in ethnic Chinese people, which are the same as the predominant genotypes of South Asia and East Asia, respectively [38].

By combining the *cagA*, *cag* right-end junction, and *vacA* genotypes of more than 1000 *H. pylori* strains collected from East Asia, Southeast Asia, South Asia, Central Asia, Europe, Africa, North America, and South America [12, 38], four major groups (East Asia type, South/Central Asia type, Iberian/Africa type, and Europe type) can be defined according to geographical associations (Table 2.2). In these groups, *cagA*-negative and/or *vacA* m2 genotypes are not taken into account, but we can predict the geographical origins of each group using available genotypes (i.e., strains with *cagA* negative, but *vacA* s1a-m1a is predicted to be of the Europe type). Overall, the genotype of the virulence genes is important, not only as a tool to track human migration but also for epidemiological studies of *H. pylori*-related gastroduodenal diseases, especially in areas where multiple genotypes coexist (e.g., virulent East Asian type and less virulent South/Central Asian type in Thailand).

The genotypes of the virulence genes have provided important information about human migration to the Americas. The Americas were populated by humans

Table 2.2 Predominant virulence genotypes on *cagA* and *vacA* genes according to geographic area

Geographical area	East Asia	South/Central Asia	Europe (except for the Iberian Peninsula)	Europe (Iberian Peninsula)/Africa
Virulence genes genotype	East Asia	South/Central Asia	Europe	Iberian/Africa
cagA gene genotype	East Asia	Western	Western	Western
cag right-end junction genotype	II	III (V)	I	I
vacA s1 subtype	s1c	s1a	s1a	s1a
vacA m1 subtype	m1b	m1c	m1a	m1a

of East Asian ancestry approximately 15,000 years ago. Over the last 500 years, Europeans and Africans have come to the Americas, leading to an increasing Mestizo (mixture of Amerindian and European ancestry) population. Our group has discovered that approximately 25 % of the *H. pylori* isolates from Native Colombians and Native Alaskans possess novel *vacA* and/or *cagA* structures that are similar, but not identical, to structures from East Asia (i.e., *vacA* s1c-m1b-like, East Asian-like *cagA*) [12]. Native Venezuelans are also reported to have a high frequency of the *vacA* s1c genotype [55]. These data confirm that *H. pylori* accompanied humans when they crossed the Bering Strait from Asia to the New World. Importantly, none of the *H. pylori* strains from Mestizo populations possess East Asian-like genotypes. Sequence analysis of *H. pylori* genomes has shown that East Asian-like Amerindian strains are the least genetically diverse, probably because of a genetic bottleneck, whereas European strains are the most diverse among Amerindian, European, African, and East Asian strains [56]. If diversity is important for the success of *H. pylori* colonization, the East Asian-like Amerindian strains may lack the needed diversity to compete with the diverse *H. pylori* population brought to the New World by non-Amerindian hosts and has therefore disappeared.

2.4 Genome-Wide for Evolutionary Study

Analysis of MLST data and virulence factors revealed much information about the pathogenicity and genealogy of *H. pylori*; however, these approaches focus on a small number of genes and may miss information conveyed by the rest of the genome. Genome-wide analyses using DNA microarray or whole-genome sequencing technology give a broad view on the genome of *H. pylori*.

Microarray analysis provides comprehensive information about gene contents of different strains and helps identify strain-specific genes as well as core genes shared by multiple strains. Salama et al. examined the genomic content of 15 clinical isolates using a whole-genome DNA microarray and defined 1281 genes as functional core genes [57]. They identified candidates of virulence genes on the basis of coinheritance with the *cag* PAI. A similar approach was used to elucidate the genomic diversity of isolates obtained from clinical patients in China [58]. The whole-genome sequencing technology is another powerful tool to study the evolution and pathogenicity of *H. pylori*.

Since the first release of the whole genome of strain 26,695 [59], the sequences of more than 20 genomes were determined by Sanger sequencing or the massively parallel sequencing technology. Accumulation of whole-genome data enables extensive sequence analyses of *H. pylori* strains. About 1200 core genes were identified by comparison of peptic ulcer strain P12 and six other *H. pylori* genomes, which were in agreement with preceding studies [60]. The authors found that the P12 genome contains three plasticity zones and that one of them is capable of self-excision and horizontal transfer by conjugation. Their result suggests that conjugative transfer of genomic islands may contribute to the genetic diversity of *H. pylori*. McClain et al. compared genome sequences of an isolate obtained from a patient with gastric cancer (strain 98–10) and an isolate from a patient with gastric ulcer (strain B128) [61]. Strain 98–10 was found to be closely related to East Asian strains, while strain B128 was related to European strains. They determined strain-specific genes of strain 98–10 as candidate genes associated with gastric cancer. East Asian strains are known for their stronger carcinogenicity compared to strains of other areas. Kawai et al. investigated the evolution of East Asian strains using 20 whole genomes of Japanese, Korean, Amerindian, European, and West African strains [62]. Phylogenetic analysis revealed a greater divergence between the East Asian strains and the European strains in genes related to virulence factors, outer membrane proteins, and lipopolysaccharide synthesis enzymes. They examined positively selected amino acid changes and mapped the identified residues on CagA, VacA, HomC, SotB, and MiaA proteins.

Currently, we took advantage of next-generation sequencers to read genomic sequences of more than 40 *H. pylori* strains mainly from Asian populations and attempted de novo assembly (unpublished observation). Although we cannot determine the whole genomes yet, we could construct a substantial size of contigs and predicted 1200–1500 genes for each strain. Using these data, we determined orthologous genes among our samples and strains whose whole genomes were released into public databases. A phylogenetic tree constructed by concatenated sequences of the orthologous genes showed more reliable results than a phylogenetic tree constructed by using MLST data. Compared with the tree based on MLST data, the tree constructed by using concatenated genes showed better branching with higher bootstrap values between hpEurope and hpAsia2, as well as between hspEAsia and hspAmerind. Data obtained by using the massively parallel sequencing technology provide valuable information on the genealogy of *H. pylori* strains, as well as on candidates of drug resistance genes and new virulence factors.

2.5 Conclusion

H. pylori typing is very useful as a tool for tracking human migrations. Serial studies of large numbers of *H. pylori* strains from all over the world, including strains isolated from aboriginal populations, have shown that MLST analysis of *H. pylori* sequences provides more detailed information on human migrations than does human genetic analysis. However, there are still a number of untapped areas in the world, including a number of isolated aboriginal populations in Siberia, Mongolia, China, Indonesia, and Japan (Ainu tribe), and it will be interesting to study *H. pylori* strains isolated from these different groups. To date, subcategorization of East Asian strains (hspEAsia) has not been possible because of high homology among East Asian strains. However, the genotyping of virulence genes has shown that the *vacA* middle region can be useful for distinguishing strains of the northern parts of East Asian countries from those of the south. Genome-wide analyses using DNA microarray or whole-genome sequencing technology give a broad view on the genome of *H. pylori*. These methods may complete the weakness of MLST and virulence factors. In particular, next-generation sequencers enabled us to efficiently investigate not only the evolution of *H. pylori*, but also novel virulence factors and genomic changes related to drug resistance.

References

1. Suerbaum S, Michetti P. *Helicobacter pylori* infection. N Engl J Med. 2002;347(15):1175–86.
2. Achtman M, Azuma T, Berg DE, Ito Y, Morelli G, Pan ZJ, Suerbaum S, Thompson SA, van der Ende A, van Doorn LJ. Recombination and clonal groupings within *Helicobacter pylori* from different geographical regions. Mol Microbiol. 1999;32(3):459–70.
3. Li WH, Sadler LA. Low nucleotide diversity in man. Genetics. 1991;129(2):513–23.
4. Suerbaum S, Josenhans C. *Helicobacter pylori* evolution and phenotypic diversification in a changing host. Nat Rev Microbiol. 2007;5(6):441–52.
5. Falush D, Wirth T, Linz B, Pritchard JK, Stephens M, Kidd M, Blaser MJ, Graham DY, Vacher S, Perez-Perez GI, et al. Traces of human migrations in *Helicobacter pylori* populations. Science. 2003;299(5612):1582–5.
6. Yamaoka Y. Mechanisms of disease: *Helicobacter pylori* virulence factors. Nat Rev Gastroenterol Hepatol. 2010;7(11):629–41.
7. Yamaoka Y, Ojo O, Fujimoto S, Odenbreit S, Haas R, Gutierrez O, El-Zimaity HM, Reddy R, Arnqvist A, Graham DY. *Helicobacter pylori* outer membrane proteins and gastroduodenal disease. Gut. 2006;55(6):775–81.
8. Yamaoka Y, Kodama T, Kita M, Imanishi J, Kashima K, Graham DY. Relationship of vacA genotypes of *Helicobacter pylori* to *cagA* status, cytotoxin production, and clinical outcome. Helicobacter. 1998;3(4):241–53.
9. Yamaoka Y, El-Zimaity HM, Gutierrez O, Figura N, Kim JG, Kodama T, Kashima K, Graham DY. Relationship between the *cagA* 3' repeat region of *Helicobacter pylori*, gastric histology, and susceptibility to low pH. Gastroenterology. 1999;117(2):342–9.
10. Yamaoka Y. Roles of *Helicobacter pylori* BabA in gastroduodenal pathogenesis. World J Gastroenterol. 2008;14(27):4265–72.

11. Jung SW, Sugimoto M, Graham DY. Yamaoka Y: *homB* status of *Helicobacter pylori* as a novel marker to distinguish gastric cancer from duodenal ulcer. J Clin Microbiol. 2009;47 (10):3241–5.
12. Yamaoka Y, Orito E, Mizokami M, Gutierrez O, Saitou N, Kodama T, Osato MS, Kim JG, Ramirez FC, Mahachai V, et al. *Helicobacter pylori* in North and South America before Columbus. FEBS Lett. 2002;517(1–3):180–4.
13. Linz B, Balloux F, Moodley Y, Manica A, Liu H, Roumagnac P, Falush D, Stamer C, Prugnolle F, van der Merwe SW, et al. An African origin for the intimate association between humans and *Helicobacter pylori*. Nature. 2007;445(7130):915–8.
14. Moodley Y, Linz B, Yamaoka Y, Windsor HM, Breurec S, Wu JY, Maady A, Bernhoft S, Thiberge JM, Phuanukoonnon S, et al. The peopling of the Pacific from a bacterial perspective. Science. 2009;323(5913):527–30.
15. Wirth T, Wang X, Linz B, Novick RP, Lum JK, Blaser M, Morelli G, Falush D, Achtman M. Distinguishing human ethnic groups by means of sequences from *Helicobacter pylori*: lessons from Ladakh. Proc Natl Acad Sci U S A. 2004;101(14):4746–51.
16. Maiden MC, Bygraves JA, Feil E, Morelli G, Russell JE, Urwin R, Zhang Q, Zhou J, Zurth K, Caugant DA, et al. Multilocus sequence typing: a portable approach to the identification of clones within populations of pathogenic microorganisms. Proc Natl Acad Sci U S A. 1998;95 (6):3140–5.
17. Covacci A, Telford JL, Del Giudice G, Parsonnet J, Rappuoli R. *Helicobacter pylori* virulence and genetic geography. Science. 1999;284(5418):1328–33.
18. Moodley Y, Linz B, Bond RP, Nieuwoudt M, Soodyall H, Schlebusch CM, Bernhoft S, Hale J, Suerbaum S, Mugisha L, et al. Age of the association between *Helicobacter pylori* and man. PLoS Pathog. 2012;8(5), e1002693.
19. Suzuki R, Shiota S, Yamaoka Y. Molecular epidemiology, population genetics, and pathogenic role of *Helicobacter pylori*. Infect Genet Evol. 2012;12(2):203–13.
20. Correa P, Piazuelo MB. Evolutionary history of the *Helicobacter pylori* genome: implications for gastric carcinogenesis. Gut Liver. 2012;6(1):21–8.
21. Devi SM, Ahmed I, Francalacci P, Hussain MA, Akhter Y, Alvi A, Sechi LA, Megraud F, Ahmed N. Ancestral European roots of *Helicobacter pylori* in India. BMC Genomics. 2007;8:184.
22. Tay CY, Mitchell H, Dong Q, Goh KL, Dawes IW, Lan R. Population structure of *Helicobacter pylori* among ethnic groups in Malaysia: recent acquisition of the bacterium by the Malay population. BMC Microbiol. 2009;9:126.
23. Latifi-Navid S, Ghorashi SA, Siavoshi F, Linz B, Massarrat S, Khegay T, Salmanian AH, Shayesteh AA, Masoodi M, Ghanadi K, et al. Ethnic and geographic differentiation of *Helicobacter pylori* within Iran. PLoS One. 2010;5(3), e9645.
24. Breurec S, Guillard B, Hem S, Brisse S, Dieye FB, Huerre M, Oung C, Raymond J, Tan TS, Thiberge JM, et al. Evolutionary history of *Helicobacter pylori* sequences reflect past human migrations in Southeast Asia. PLoS One. 2011;6(7), e22058.
25. Gressmann H, Linz B, Ghai R, Pleissner KP, Schlapbach R, Yamaoka Y, Kraft C, Suerbaum S, Meyer TF, Achtman M. Gain and loss of multiple genes during the evolution of *Helicobacter pylori*. PLoS Genet. 2005;1(4), e43.
26. Ishida T, Hinuma Y. The origin of Japanese HTLV-I. Nature. 1986;322(6079):504.
27. Miftahussurur M, Tuda J, Suzuki R, Kido Y, Kawamoto F, Matsuda M, Tantular IS, Pusarawati S, Nasronudin, Harijanto PN, et al. Extremely low *Helicobacter pylori* prevalence in North Sulawesi, Indonesia and identification of a Maori-tribe type strain: a cross sectional study. Gut Pathog. 2014;6(1):42.
28. Gangwer KA, Shaffer CL, Suerbaum S, Lacy DB, Cover TL, Bordenstein SR. Molecular evolution of the *Helicobacter pylori* vacuolating toxin gene vacA. J Bacteriol. 2010;192 (23):6126–35.

29. Olbermann P, Josenhans C, Moodley Y, Uhr M, Stamer C, Vauterin M, Suerbaum S, Achtman M, Linz B. A global overview of the genetic and functional diversity in the *Helicobacter pylori* cag pathogenicity island. PLoS Genet. 2010;6(8), e1001069.
30. Censini S, Lange C, Xiang Z, Crabtree JE, Ghiara P, Borodovsky M, Rappuoli R, Covacci A. *cag*, a pathogenicity island of *Helicobacter pylori*, encodes type I-specific and disease-associated virulence factors. Proc Natl Acad Sci U S A. 1996;93(25):14648–53.
31. Asahi M, Azuma T, Ito S, Ito Y, Suto H, Nagai Y, Tsubokawa M, Tohyama Y, Maeda S, Omata M, et al. *Helicobacter pylori* CagA protein can be tyrosine phosphorylated in gastric epithelial cells. J Exp Med. 2000;191(4):593–602.
32. Odenbreit S, Puls J, Sedlmaier B, Gerland E, Fischer W, Haas R. Translocation of *Helicobacter pylori* CagA into gastric epithelial cells by type IV secretion. Science. 2000;287(5457):1497–500.
33. Segal ED, Cha J, Lo J, Falkow S, Tompkins LS. Altered states: involvement of phosphorylated CagA in the induction of host cellular growth changes by *Helicobacter pylori*. Proc Natl Acad Sci U S A. 1999;96(25):14559–64.
34. Backert S, Selbach M. Role of type IV secretion in *Helicobacter pylori* pathogenesis. Cell Microbiol. 2008;10(8):1573–81.
35. Yamaoka Y, Kodama T, Kashima K, Graham DY, Sepulveda AR. Variants of the 3′ region of the *cagA* gene in *Helicobacter pylori* isolates from patients with different *H. pylori*-associated diseases. J Clin Microbiol. 1998;36(8):2258–63.
36. Yamaoka Y, Osato MS, Sepulveda AR, Gutierrez O, Figura N, Kim JG, Kodama T, Kashima K, Graham DY. Molecular epidemiology of *Helicobacter pylori*: separation of *H. pylori* from East Asian and non-Asian countries. Epidemiol Infect. 2000;124(1):91–6.
37. van Doorn LJ, Figueiredo C, Sanna R, Blaser MJ, Quint WG. Distinct variants of *Helicobacter pylori cagA* are associated with vacA subtypes. J Clin Microbiol. 1999;37(7):2306–11.
38. Vilaichone RK, Mahachai V, Tumwasorn S, Wu JY, Graham DY, Yamaoka Y. Molecular epidemiology and outcome of *Helicobacter pylori* infection in Thailand: a cultural cross roads. Helicobacter. 2004;9(5):453–9.
39. Truong BX, Mai VT, Tanaka H, le Ly T, Thong TM, Hai HH, Van Long D, Furumatsu K, Yoshida M, Kutsumi H, et al. Diverse characteristics of the CagA gene of *Helicobacter pylori* strains collected from patients from southern Vietnam with gastric cancer and peptic ulcer. J Clin Microbiol. 2009;47(12):4021–8.
40. Uchida T, Nguyen LT, Takayama A, Okimoto T, Kodama M, Murakami K, Matsuhisa T, Trinh TD, Ta L, Ho DQ, et al. Analysis of virulence factors of *Helicobacter pylori* isolated from a Vietnamese population. BMC Microbiol. 2009;9:175.
41. Xia Y, Yamaoka Y, Zhu Q, Matha I, Gao X. A comprehensive sequence and disease correlation analyses for the C-terminal region of CagA protein of *Helicobacter pylori*. PLoS One. 2009;4(11), e7736.
42. Suzuki M, Kiga K, Kersulyte D, Cok J, Hooper CC, Mimuro H, Sanada T, Suzuki S, Oyama M, Kozuka-Hata H, et al. Attenuated CagA oncoprotein in *Helicobacter pylori* from Amerindians in Peruvian Amazon. J Biol Chem. 2011;286(34):29964–72.
43. Furuta Y, Yahara K, Hatakeyama M, Kobayashi I. Evolution of *cagA* oncogene of *Helicobacter pylori* through recombination. PLoS One. 2011;6(8), e23499.
44. Kersulyte D, Mukhopadhyay AK, Velapatino B, Su W, Pan Z, Garcia C, Hernandez V, Valdez Y, Mistry RS, Gilman RH, et al. Differences in genotypes of *Helicobacter pylori* from different human populations. J Bacteriol. 2000;182(11):3210–8.
45. Cover TL, Blaser MJ. Purification and characterization of the vacuolating toxin from *Helicobacter pylori*. J Biol Chem. 1992;267(15):10570–5.
46. Leunk RD. Production of a cytotoxin by *Helicobacter pylori*. Rev Infect Dis. 1991;13 Suppl 8: S686–9.
47. Atherton JC, Cao P, Peek Jr RM, Tummuru MK, Blaser MJ, Cover TL. Mosaicism in vacuolating cytotoxin alleles of *Helicobacter pylori*. Association of specific *vacA* types with cytotoxin production and peptic ulceration. J Biol Chem. 1995;270(30):17771–7.

48. Atherton JC, Peek Jr RM, Tham KT, Cover TL, Blaser MJ. Clinical and pathological importance of heterogeneity in *vacA*, the vacuolating cytotoxin gene of *Helicobacter pylori*. Gastroenterology. 1997;112(1):92–9.
49. Atherton JC. The pathogenesis of *Helicobacter pylori*-induced gastro-duodenal diseases. Annu Rev Pathol. 2006;1:63–96.
50. Letley DP, Lastovica A, Louw JA, Hawkey CJ, Atherton JC. Allelic diversity of the *Helicobacter pylori* vacuolating cytotoxin gene in South Africa: rarity of the *vacA* s1a genotype and natural occurrence of an s2/m1 allele. J Clin Microbiol. 1999;37(4):1203–5.
51. Mukhopadhyay AK, Kersulyte D, Jeong JY, Datta S, Ito Y, Chowdhury A, Chowdhury S, Santra A, Bhattacharya SK, Azuma T, et al. Distinctiveness of genotypes of *Helicobacter pylori* in Calcutta, India. J Bacteriol. 2000;182(11):3219–27.
52. Hisatsune J, Nakayama M, Isomoto H, Kurazono H, Mukaida N, Mukhopadhyay AK, Azuma T, Yamaoka Y, Sap J, Yamasaki E, et al. Molecular characterization of *Helicobacter pylori* VacA induction of IL-8 in U937 cells reveals a prominent role for p38MAPK in activating transcription factor-2, cAMP response element binding protein, and NF-kappaB activation. J Immunol. 2008;180(7):5017–27.
53. Sugimoto M, Yamaoka Y. The association of *vacA* genotype and *Helicobacter pylori*-related disease in Latin American and African populations. Clin Microbiol Infect. 2009;15(9):835–42.
54. Yamaoka Y, Malaty HM, Osato MS, Graham DY. Conservation of *Helicobacter pylori* genotypes in different ethnic groups in Houston, Texas. J Infect Dis. 2000;181(6):2083–6.
55. Ghose C, Perez-Perez GI, Dominguez-Bello MG, Pride DT, Bravi CM, Blaser MJ. East Asian genotypes of *Helicobacter pylori* strains in Amerindians provide evidence for its ancient human carriage. Proc Natl Acad Sci U S A. 2002;99(23):15107–11.
56. Dominguez-Bello MG, Perez ME, Bortolini MC, Salzano FM, Pericchi LR, Zambrano-Guzman O, Linz B. Amerindian *Helicobacter pylori* strains go extinct, as european strains expand their host range. PLoS One. 2008;3(10), e3307.
57. Salama N, Guillemin K, McDaniel TK, Sherlock G, Tompkins L, Falkow S. A whole-genome microarray reveals genetic diversity among *Helicobacter pylori* strains. Proc Natl Acad Sci U S A. 2000;97(26):14668–73.
58. Han YH, Liu WZ, Shi YZ, Lu LQ, Xiao S, Zhang QH, Zhao GP. Comparative genomics profiling of clinical isolates of *Helicobacter pylori* in Chinese populations using DNA microarray. J Microbiol. 2007;45(1):21–8.
59. Tomb JF, White O, Kerlavage AR, Clayton RA, Sutton GG, Fleischmann RD, Ketchum KA, Klenk HP, Gill S, Dougherty BA, et al. The complete genome sequence of the gastric pathogen *Helicobacter pylori*. Nature. 1997;388(6642):539–47.
60. Fischer W, Windhager L, Rohrer S, Zeiller M, Karnholz A, Hoffmann R, Zimmer R, Haas R. Strain-specific genes of *Helicobacter pylori*: genome evolution driven by a novel type IV secretion system and genomic island transfer. Nucleic Acids Res. 2010;38(18):6089–101.
61. McClain MS, Shaffer CL, Israel DA, Peek Jr RM, Cover TL. Genome sequence analysis of *Helicobacter pylori* strains associated with gastric ulceration and gastric cancer. BMC Genomics. 2009;10:3.
62. Kawai M, Furuta Y, Yahara K, Tsuru T, Oshima K, Handa N, Takahashi N, Yoshida M, Azuma T, Hattori M, et al. Evolution in an oncogenic bacterial species with extreme genome plasticity: *Helicobacter pylori* East Asian genomes. BMC Microbiol. 2011;11:104.

Chapter 3
CagA

Yoshie Senda and Masanori Hatakeyama

Abstract Chronic infection with *Helicobacter pylori cagA*-positive strains is the strongest risk factor for gastric cancer. *H. pylori* injects the *cagA*-encoded CagA protein into the host gastric epithelial cells. Recent studies revealed that CagA acts as a pathogenic/oncogenic scaffold, which promotes oncogenic signaling in the delivered host cells. Indeed, CagA interacts with a variety of cellular proteins and deregulates their functions. More specifically, the N-terminal structured region of CagA associates with ASPP2 and RUNX3. The C-terminal disordered region of CagA possesses multiple segments containing the Glu-Pro-Ile-Tyr-Ala (EPIYA) motif, which undergoes tyrosine phosphorylation by Src family kinases (SFKs), and the CagA-multimerization (CM) sequence. The EPIYA segments interact with SH2 domain-containing proteins such as SHP2 and Csk in a tyrosine phosphorylation-dependent manner, whereas the CM sequence binds to the polarity-regulating kinase PAR1 in a tyrosine phosphorylation-independent manner. We propose a hypothesis that mammalian proteome contains a distinct class of proteins carrying an EPIYA or EPIYA-like sequence, which are imitated by bacterial EPIYA effectors such as CagA to perturb intracellular signaling in the host cells.

Keywords CagA • Pathogenic/oncogenic scaffold • EPIYA motif/segment

3.1 Introduction

Gastric cancer is one of the most common neoplasms and the third leading cause of cancer mortality in the world. East Asian countries (Japan, Korea, and China) have the highest incidence rates of gastric cancer [1]. Chronic infection with *Helicobacter pylori* is associated with gastric diseases including cancer [2]. *H. pylori* strains that possess cytotoxin-associated gene pathogenicity island (*cag* PAI) in their genome are more virulent strains than those without *cag* PAI and are associated with more severe gastric mucosal damages. The *cag* PAI region, a ~37 kb genomic region acquired by horizontal transmission from unknown origin,

Y. Senda • M. Hatakeyama (✉)
Division of Microbiology, Graduate School of Medicine, The University of Tokyo, 7-3-1 Hongo, Bunkyo-Ku, Tokyo 113-0033, Japan
e-mail: mhata@m.u-tokyo.ac.jp

H. Suzuki et al. (eds.), *Helicobacter pylori*, DOI 10.1007/978-4-431-55705-0_3

contains ~30 genes encoding a type IV secretion system (T4SS) as well as cytotoxin-associated gene A (CagA). T4SS is a needle-shaped apparatus formed by supramolecular assembly of multiple Cag proteins [3, 4], and CagA, a 120–145 kDa protein encoded by the *cagA* gene, is delivered into the host gastric epithelial cells via T4SS [5]. Epidemiological studies revealed the increased risk for gastric carcinoma in individuals infected with *cagA*-positive *H. pylori* strains compared with those infected with *cagA*-negative strains [6–8].

3.2 Invasion of CagA into the Gastric Epithelial Cells

In normal eukaryotic cells, the outer leaflet of plasma membrane is enriched with choline-containing phospholipids and glycosphingolipids, whereas the inner leaflet is enriched with amine-containing phospholipids such as phosphatidylserine (PS), phosphatidylethanolamine (PE), phosphatidic acid (PA), and phosphatidylinositol (PI) [9]. The direct attachment of *H. pylori* to gastric epithelial cells induces the transition of the plasma membrane PS to the outer leaflet. The externalization of PS upon *H. pylori* infection is rapid, transient, and independent of apoptotic process. At the interface between T4SS and host cell surface, CagA directly binds to the externalized membrane PS via the electrostatic interaction, which is critical for the delivery of CagA into host cells. Delivered CagA again binds to the membrane PS and thereby localizes to the inner face of the plasma membrane [10]. Inside the cells, CagA undergoes tyrosine phosphorylation at the Glu-Pro-Ile-Tyr-Ala (EPIYA) motifs, which are present in a variable number in its C-terminal region, by Src family kinases (SFKs) and c-Abl kinase [7, 11, 12]. CagA promiscuously interacts with multiple cellular proteins via its C-terminal region in both tyrosine-phosphorylation-dependent and tyrosine-phosphorylation-independent manners and thereby perturbs the host cell signaling. As such, CagA is considered to act as a bacterium-derived scaffold/hub protein that potentiates oncogenic signaling in delivered gastric epithelial cells [13].

3.3 Structure of CagA

CagA comprises more than 1200 amino acid residues with sequence diversity in its C-terminal region among distinct *H. pylori* isolates. Crystal structure analysis revealed that the CagA N-terminal region (~70 % of the entire protein) has a solid structure, whereas the C-terminal region (~30 % of the entire protein) containing the EPIYA motifs is intrinsically disordered (Fig. 3.1) [14, 15]. The N-terminal region of CagA is composed of three distinct domains, domains I, II, and III. Domain I, a mobile module with α-helical structure, mediates the interaction of CagA with ASPP2, a tumor suppressor apoptosis-stimulating protein of p53,

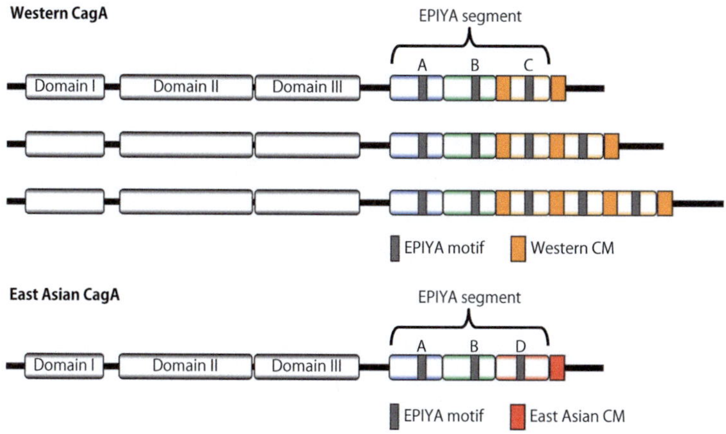

Fig. 3.1 Structure of *H. pylori* CagA. The N-terminal region of CagA contains domains I, II, and III. Based on the structural diversity of the C-terminal region, CagA is classified into Western CagA and East Asian CagA. The C-terminal region of Western CagA contains the EPIYA-A, EPIYA-B, and EPIYA-C segments. Western CagA contains at least two CM motifs. The C-terminal region of East Asian CagA contains the EPIYA-A, EPIYA-B, and EPIYA-D segments. East Asian CagA has a single CM motif

that induce the degradation of p53 (Fig. 3.2) [16, 17]. Domain I also binds to a tumor suppressor RUNX3 and thereby inactivates the tumor suppressor [18].

Domain II forms a structural core of CagA and contains the basic patch, a cluster of basic residues, with which CagA binds to the membrane PS. Domain II also contains a large antiparallel β-sheet, through which CagA associates with β1-integrin. Domain III intramolecularly interacts with the disordered C-terminal region, stabilizing the complex of CagA with multiple cellular molecules [14].

In the disordered C-terminal region, four EPIYA segments, EPIYA-A, EPIYA-B, EPIYA-C, and EPIYA-D, are classified based on the amino acid sequence surrounding each of the EPIYA motifs. The diversity of the C-terminal region is due to combination and/or order of the four distinct EPIYA segments, which serve as tyrosine phosphorylation sites [19, 20]. The EPIYA-A and EPIYA-B segments are common in almost all CagA species. The EPIYA-C segment is specific to "Western CagA," which is distributed in Europe, North America, and Australia, whereas the EPIYA-D segment is specific to "East Asian CagA," which is distributed in East Asian countries. Furthermore, the EPIYA-C segment duplicates in variable numbers (from 0 to 5) among Western CagA species [19].

The CagA-multimerization (CM) sequence, which was originally identified as a CagA sequence that is responsible for CagA multimerization (dimerization) in host cells, is a 16 amino acid stretch present within the EPIYA-C segment and in also the end of EPIYA-repeat region. Accordingly, Western CagA proteins with ABC, ABCC, and ABCCC types contain two, three, and four CM motifs, respectively. East Asian CagA has a single CM motif that is located immediately following EPIYA-D segment (Fig. 3.1) [19, 21, 22].

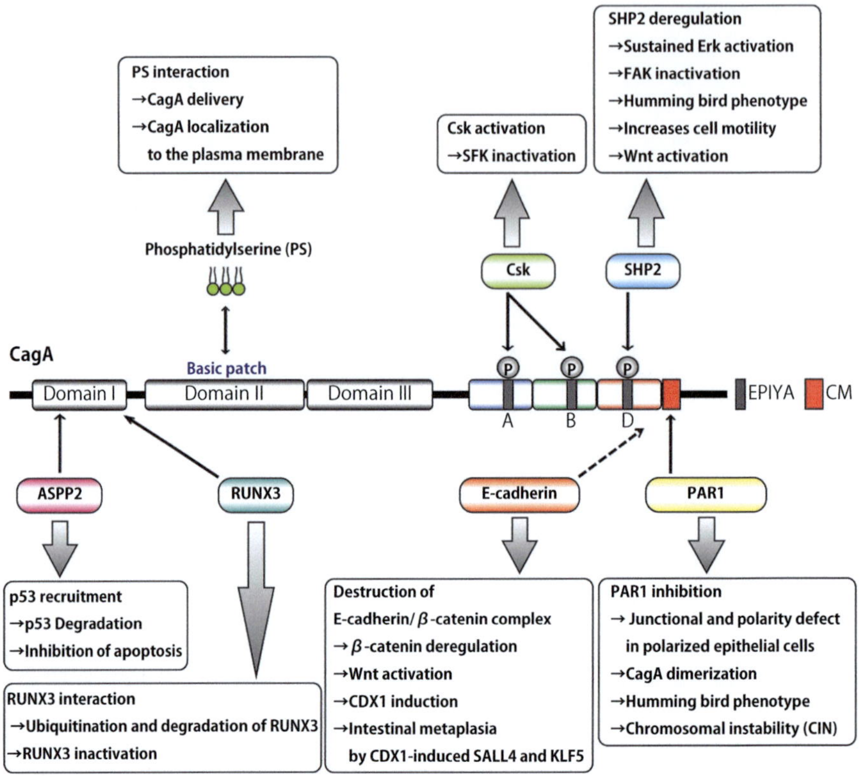

Fig. 3.2 *H. pylori* CagA as a pathogenic/oncogenic scaffold. Domain I of CagA binds to ASPP2. CagA-ASPP2 complex recruits p53 and induces degradation of p53 to inhibit apoptosis. Domain I also binds to RUNX3 and thus inactivates its tumor-suppressive function via degradation. Domain II of CagA contains basic patch, which is required for CagA delivery via T4SS and localization to the plasma membrane. Tyrosine-phosphorylated EPIYA motifs can bind to Csk and SHP2. The CagA-Csk complex activates Csk, which in turn inhibits SFK activity. The CagA-SHP2 complex deregulates the pro-oncogenic SHP2 phosphatase activity. CagA associates with E-cadherin in a phosphorylation-independent manner, thereby inducing aberrant Wnt activation that leads to the development of intestinal metaplasia. The CM motif binds to and inhibits PAR1. This interaction causes junctional and polarity defects while inducing chromosomal instability

3.4 EPIYA-Dependent CagA Function

CagA is phosphorylated at the EPIYA segments in the C-terminal region by SFKs and c-Abl [7, 23]. Upon tyrosine phosphorylation, CagA interacts with the SHP2 protein tyrosine phosphatase via the EPIYA-C or EPIYA-D segment (Fig. 3.2) [20, 24, 25]. SHP2 possesses two SH2 domains, which can interact with tyrosine-phosphorylated peptides, in the N-terminal region, followed by a protein tyrosine phosphatase (PTP) domain. SHP2 is required for the full activation of the Ras-Erk pathway and is known as a *bona fide* oncoprotein because gain-of-function mutations in SHP2 are associated with various human malignancies [26, 27]. CagA deregulates SHP2 phosphatase activity through complex formation and aberrantly activates Erk [24, 25]. Also, CagA-deregulated SHP2 dephosphorylates to inactivate focal adhesion kinase (FAK) [28]. FAK is a tyrosine kinase that regulates the turnover of focal adhesion spots [29]. Inactivation of FAK by CagA induces elongated cell shape known as the hummingbird phenotype and increases cell motility [28, 30].

Approximately 70 % of Western CagA species have a single EPIYA-C segment, and some have two or three [19, 31]. CagA with an increased number of EPIYA-C segments shows an increased ability to bind to and thereby deregulate SHP2 in a tyrosine phosphorylation-dependent manner. In contrast to the molecular polymorphism in Western CagA, almost all East Asian CagA have a single EPIYA-D segment [19, 31]. Notably, the EPIYA-D segment binds to SHP2 more robustly than the EPIYA-C segment does [20].

On the other hand, tyrosine-phosphorylated EPIYA-A or EPIYA-B segment serves a binding site for C-terminal Src kinase (Csk) (Fig. 3.2) [7, 32]. Csk is a non-receptor protein tyrosine kinase that phosphorylates the negative regulatory site of SFKs to inhibit their kinase activities [33]. CagA recruits Csk to the plasma membrane, where it activates Csk to inhibit membrane-anchored SFKs. As CagA is phosphorylated by SFKs, Csk activation at the membrane causes reduction of CagA tyrosine phosphorylation. CagA-mediated Csk activation therefore attenuates CagA-SHP2 signaling. This negative feedback loop may enable long-term colonization of *H. pylori* in the stomach without eliciting fatal damages to the host [32].

In addition, tyrosine-phosphorylated CagA promiscuously binds to a number of host proteins possessing SH2 domain(s) such as CrkI, CrkII, CrkL, Grb7, PI3-kinase, and Ras-GAP. The CagA-Crk complex also enhances cell scattering, motility, and proliferation. Whereas the functions and biological significances of these complexes have yet to be elucidated, it appears that CagA mimics tyrosine-phosphorylated host proteins to perturb intracellular signaling in the host gastric epithelial cells [13, 34].

The adaptor protein Grb2 binds to CagA via EPIYA region in a tyrosine phosphorylation-independent manner. Grb2 recruits Grb2-associated Sos, a guanine-exchange factor of the small GTPase Ras. Accordingly, the CagA-Grb2-Sos complex stimulates Ras-Erk pathway, which also promotes cell scattering and cell proliferation [35].

3.5 CM Motif-Dependent CagA Function

CagA directly interacts with a polarity-regulating serine/threonine kinase, partitioning-defective 1 (PAR1)/microtubule affinity-regulating kinase (MARK) via the CM motif in a tyrosine phosphorylation-independent manner (Fig. 3.2) [36]. CagA inhibits PAR1 kinase activity by binding to the catalytic domain of PAR1. Since PAR1 is a master regulator of cell polarity [37], the inhibition of

PAR1 activity by CagA causes junctional and polarity defects in polarized epithelial cells [36].

In mammals, PAR1 comprises four homologues (PAR1a/MARK3, PAR1b/MARK2, PAR1c/MARK1, and PAR1d/MARK4). CagA interacts with all the members of the PAR1 family kinases, among which PAR1b is the strongest binding partner [38]. Because PAR1 exists as a homodimer in cells, two CagA molecules passively dimerize through binding to PAR1. CagA dimerization stabilizes CagA-SHP2 complex and thereby potentiates SHP2 deregulation, causing enhancement of the hummingbird phenotype by CagA [39].

PAR1 phosphorylates microtubule-associated proteins (MAPs) on their tubulin-binding repeats to destabilize microtubules [40]. Microtubules are required for not only polarity regulation but also the formation of mitotic spindles. Inhibition of PAR1 kinase activity by CagA perturbs microtubule stability and causes microtubule-based spindle dysfunction. Accordingly, CagA-expressing cells display a delay in the transition from prophase to metaphase during mitosis, showing spindle misorientation. Thus, CagA gives rise to chromosomal instability (CIN) [41].

In non-polarized epithelial cells, CagA deregulates Erk signaling and induces senescence-like proliferation arrest via accumulation of p21 cyclin-dependent kinase (CDK) inhibitor [42]. In polarized epithelial cells, CagA causes disruption of tight junctions and loss of epithelial polarity [36, 43]. CagA-expressing cells are extruded from the polarized monolayer and undergo multiple rounds of cell divisions while overcoming senescence. In polarized epithelial cells, CagA-induced Erk signals prevent accumulation of p21 by activating a guanine nucleotide exchange factor-H1 (GEF-H1)-RhoA-RhoA-associated kinase (ROCK)-c-Myc pathway [42, 44]. CagA-expressing cells also show morphological changes resembling the epithelial-mesenchymal transition (EMT) [42, 45]. Recent study reported that CagA-induced EMT is caused by stabilization of Snail, a transcriptional repressor of E-cadherin [46]. Mechanistically, CagA binds to GSK-3, which phosphorylates and destabilizes Snail, at the CagA N-terminal region, causing GSK-3 translocation and inactivation.

c-Met receptor tyrosine kinase also interacts with CagA via the CM motif. This interaction causes deregulation of c-Met that aberrantly stimulates PI3-kinase/Akt kinase signaling. Hence, CagA-c-Met interaction activates both Wnt signaling and NF-κB [13, 34].

3.6 Perturbation of β-Catenin Signal by CagA

β-catenin is well known to be involved in canonical Wnt signaling, which is critical for growth and differentiation in various types of cells [47]. β-catenin is localized to adherence junction by interacting with E-cadherin. The E-cadherin/β-catenin complex plays an important role as a major component of the adherence junctions [48, 49]. The oncogenic *H. pylori* strain activates β-catenin in gastric epithelial cells in Mongolian gerbils [50]. CagA interacts with E-cadherin in a tyrosine phosphorylation-independent manner (Fig. 3.2) [51, 52]. The interaction of CagA with E-cadherin destabilizes the E-cadherin/β-catenin complex and thereby promotes cytoplasmic/nuclear accumulation of β-catenin [51]. The CagA CM motif is required for deregulation of β-catenin. Deregulated β-catenin by CagA aberrantly activates β-catenin-dependent Wnt target genes including *CDX1*. *CDX1* encodes caudal-related homeobox transcription factor CDX1, which is specifically expressed in the intestine and plays a crucial role in development and maintenance of intestinal epithelia.

Intestinal metaplasia of gastric epithelium is an *H. pylori*-associated precancerous lesion. Although CDX1 is not expressed in normal gastric epithelium, chronic infection with *H. pylori* induces CDX1 expression in the gastric epithelial cells via nonphysiological Wnt activation by CagA [51]. Ectopically expressed CDX1 then induces the expression of *SALL4*, a zinc finger transcription factor playing an important role in maintaining self-renewal and pluripotency. CDX1 also upregulates *KLF5*, a member of the KLF family of transcription factors that are associated with stemness [53]. CagA-expressing gastric epithelial cells therefore dedifferentiate into tissue stemlike progenitor cells by CDX1-induced SALL4 and KLF5 and then transdifferentiate into intestinal epithelial cells. This provides a mechanism that underlies the development of intestinal metaplasia in stomach [53].

β-catenin controls embryogenesis and homeostasis as an effector of the canonical Wnt pathway, where Wnt ligand binds to Frizzled receptor to stabilize β-catenin. Stabilized β-catenin then translocalizes into the nucleus, where it interacts with transcription factors such as T-cell transcription factor (TCF)/lymphocyte enhancer factor (LEF) to upregulate Wnt target genes [47, 54]. Recent studies reported that SHP2, a primary target of CagA, is present not only in the cytoplasm but also in the nucleus. In the nucleus, SHP2 dephosphorylate parafibromin/Cdc73, a core component of the RNA polymerase II-associated factor (PAF) complex. Dephosphorylated parafibromin can bind to β-catenin, and the parafibromin/β-catenin complex induces expression of Wnt target genes such as *cyclin D1* and *c-myc* [55]. SHP2 also binds to transcriptional coactivators YAP and TAZ, which are targeted by the mammalian Hippo pathway that controls cell proliferation and apoptosis. At low cell density, YAP/TAZ localizes to and recruits SHP2 to the nucleus [56]. Nuclear SHP2 stimulates transcription of Wnt target genes by dephosphorylating parafibromin.

3.7 In Vivo Oncogenic Activity of CagA

In Mongolian gerbils, gastric adenocarcinoma can occur upon long-term infection with *H. pylori* [57]. The production of CagA and T4SS potentiates colonization of *H. pylori*, epithelial cell proliferation, significant atrophy, and mucous gland metaplasia in the stomach body [58]. Moreover, a particular *H. pylori* adapted to host

Mongolian gerbils can rapidly induce the development of gastric dysplasia and adenocarcinoma [50].

Oncogenic potential of CagA in vivo was directly investigated by employing mice transgenically expressing CagA throughout the body [59]. The CagA transgenic mice develop gastric carcinoma, small intestinal carcinoma, or hematological malignancy. However, these neoplastic changes do not occur in transgenic mice expressing a CagA mutant that does not undergo tyrosine phosphorylation. These observations indicate that tyrosine phosphorylation of CagA is critical for the development of malignant neoplasias in CagA transgenic mice.

In zebrafish model, transgenic expression of CagA induced hyperproliferation of intestinal epithelial cells and activates Wnt target genes *cyclinD1*, *axin2*, and zebrafish *c-myc* ortholog *myca* in a tyrosine phosphorylation-independent manner. Long-term expression of CagA causes hyperplasia in a phosphorylation-dependent manner [60]. These in vivo models clearly demonstrate the oncogenic potential of the bacterial protein in metazoans.

3.8 Bacterial EPIYA-Containing Effectors and Mammalian EPIYA-Containing Proteins

It has been reported that several pathogenic bacteria produce the effector proteins that contain multiple EPIYA or EPIYA-like motifs as is the case of *H. pylori* CagA (Table 3.1). These EPIYA effectors include Tarp of *Chlamydia trachomatis*, which causes sexually transmitted diseases and blindness; BepD, BepE, and BepF of *Bartonella henselae*, which causes human diseases such as cat-scratch disease, bacillary angiomatosis, bacillary peliosis hepatis, and neuroretinitis; AnkA of *Anaplasma phagocytophilum*, which causes human granulocytic anaplasmosis; LspA1 and LspA2 of *Haemophilus ducreyi*, which causes the sexually transmitted disease chancroid; and Tir of enteropathogenic *Escherichia coli* (EPEC). These bacterial EPIYA effectors are delivered into host cells by type III or IV secretion system. Inside the cells, these effectors undergo tyrosine phosphorylation at the EPIYA motifs by host cell kinases to acquire the ability to interact with multiple SH2 domain-containing cellular proteins and thereby to perturb host cell signaling [13, 61]. These bacterial EPIYA effectors share no sequence homology with each other except for the EPIYA motifs, indicating that they didn't molecularly evolve from a common ancestor. As the bacterial EPIYA effectors do not undergo tyrosine phosphorylation inside the bacteria and the bacterial proteome does not have SH2 domain-containing proteins, it was reasonable to assume the presence of a mammalian EPIYA-containing protein(s), the function of which is mimicked by the bacterial effectors [61].

A proteome analysis revealed that some mammalian proteins such as Pragmin/Sgk223, p140Cap/SRC kinase signaling inhibitor 1, and Partitioning defective 3 homolog B/PAR3β have the EPIYA or EPIYA-like motifs (Table 3.2)

Table 3.1 Bacterial effectors that contain EPIYA (or like) motifs

Pathogen	Effector protein	EPIYA (or like) motifs
H. pylori	CagA	EPIYA
C. trachomatis	Tarp	ENIYE
B. henselae	Bep	EPLYA
		EVVYA
		TPLYA
		EPLYA
A. phagocytophilum	AnkA	ESIYE
		EDLYA
		ESIYA
		EPIYA
H. ducreyi	LspA	EPIYG
		EPVYA
EPEC	Tir	VNPYA
		EHIYD
		EPIYD

[62]. A pseudokinase Pragmin/Sgk223, which was originally identified as an effector of Rnd2 GTPase, contains a single EPIYA motif at the N-terminal region. Pragmin binds to Rnd2 to stimulate RhoA activity [63]. Pragmin also undergoes tyrosine phosphorylation on the EPIYA motif by SFKs [64]. Tyrosine-phosphorylated Pragmin binds to and sequestrates Csk in the cytoplasm, thereby maintaining SFKs active [64]. A recent study demonstrated that Pragmin acts as a Notch activation complex kinase (NACK) and a Notch transcriptional coactivator and that homozygous loss of *Pragmin* causes embryonic lethality [65]. p140Cap/SRC kinase signaling inhibitor 1 has two EPIYA-like motifs (EPLYA and EGLYA), where it is tyrosine phosphorylated [66]. p140Cap binds to and activates Csk dependently of tyrosine phosphorylation at the EPIYA-like motifs. p140Cap regulates integrin signaling via p140Cap-Csk complex formation [67]. Partitioning defective 3 homologue B/PAR3β has a single EPIYA-like motif (EGLYA). PAR3β was originally identified as a homologue of the cell polarity protein PAR3 [68] and is essential for mammary gland stem cell maintenance [69]. Although the EPIYA-like motif of PAR3β is known to undergo tyrosine phosphorylation, its functional significance has yet to be elucidated [62].

Although each mammalian EPIYA-containing protein specifically interacts with particular SH2 domain-containing protein(s) to control the intracellular molecular network, the bacterial EPIYA effectors promiscuously interact with a number of host proteins. It is therefore suggested that bacterial EPIYA effectors act as "a master key" that picks the host protein-protein interactions and perturb multiple host signaling pathways involved in a broad range of physiological functions. In this regard, *H. pylori* CagA is the most investigated and archetypal bacterial EPIYA

Table 3.2 Mammalian proteins that contain EPIYA (or like) motifs

EPIYA (or like) motifs	Protein name
EPIYA	Pragmin/Sgk223
	General transcription factor TFIIE, α subunit
	Solute carrier family 2
	Transmembrane protein 218
	Coiled-coil domain-containing protein 146
EGLYA	p140Cap/SRC kinase signaling inhibitor 1
	Oxidative stress-induced growth inhibitor 1
	Short transient receptor potential channel 3, 6, 7
	Zinc finger SWIM domain-containing protein 4
	Zinc finger SWIM domain-containing protein 5
	Partitioning defective 3 homologue B/PAR3β
	Citron Rho-interacting kinase
	Filamin B
	Beta/gamma crystallin domain-containing protein 3
	Striated muscle preferentially expressed protein kinase
	WD repeat and FYVE domain-containing protein 3
	Malonyl-CoA-acyl carrier protein transacylase, mitochondrial
EPLYA	p140Cap/SRC kinase signaling inhibitor 1
	Putative fidgetin-like protein 2
ESIYE	Retinal dehydrogenase 2
	Carcinoembryonic antigen-related cell adhesion molecule 20
	Trafficking protein particle complex subunit 8
EDLYA	Myotubularin-related protein 6
	Myotubularin-related protein 8
	Early endosome antigen 1
	Pre-mRNA-processing-splicing factor 8
	Piezo-type mechanosensitive ion channel component 2
ESIYA	Solute carrier family 25 member 38
EHIYD	Protein unc-119 homologue A
	Occludin
	Clathrin interactor 1
EPIYD	Nuclear receptor ROR-beta
	CUB and sushi domain-containing protein 3
ENIYE	Torsin-1A
	Zinc finger protein 674
	Zinc finger protein 699
	Zinc finger protein 568
	Sodium/hydrogen exchanger 9
	MICAL-like protein 2
	Putative ATP-dependent RNA helicase DHX30
	Testis- and ovary-specific PAZ domain-containing protein 1

(continued)

Table 3.2 (continued)

EPIYA (or like) motifs	Protein name
EPVYA	Palmdelphin
	Tripartite motif-containing protein 1b Tripartite motif-containing protein 1b-like protein
	Rho GTPase-activating protein 27 isoform
EVVYA	Probable ubiquitin carboxyl-terminal hydrolase FAF-Y
TPLYA	39S ribosomal protein L3, mitochondrial
	Zinc finger MIZ domain-containing protein 2
	Bile salt export pump
	Xin actin-binding repeat-containing protein 2
	Zinc finger protein 469

effector [34]. Possibly, pathogenic bacteria exploit the EPIYA effectors to achieve a successful colonization in the host.

3.9 Conclusion

The chronic infection with *H. pylori cagA*-positive strains is the strongest risk factor of gastric cancer. *H. pylori* delivers CagA into host gastric epithelial cells via T4SS. Upon delivery, CagA promiscuously interacts with a number of host cell proteins in both phosphorylation-dependent and phosphorylation-independent manners to subvert physiological cell functions. Carcinogenesis requires two major events. One is the activation of oncoprotein and the other is the inactivation of tumor suppressor. *H. pylori* CagA interacts with both of them and successfully disturbs their functions. Human gastric organoid (hGO) method has recently been reported [70]. This newly developed experimental approach may provide additional insights into our understanding of gastric carcinogenesis mediated by *H. pylori* CagA.

References

1. Torre LA, Bray F, Siegel RL, Ferlay J, Lortet-Tieulent J, Jemal A. Global cancer statistics, 2012. CA Cancer J Clin. 2015;65(2):87–108. doi:10.3322/caac.21262.
2. Shanks AM, El-Omar EM. *Helicobacter pylori* infection, host genetics and gastric cancer. J Dig Dis. 2009;10(3):157–64. doi:10.1111/j.1751-2980.2009.00380.x.
3. Odenbreit S, Puls J, Sedlmaier B, Gerland E, Fischer W, Haas R. Translocation of *Helicobacter pylori* CagA into gastric epithelial cells by type IV secretion. Science. 2000;287(5457):1497–500.

4. Rohde M, Püls J, Buhrdorf R, Fischer W, Haas R. A novel sheathed surface organelle of the *Helicobacter pylori cag* type IV secretion system. Mol Microbiol. 2003;49(1):219–34. doi:10.1046/j.1365-2958.2003.03549.x.
5. Fischer W. Assembly and molecular mode of action of the *Helicobacter pylori* Cag type IV secretion apparatus. FEBS J. 2011;278(8):1203–12. doi:10.1111/j.1742-4658.2011.08036.x.
6. Ekström AM, Held M, Hansson L-E, Engstrand L, Nyrén O. *Helicobacter pylori* in gastric cancer established by CagA immunoblot as a marker of past infection. Gastroenterology. 2001;121(4):784–91. doi:10.1053/gast.2001.27999.
7. Hatakeyama M. Oncogenic mechanisms of the *Helicobacter pylori* CagA protein. Nat Rev Cancer. 2004;4(9):688–94. doi:10.1038/nrc1433.
8. Parsonnet J, Friedman GD, Orentreich N, Vogelman H. Risk for gastric cancer in people with CagA positive or CagA negative *Helicobacter pylori* infection. Gut. 1997;40(3):297–301.
9. Daleke DL, Lyles JV. Identification and purification of aminophospholipid flippases. Biochim Biophys Acta. 2000;1486(1):108–27.
10. Murata-Kamiya N, Kikuchi K, Hayashi T, Higashi H, Hatakeyama M. *Helicobacter pylori* exploits host membrane phosphatidylserine for delivery, localization, and pathophysiological action of the CagA oncoprotein. Cell Host Microbe. 2010;7(5):399–411. doi:10.1016/j.chom.2010.04.005.
11. Backert S, Feller SM, Wessler S. Emerging roles of Abl family tyrosine kinases in microbial pathogenesis. Trends Biochem Sci. 2008;33(2):80–90. doi:10.1016/j.tibs.2007.10.006.
12. Mueller D, Tegtmeyer N, Brandt S, Yamaoka Y, De Poire E, Sgouras D, et al. c-Src and c-Abl kinases control hierarchic phosphorylation and function of the CagA effector protein in Western and East Asian *Helicobacter pylori* strains. J Clin Invest. 2012;122(4):1553–66. doi:10.1172/JCI61143.
13. Hatakeyama M. *Helicobacter pylori* CagA and gastric cancer: a paradigm for hit-and-run carcinogenesis. Cell Host Microbe. 2014;15(3):306–16. doi:10.1016/j.chom.2014.02.008.
14. Hayashi T, Senda M, Morohashi H, Higashi H, Horio M, Kashiba Y, et al. Tertiary structure-function analysis reveals the pathogenic signaling potentiation mechanism of *Helicobacter pylori* oncogenic effector CagA. Cell Host Microbe. 2012;12(1):20–33. doi:10.1016/j.chom.2012.05.010.
15. Kaplan-Turkoz B, Jimenez-Soto LF, Dian C, Ertl C, Remaut H, Louche A, et al. Structural insights into *Helicobacter pylori* oncoprotein CagA interaction with β1 integrin. Proc Natl Acad Sci USA. 2012;109(36):14640–5. doi:10.1073/pnas.1206098109.
16. Buti L, Spooner E, Van der Veen AG, Rappuoli R, Covacci A, Ploegh HL. *Helicobacter pylori* cytotoxin-associated gene A (CagA) subverts the apoptosis-stimulating protein of p53 (ASPP2) tumor suppressor pathway of the host. Proc Natl Acad Sci USA. 2011;108(22):9238–43. doi:10.1073/pnas.1106200108.
17. Nesic D, Buti L, Lu X, Stebbins CE. Structure of the *Helicobacter pylori* CagA oncoprotein bound to the human tumor suppressor ASPP2. Proc Natl Acad Sci USA. 2014;111(4):1562–7. doi:10.1073/pnas.1320631111.
18. Tsang YH, Lamb A, Romero-Gallo J, Huang B, Ito K, Peek Jr RM, et al. *Helicobacter pylori* CagA targets gastric tumor suppressor RUNX3 for proteasome-mediated degradation. Oncogene. 2010;29(41):5643–50. doi:10.1038/onc.2010.304.
19. Hatakeyama M. Anthropological and clinical implications for the structural diversity of the *Helicobacter pylori* CagA oncoprotein. Cancer Sci. 2011;102(1):36–43. doi:10.1111/j.1349-7006.2010.01743.x.
20. Higashi H, Tsutsumi R, Fujita A, Yamazaki S, Asaka M, Azuma T, et al. Biological activity of the *Helicobacter pylori* virulence factor CagA is determined by variation in the tyrosine phosphorylation sites. Proc Natl Acad Sci USA. 2002;99(22):14428–33. doi:10.1073/pnas.222375399.
21. Ren S, Higashi H, Lu H, Azuma T, Hatakeyama M. Structural basis and functional consequence of *Helicobacter pylori* CagA multimerization in cells. J Biol Chem. 2006;281(43):32344–52. doi:10.1074/jbc.M606172200.

22. Lu HS, Saito Y, Umeda M, Murata-Kamiya N, Zhang HM, Higashi H, et al. Structural and functional diversity in the PAR1b/MARK2-binding region of *Helicobacter pylori* CagA. Cancer Sci. 2008;99(10):2004–11. doi:10.1111/j.1349-7006.2008.00950.x.
23. Muller A. Multistep activation of the *Helicobacter pylori* effector CagA. J Clin Invest. 2012;122(4):1192–5. doi:10.1172/JCI61578.
24. Higashi H, Nakaya A, Tsutsumi R, Yokoyama K, Fujii Y, Ishikawa S, et al. *Helicobacter pylori* CagA induces Ras-independent morphogenetic response through SHP-2 recruitment and activation. J Biol Chem. 2004;279(17):17205–16. doi:10.1074/jbc.M309964200.
25. Higashi H, Tsutsumi R, Muto S, Sugiyama T, Azuma T, Asaka M, et al. SHP-2 tyrosine phosphatase as an intracellular target of *Helicobacter pylori* CagA protein. Science. 2002;295 (5555):683–6. doi:10.1126/science.1067147.
26. Chan RJ, Feng GS. *PTPN11* is the first identified proto-oncogene that encodes a tyrosine phosphatase. Blood. 2007;109(3):862–7. doi:10.1182/blood-2006-07-028829.
27. Mohi MG, Neel BG. The role of Shp2 (*PTPN11*) in cancer. Curr Opin Genet Dev. 2007;17 (1):23–30. doi:10.1016/j.gde.2006.12.011.
28. Tsutsumi R, Takahashi A, Azuma T, Higashi H, Hatakeyama M. Focal adhesion kinase is a substrate and downstream effector of SHP-2 complexed with *Helicobacter pylori* CagA. Mol Cell Biol. 2006;26(1):261–76. doi:10.1128/MCB.26.1.261-276.2006.
29. Parsons JT. Focal adhesion kinase: the first ten years. J Cell Sci. 2003;116(8):1409–16. doi:10.1242/jcs.00373.
30. Segal ED, Cha J, Lo J, Falkow S, Tompkins LS. Altered states: involvement of phosphorylated CagA in the induction of host cellular growth changes by *Helicobacter pylori*. Proc Natl Acad Sci USA. 1999;96(25):14559–64.
31. Xia Y, Yamaoka Y, Zhu Q, Matha I, Gao X. A comprehensive sequence and disease correlation analyses for the C-terminal region of CagA protein of *Helicobacter pylori*. PLoS One. 2009;4(11):e7736. doi:10.1371/journal.pone.0007736.
32. Tsutsumi R, Higashi H, Higuchi M, Okada M, Hatakeyama M. Attenuation of *Helicobacter pylori* CagA × SHP-2 signaling by interaction between CagA and C-terminal Src kinase. J Biol Chem. 2003;278(6):3664–70. doi:10.1074/jbc.M208155200.
33. Okada M. Regulation of the SRC family kinases by Csk. Int J Biol Sci. 2012;8(10):1385–97. doi:10.7150/ijbs.5141.
34. Backert S, Tegtmeyer N, Selbach M. The versatility of *Helicobacter pylori* CagA effector protein functions: the master key hypothesis. Helicobacter. 2010;15(3):163–76. doi:10.1111/j.1523-5378.2010.00759.x.
35. Mimuro H, Suzuki T, Tanaka J, Asahi M, Haas R, Sasakawa C. Grb2 is a key mediator of *Helicobacter pylori* CagA protein activities. Mol Cell. 2002;10(4):745–55.
36. Saadat I, Higashi H, Obuse C, Umeda M, Murata-Kamiya N, Saito Y, et al. *Helicobacter pylori* CagA targets PAR1/MARK kinase to disrupt epithelial cell polarity. Nature. 2007;447 (7142):330–3. doi:10.1038/nature05765.
37. Suzuki A, Ohno S. The PAR-aPKC system: lessons in polarity. J Cell Sci. 2006;119 (Pt 6):979–87. doi:10.1242/jcs.02898.
38. Lu H, Murata-Kamiya N, Saito Y, Hatakeyama M. Role of partitioning-defective 1/microtubule affinity-regulating kinases in the morphogenetic activity of *Helicobacter pylori* CagA. J Biol Chem. 2009;284(34):23024–36. doi:10.1074/jbc.M109.001008.
39. Nagase L, Murata-Kamiya N, Hatakeyama M. Potentiation of *Helicobacter pylori* CagA protein virulence through homodimerization. J Biol Chem. 2011;286(38):33622–31. doi:10.1074/jbc.M111.258673.
40. Drewes G, Ebneth A, Preuss U, Mandelkow EM, Mandelkow E. MARK, a novel family of protein kinases that phosphorylate microtubule-associated proteins and trigger microtubule disruption. Cell. 1997;89(2):297–308.
41. Umeda M, Murata-Kamiya N, Saito Y, Ohba Y, Takahashi M, Hatakeyama M. *Helicobacter pylori* CagA causes mitotic impairment and induces chromosomal instability. J Biol Chem. 2009;284(33):22166–72. doi:10.1074/jbc.M109.035766.

42. Saito Y, Murata-Kamiya N, Hirayama T, Ohba Y, Hatakeyama M. Conversion of *Helicobacter pylori* CagA from senescence inducer to oncogenic driver through polarity-dependent regulation of p21. J Exp Med. 2010;207(10):2157–74. doi:10.1084/jem.20100602.
43. Amieva MR, Vogelmann R, Covacci A, Tompkins LS, Nelson WJ, Falkow S. Disruption of the epithelial apical-junctional complex by *Helicobacter pylori* CagA. Science. 2003;300 (5624):1430–4. doi:10.1126/science.1081919.
44. Yamahashi Y, Saito Y, Murata-Kamiya N, Hatakeyama M. Polarity-regulating kinase partitioning-defective 1b (PAR1b) phosphorylates guanine nucleotide exchange factor H1 (GEF-H1) to regulate RhoA-dependent actin cytoskeletal reorganization. J Biol Chem. 2011;286(52):44576–84. doi:10.1074/jbc.M111.267021.
45. Bagnoli F, Buti L, Tompkins L, Covacci A, Amieva MR. *Helicobacter pylori* CagA induces a transition from polarized to invasive phenotypes in MDCK cells. Proc Natl Acad Sci USA. 2005;102(45):16339–44. doi:10.1073/pnas.0502598102.
46. Lee DG, Kim HS, Lee YS, Kim S, Cha SY, Ota I, et al. *Helicobacter pylori* CagA promotes Snail-mediated epithelial-mesenchymal transition by reducing GSK-3 activity. Nat Commun. 2014;5:4423. doi:10.1038/ncomms5423.
47. Wang J, Wynshaw-Boris A. The canonical Wnt pathway in early mammalian embryogenesis and stem cell maintenance/differentiation. Curr Opin Genet Dev. 2004;14(5):533–9. doi:10.1016/j.gde.2004.07.013.
48. Takeichi M. Cadherin cell adhesion receptors as a morphogenetic regulator. Science. 1991;251 (5000):1451–5.
49. Murata-Kamiya N. Pathophysiological functions of the CagA oncoprotein during infection by *Helicobacter pylori*. Microbes Infect. 2011;13(10):799–807. doi:10.1016/j.micinf.2011.03.011.
50. Franco AT, Israel DA, Washington MK, Krishna U, Fox JG, Rogers AB, et al. Activation of β-catenin by carcinogenic *Helicobacter pylori*. Proc Natl Acad Sci USA. 2005;102 (30):10646–51. doi:10.1073/pnas.0504927102.
51. Murata-Kamiya N, Kurashima Y, Teishikata Y, Yamahashi Y, Saito Y, Higashi H, et al. *Helicobacter pylori* CagA interacts with E-cadherin and deregulates the beta-catenin signal that promotes intestinal transdifferentiation in gastric epithelial cells. Oncogene. 2007;26(32):4617–26. doi:10.1038/sj.onc.1210251.
52. Kurashima Y, Murata-Kamiya N, Kikuchi K, Higashi H, Azuma T, Kondo S, et al. Deregulation of β-catenin signal by *Helicobacter pylori* CagA requires the CagA-multimerization sequence. Int J Cancer. 2008;122(4):823–31. doi:10.1002/ijc.23190.
53. Fujii Y, Yoshihashi K, Suzuki H, Tsutsumi S, Mutoh H, Maeda S, et al. CDX1 confers intestinal phenotype on gastric epithelial cells via induction of stemness-associated reprogramming factors SALL4 and KLF5. Proc Natl Acad Sci USA. 2012;109(50):20584–9. doi:10.1073/pnas.1208651109.
54. Clevers H, Nusse R. Wnt/β-catenin signaling and disease. Cell. 2012;149(6):1192–205. doi:10.1016/j.cell.2012.05.012.
55. Takahashi A, Tsutsumi R, Kikuchi I, Obuse C, Saito Y, Seidi A, et al. SHP2 tyrosine phosphatase converts parafibromin/Cdc73 from a tumor suppressor to an oncogenic driver. Mol Cell. 2011;43(1):45–56. doi:10.1016/j.molcel.2011.05.014.
56. Tsutsumi R, Masoudi M, Takahashi A, Fujii Y, Hayashi T, Kikuchi I, et al. YAP and TAZ, Hippo signaling targets, act as a rheostat for nuclear SHP2 function. Dev Cell. 2013;26 (6):658–65. doi:10.1016/j.devcel.2013.08.013.
57. Watanabe T, Tada M, Nagai H, Sasaki S, Nakao M. *Helicobacter pylori* infection induces gastric cancer in mongolian gerbils. Gastroenterology. 1998;115(3):642–8.
58. Rieder G, Merchant JL, Haas R. *Helicobacter pylori cag*-Type IV secretion system facilitates corpus colonization to induce precancerous conditions in mongolian gerbils. Gastroenterology. 2005;128(5):1229–42. doi:10.1053/j.gastro.2005.02.064.

59. Ohnishi N, Yuasa H, Tanaka S, Sawa H, Miura M, Matsui A, et al. Transgenic expression of *Helicobacter pylori* CagA induces gastrointestinal and hematopoietic neoplasms in mouse. Proc Natl Acad Sci USA. 2008;105(3):1003–8. doi:10.1073/pnas.0711183105.
60. Neal JT, Peterson TS, Kent ML, Guillemin K. *H. pylori* virulence factor CagA increases intestinal cell proliferation by Wnt pathway activation in a transgenic zebrafish model. Dis Model Mech. 2013;6(3):802–10. doi:10.1242/dmm.011163.
61. Hayashi T, Morohashi H, Hatakeyama M. Bacterial EPIYA effectors – where do they come from? What are they? Where are they going? Cell Microbiol. 2013;15(3):377–85. doi:10.1111/cmi.12040.
62. Safari F. EPIYA (or -like) motifs in mammalian proteins. J King Saud Univ Sci. 2014;26(4):276–84. doi:10.1016/j.jksus.2014.05.001.
63. Tanaka H, Katoh H, Negishi M. Pragmin, a novel effector of Rnd2 GTPase, stimulates RhoA activity. J Biol Chem. 2006;281(15):10355–64. doi:10.1074/jbc.M511314200.
64. Safari F, Murata-Kamiya N, Saito Y, Hatakeyama M. Mammalian Pragmin regulates Src family kinases via the Glu-Pro-Ile-Tyr-Ala (EPIYA) motif that is exploited by bacterial effectors. Proc Natl Acad Sci USA. 2011;108(36):14938–43. doi:10.1073/pnas.1107740108.
65. Weaver KL, Alves-Guerra MC, Jin K, Wang Z, Han X, Ranganathan P, et al. NACK is an integral component of the Notch transcriptional activation complex and is critical for development and tumorigenesis. Cancer Res. 2014;74(17):4741–51. doi:10.1158/0008-5472.CAN-14-1547.
66. Repetto D, Aramu S, Boeri Erba E, Sharma N, Grasso S, Russo I, et al. Mapping of p140Cap phosphorylation sites: the EPLYA and EGLYA motifs have a key role in tyrosine phosphorylation and Csk binding, and are substrates of the Abl kinase. PLoS One. 2013;8(1):e54931. doi:10.1371/journal.pone.0054931.
67. Di Stefano P, Damiano L, Cabodi S, Aramu S, Tordella L, Praduroux A, et al. p140Cap protein suppresses tumour cell properties, regulating Csk and Src kinase activity. EMBO J. 2007;26(12):2843–55. doi:10.1038/sj.emboj.7601724.
68. Kohjima M, Noda Y, Takeya R, Saito N, Takeuchi K, Sumimoto H. PAR3β, a novel homologue of the cell polarity protein PAR3, localizes to tight junctions. Biochem Biophys Res Commun. 2002;299(4):641–6.
69. Huo Y, Macara IG. The Par3-like polarity protein Par3L is essential for mammary stem cell maintenance. Nat Cell Biol. 2014;16(6):529–37. doi:10.1038/ncb2969.
70. McCracken KW, Cata EM, Crawford CM, Sinagoga KL, Schumacher M, Rockich BE, et al. Modelling human development and disease in pluripotent stem-cell-derived gastric organoids. Nature. 2014;516(7531):400–4. doi:10.1038/nature13863.

Chapter 4
Helicobacter pylori VacA Exhibits Pleiotropic Actions in Host Cells

Masayuki Nakano, Toshiya Hirayama, Joel Moss, and Kinnosuke Yahiro

Abstract *Helicobacter pylori* vacuolating cytotoxin (VacA) is a major virulence factor, with pleiotropic actions on target cells including induction of vacuole formation, mitochondrial dysfunction leading to apoptosis, modulation of signal transduction pathways associated with autophagy, inhibition of T cell proliferation, and production of inflammatory cytokines. Numerous epidemiological studies have indicated that the allelic diversity within four variable regions of the *vacA* gene might be associated with cell type-specific binding as well as specific clinical outcomes in *H. pylori* infection. VacA binds to receptors such as receptor protein tyrosine phosphatases (RPTPα and RPTPβ), low-density lipoprotein receptor-related protein-1 (LRP1), fibronectin, CD18, and sphingomyelin to facilitate its action, suggesting the involvement of these receptors in the pathogenesis of *H. pylori* infection. RPTPβ contributes to ulceration in gastric epithelial cells and LRP1 is involved in the induction of autophagy. Interestingly, it has been suggested that CagA is degraded by VacA-induced autophagy and that the interaction between these two molecules is associated with the pathogenesis of gastric diseases. Therefore, better understanding of the mechanism of VacA toxicity may provide valuable information regarding appropriate medical care for gastroduodenal diseases caused by *H. pylori* infection.

Keywords *Helicobacter pylori* • VacA • Vacuolation • Autophagy

---

M. Nakano (✉) • T. Hirayama
Department of Bacteriology, Institute of Tropical Medicine, Nagasaki University, 1-12-4, Sakamoto, Nagasaki 852-8523, Japan
e-mail: manakano@nagasaki-u.ac.jp

J. Moss
Cardiovascular and Pulmonary Branch, NHLBI, NIH, Bethesda, MD, USA

K. Yahiro
Department of Molecular Infectiology, Graduate School of Medicine, Chiba University, Chiba, Japan

H. Suzuki et al. (eds.), *Helicobacter pylori*, DOI 10.1007/978-4-431-55705-0_4

4.1 Introduction

Helicobacter pylori is a Gram-negative bacterium that causes gastroduodenal diseases including chronic gastritis, peptic ulcers, gastric adenocarcinoma, and gastric lymphoma. A number of virulence factors are involved in the pathogenesis of *H. pylori* infection. During colonization of the gastric epithelium, *H. pylori* secretes a potent protein toxin, vacuolating cytotoxin, termed VacA, which is considered to be an important virulence factor associated with peptic ulcer disease, indicating that VacA is critical to the pathogenicity of *H. pylori* [1–4]. Although vacuolation, a hallmark of VacA function (Fig. 4.1), is readily observed in VacA-intoxicated cells in vitro, its function in vivo in *H. pylori* infection is unclear.

In 1988, Leunk et al. reported that the broth culture filtrates of *H. pylori* caused the generation of large vacuoles in the cytoplasm of cultured mammalian cells [5]. After Cover et al. purified an 87-kDa VacA from culture supernatant [6], the entire nucleotide sequence of the *vacA* gene and its deduced amino acid sequence were determined, revealing the presence of a protoxin of about 140 kDa that lacked homology to any other protein [7]. Vacuolating activity was enhanced by acidic (<pH 5.5) or alkaline (>pH 9.5) conditions due to conformational changes in the protein [8, 9]. Further, vacuolating activity was also potentiated by weak bases, including ammonium chloride and nicotine [10]. After binding to the cell surface receptors, VacA internalizes into epithelial cells by endocytosis and then induces the formation of vacuoles, which are derived from late endosomes. Maturation of vacuoles involves several host molecules including a small GTPase, Rab7, and vacuolar-type ATPase (V-ATPase) proton pump. Rab7 is essential for the efficient

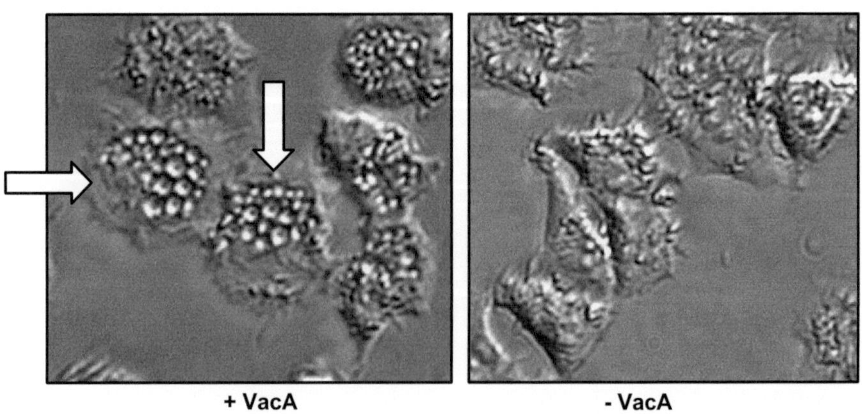

Fig. 4.1 Vacuolation in the cytoplasm of mammalian cells. Purified VacA (*left*) or vehicle (*right*) was added, and cultured mammalian cells were incubated at 37 °C, 5 % CO_2 for several hours. Many vacuoles (*arrows*) were observed in the cytoplasm of cells incubated in the presence of VacA, while vehicle did not cause vacuolation. Thus, VacA is the responsible protein for vacuole formation

membrane flow from early to late endosome, and V-ATPase is involved in osmotic swelling of the endosomal compartment, resulting in generation of VacA-dependent vacuolation [11–13].

VacA also causes mitochondrial dysfunction, leading eventually to apoptotic cell death through cytochrome *c* release and caspase activation [14–16]. Furthermore, VacA also modulates membrane potential and signal transduction pathways associated with immune responses, induction of autophagy, and inhibition of T cell proliferation [17–21]. There is some evidence of an antagonistic relationship between VacA and cytotoxin-associated gene A (CagA), another important *H. pylori* virulence factor [22–24], suggesting that both toxins contribute to the development of gastric disorders in *H. pylori* infection. In this review, we will discuss the multiple functions of VacA on target host cells and also discuss the latest research on VacA function.

4.2 Protein Structure and Domains of VacA

All isolated strains of *H. pylori* contain the *vacA* gene encoding a 140-kDa precursor protein, which has three major domains, i.e., signal sequence, mature toxin, and autotransporter domains. Proteolytic cleavage of the signal sequence and autotransporter domain from a VacA protoxin results in a mature VacA toxin of about 90-kDa molecular mass, which is secreted into the extracellular space by a type V secretion system (Fig. 4.2a) [1, 6, 7, 25]. Secreted mature VacA forms an oligomeric structure, and the assembled VacA has an expected molecular mass of around 1000 kDa under neutral pH or non-denaturing conditions [6]. Upon imaging analysis by electron microscopy, VacA oligomer is composed of six or seven VacA monomers that form a flower-shaped structure with a central ring. VacA oligomerization leads to the assembly of a double-layered form of a single VacA oligomer (consisting of 12 or 14 VacA monomers), rather than the single-layered oligomer [26–29]. VacA oligomers observed in the lipid bilayer of cell or mitochondrial membrane seem to contribute to anion-selective membrane channel formation [26, 30]. The mature VacA is further cleaved into two fragments by proteolytic digestion and consists of 33-kDa N-terminal (termed p33) and 55-kDa C-terminal fragments (termed p55) [1, 27, 31]. It has been proposed that in the three-dimensional structures of VacA oligomers, p55 represents the peripheral arms and p33, the central core of the complexes [29].

Although many studies about the cellular functions of both p33 and p55 have been performed, their functions are still debated. Several lines of evidence have indicated that p55 is involved in binding to host cells, whereas p33 plays a role in the formation of anion channels in the lipid membrane and is targeted to the inner mitochondrial membrane, leading to the loss of membrane potential [14]. In addition, it has been reported that p33 is crucial in pore formation in lipid bilayers [32]. The N-terminal of p33 is predicted to contain a hydrophobic region, which is

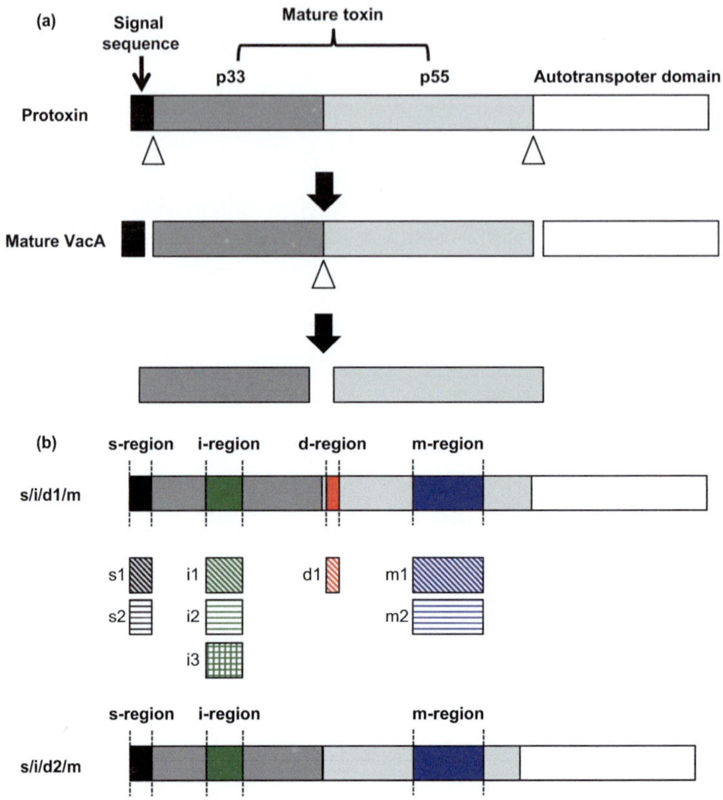

Fig. 4.2 *vacA* gene allelic and protein structures. (**a**) Precursor VacA is synthesized in *H. pylori* with three functional domains (signal sequence, mature toxin, and autotransporter domain). Mature VacA protein, which lacks two fragments (signal sequence and autotransporter domain), is generated by proteolytic digestion and is then secreted by a type V secretion system. The p33 and p55 fragments might be generated from the mature form of VacA. *Arrowheads* indicate cleavage sites. (**b**) The *vacA* gene as four variable regions: s-region (*black*, s1 and s2), i-region (*green*, i1, i2, and i3), d-region (*red*, d1), and m-region (*blue*, m1 and m2). The *vacA* gene allele d2 is missing the d1 region (*lower figure*)

required for not only channel formation but also vacuolating activity and trafficking to the inner mitochondrial membrane [33, 34]. Furthermore, the proline residue at position 9 and glycine residue at position 14 in p33 play key roles in these activities [35]. In addition, GXXXG motif at residues 14–18 in p33 is also important for channel formation [34]. However, mutant VacA lacking a part of the hydrophobic region retained channel activity in planar lipid bilayers [32]. Other studies have shown that p33 and p55 are involved in cell binding and channel formation [36, 37]. Thus, further studies to define the molecular basis underlying p33 and p55 activities are required.

4.3 Gene Structure of *vacA*

4.3.1 Polymorphism of the vacA *Gene*

Although all *H. pylori* strains isolated from patients possess a *vacA* gene, only approximately 50 % of the isolates show vacuolating activity because of allelic diversity within the *vacA* gene [2, 6]. Nucleotide sequence analyses of the *vacA* gene have revealed that polymorphisms exist in four variable regions (Fig. 4.2b) [38]. The most characterized allelic diversities are observed within the signal sequence region (s-region) and the mid-region (m-region). The s-region is located at the 5′ end of the *vacA* gene and includes the signal sequence of VacA (designated s1 and s2). Interestingly, epidemiological studies have shown that strains possessing the s1 type of *vacA* are linked to the presence of the *cagA* gene [39, 40]. The s1-type allele is further classified into three subtypes s1a, s1b, and s1c, respectively [41]. Strains possessing the s2 type of *vacA* allele fail to induce vacuolation in cultured mammalian cells [2]. A hydrophilic segment consisting of 12-amino acids in the s2 type, which the s1 type lacks, suppresses VacA-induced vacuolating activity without any effect on its production [42–44]. Two m-regions (classified as m1 and m2) are located within p55; m2 is further divided into two subtypes, m2a and m2b [2, 41]. The entire m-region composed of m1 and m2 alleles is required for vacuolating activity [45, 46]. However, there is a clear difference in cell specificity between m1 and m2 [47], e.g., m2-type VacA, in contrast to m1-type VacA does not induce vacuole formation in HeLa cells irrespective of acid or alkaline activation [48]. De Guzman et al. have indicated that m2-type VacA does not bind to receptor protein tyrosine phosphatase (RPTP) α, one of the VacA-binding receptors, on membranes of HeLa cells, suggesting that posttranslational modifications of RPTPα of HeLa cells may determine sensitivity to m2 VacA [48]. Combinations of *vacA* alleles (s1/m1, s1/m2, and s2/m2) are commonly found in clinically isolated strains of *H. pylori*, but s2/m1 type of *vacA* gene is rare [2, 49].

Two other variable regions in the *vacA* gene have been identified. The intermediate region (i-region) has been identified as a third polymorphism region and is located between s- and m-regions (noted as i1, i2, and i3). Rhead et al. has indicated that the s1/m1 allele strains were predominantly i1 type, whereas all s2/m2 allele strains were i2 type and s1/m2 allele strains were variable in the i-region [50]. In addition, there is some evidence that the i-region affects human T cell functions, e.g., i2 type had a diminished capacity to inhibit the activation of nuclear factor of activated T cells (NFAT) and bound to Jurkat cells less avidly than did i1 type [51]. The fourth polymorphism is observed in the deletion region (d-region), which is located between i- and m-regions (noted as d1 and d2) [52]. It has been demonstrated that there is no deletion in the d1 genotype of the *vacA* gene, but d2 genotype contains a deleted region. However, the effects of the d-regions are still unclear.

4.3.2 *Association Between* vacA *Gene Alleles and Clinical Outcomes*

Many epidemiological studies indicate that a clear functional association between *vacA* gene alleles and clinical outcome of gastroduodenal diseases has not been found. Generally, s1/m1 *vacA* is a more toxic genotype than are s1/m2 and s2 types of the *vacA* alleles [44]. The s1/m1 *vacA* allele is predominantly isolated from patients, suggesting that strains possessing the s1/m1 *vacA* allele are more likely to be associated with gastroduodenal disease than other *vacA* alleles, whereas strains possessing s2 type of *vacA* are rarely associated with disease. The s1 or s1/m1 *vacA* allele is associated with duodenal ulceration and gastric cancer in the United States and Western Europe. On the other hand, most of the clinical isolates possess the s1/m1 *vacA* allele in East Asia including Japan and South Korea, where there is a high incidence of gastric cancer [2, 39, 53–56]. In addition, it has been proposed that i1 genotype is associated with gastric cancer and gastric ulcer in Italian populations, and therefore, the i-region might be a good indicator of the carcinogenic ability of *H. pylori* strains, while any combinations of three variable regions (s-, m-, and i-regions) in the *vacA* gene were not disease determinants in East Asian and Southeast Asian countries [50, 57, 58]. From these observations, relationships between *vacA* genotypes and specific clinical outcomes may be different in different geographic areas.

4.4 VacA Receptors on Target Cells

It is important to identify specific bacterial toxin receptors on the cell surface. These receptors play a critical role in bacterial toxin binding and entry into cells, followed by intoxication. Thus, interaction between toxin and receptors induce unique signal transduction pathways, leading to effects (e.g., morphological changes, cell damage) on the target cell.

The C-terminal domain of VacA is responsible for binding to cell surface receptors [14, 47]. We identified toxin-binding cultured cell compartments by immunoprecipitation analysis with purified VacA and biotin-labeled cultured cell lysate. In VacA-sensitive cells, we found that biotinylated cell surface proteins of 140 kDa (p140) and 250 kDa (p250) were precipitated with VacA but not inactivated VacA [59]. On the other hand, Seto and colleagues reported that the EGF receptor might be involved in VacA endocytosis, leading to induction of vacuolating activity [60]. Since these VacA-binding proteins, p140 and p250, were modified by *N*-linked sugar, we purified these proteins using lectin agarose. By LC-MS/MS analysis, RPTPα and RPTPβ were identified as VacA-binding proteins [9, 61].

De Bernard et al. reported that phorbol 12-myristate 13-acetate (PMA) induces differentiation of HL-60 cells into macrophage-like cells and induced the susceptibility to VacA [62]. It was found that expression of RPTPβ mRNA and protein was significantly induced during differentiation of HL-60 cells by PMA [63]. Consistently, knockdown of RPTPβ by antisense oligonucleotide resulted in reduction of VacA-induced vacuolation in parallel with suppression of RPTPβ expression [63]. In addition, it is known that acid or alkaline treatment induced conformational changes in VacA, which increased binding, internalization, and cytotoxicity [9, 64]. The activated VacA showed increased binding to RPTPβ, suggesting that conformational changes of VacA promote its interaction with cell surface RPTPβ [9]. Fujikawa et al. demonstrated that oral administration of VacA to wild-type mice, but not RPTPβ knockout mice, resulted in gastric ulcers, indicating that RPTPβ is essential for intoxication by VacA in gastric tissue [65]. These results suggest that VacA binding to RPTPβ has functional effects.

With G401 cells, a human kidney tumor cell line that lacks RPTPβ but forms vacuoles in the presence of VacA, we also identified p140 as RPTPα. The finding that silencing of RPTPα gene by antisense oligonucleotide in G401 cells inhibited VacA binding and induction of vacuolation supports the hypothesis that RPTPα mediates VacA intoxication [61].

Other surface factors appear to interact with VacA and serve as cell surface receptors. By surface plasmon resonance-based biosensor studies, it has been proven that VacA binds to heparin sulfate, a component of the extracellular matrix [66]. Hennig et al. demonstrated that VacA binds to fibronectin of HeLa cells, resulting in inhibition of HeLa cell adhesion, suggesting that VacA affects cytoskeleton organization and cell adhesion via interaction with fibronectin [67]. Further, Gupta et al. also reported that VacA-induced vacuolation in HeLa cells is reduced in the presence of sphingomyelinase [68]. Sewald et al. have reported that β2-integrin subunit CD18 plays an important role in VacA uptake into human T lymphocytes [69]. Through this mechanism, VacA has immunosuppressive effects on cells by inhibiting cell growth and interleukin-2 (IL-2) secretion [69]. They also demonstrated that VacA endocytosis is PKC-dependent and clathrin-independent in primary T cells [70]. Recently, we identified a new VacA receptor, low-density lipoprotein receptor-related protein-1 (LRP1), which is essential for VacA-induced autophagy and apoptosis [19].

4.5 VacA Uptake Pathway

After binding to its receptors on target cells, VacA is internalized and found in endocytic vesicles. To enter into epithelial cells, VacA associates with lipid rafts [71, 72] and then internalizes via the Cdc42-dependent pinocytic pathway without requirement for dynamin 2, ADP-ribosylation factor 6, or RhoA GTPase. In this process, VacA was associated with detergent-resistant membranes [73]. Gupta et al. also reported that sphingomyelin is important in VacA uptake and intracellular translocation [74]. Furthermore, there is some evidence that VacA is endocytosed via a GPI-anchored protein-enriched early endosomal compartment (GEEC)-

dependent pathway. Filamentous actin is involved in VacA translocation from GEECs to late endosomes [75]. In primary T lymphocytes, VacA endocytosis is regulated by PKC-, Cdc42- and Rac1-dependent pathways [70].

4.6 Biological Activities of VacA

4.6.1 Vacuolation and Autophagy

The vacuoles induced by interfering with intracellular membrane fusion in VacA-intoxicated cells contain markers of a pre-lysosomal compartment, including Rab7, Lgp110, and LAMP1 [76–78]. Further, various host factors, including dynamin, Rac1 and PIKfyve, are involved in VacA-induced vacuole formation [79–82]. The facts that VacA-induced vacuolation is suppressed by a V-ATPase inhibitor, bafilomycin A1, and V-ATPase colocalized with Rab7 in VacA-induced vacuoles suggest that a pH gradient generated through the activity of V-ATPase proton pump is required for this process (Fig. 4.3a) [83, 84].

Recent studies have reported that VacA induces autophagy, which is different from the large vacuoles formed in VacA-intoxicated cells [85–87]. VacA-induced autophagosomes and autophagolysosomes are dependent on functions of Atg family proteins (e.g., Atg5, Atg12, Atg16L1), with LC3-positive vesicles observed by confocal microscopy [85]. The internalized VacA partially colocalized with LC3 and LRP1 but not mitochondria. Additionally, the channel activity of VacA was also required for autophagosome formation [19, 85]. These findings indicate that VacA associates with autophagosomes and autophagolysosomes. Tsugawa et al. recently demonstrated that CagA is degraded by VacA-induced autophagy, which is regulated by a reactive oxygen species (ROS)-Akt-p53 signaling pathway (Fig. 4.3b) [24]. Interestingly, they found that CD44v9-expressing gastric cancer stem-like cells accumulated CagA by inhibition of VacA-induced autophagy through a mechanism involving cellular increase of glutathione due to stabilization of cystine transporter. A recent study showed that *H. pylori* have a strategy to avoid clearance by autophagy, i.e., *H. pylori* infection increased miR30BA, which directly downregulates autophagy regulatory proteins, Atg12 and Beclin1 [88]. Thus, *H. pylori* use several mechanisms to escape host autophagy, and then the persistent pathogen can produced various virulence factors, which trigger the pathological processes involved in diseases.

4.6.2 Apoptosis

VacA also induces apoptosis of various cells [11, 89]. It has been suggested that p33 N-terminal fragment of VacA has a unique sequence, which translocates into

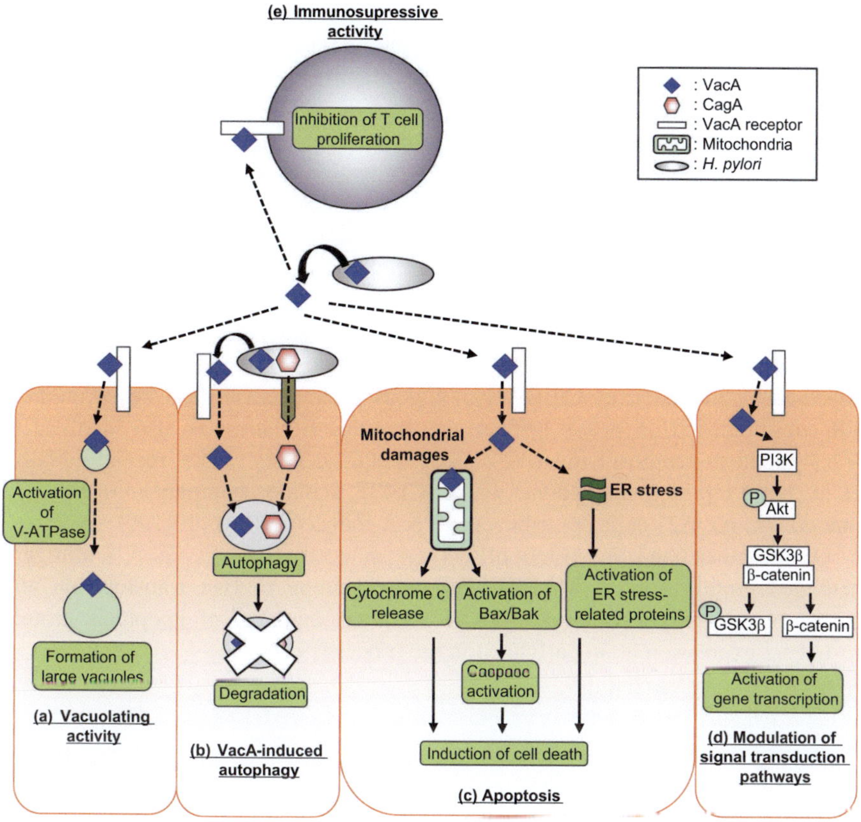

Fig. 4.3 Pleiotropic functions of VacA. During *H. pylori* infection, VacA is internalized into target host cells via its receptors. (**a**) VacA generates the formation of large vacuoles in the cytoplasm by activation of V-ATPase. (**b**) After translocation of CagA into gastric epithelial cells, CagA is degraded by VacA-induced autophagy. (**c**) VacA localizes to the mitochondria and modulates the mitochondrial functions leading to cytochrome *c* release or caspase activation, resulting in apoptosis. In addition, ER stress is also involved in VacA-induced apoptosis. (**d**) VacA modulates signal transduction pathways, e.g., VacA activates PI3K/Akt signaling pathway, resulting in phosphorylation of GSK3β, associated with induction of β-catenin release from the GSK3β/β-catenin complex. Resulting β-catenin can translocate from cytoplasm to nucleus to induce expression of genes such as cyclin D1. (**e**) VacA binds to CD18 and modulates signal transduction pathways, resulting in inhibition of T cell proliferation

mitochondria to facilitate intoxication [90]. Apoptosis induced by VacA may proceed by two mechanisms (Fig. 4.3c). After internalization into cytosol, VacA directly moves to mitochondria, followed by modulation of mitochondrial membrane permeability by its channel activity, leading to cytochrome *c* release and apoptosis [14, 91–93]. Another pathway involves the proapoptotic Bcl-2 family protein in VacA-induced apoptosis, i.e., VacA causes caspase activation and PARP cleavage, subsequent to activation of Bax/Bak on the mitochondria [16, 94, 95]. Calore et al. have demonstrated that VacA-induced apoptosis requires Bax/

Bak-dependent juxtaposition of endosomes and mitochondria and accumulation of VacA and Bax in mitochondria [96]. These findings suggest that movements of both VacA and Bax/Bak into mitochondria are important in the apoptotic pathway. Matsumoto et al. reported that VacA downregulates STAT3 expression, followed by reduction of the amounts of Bcl-2 and Bcl-xL, which are anti-apoptotic proteins [97]. Jain et al. reported that VacA-induced activation of dynamin-related protein 1 (Drp1), a regulator of mitochondrial fission, is critical to apoptosis and that activation of Drp1-dependent mitochondrial fission requires channel activity of VacA, but not Bax activation, suggesting that cross talk between Drp1 and Bax may be unidirectional in VacA-treated cells [98].

Recently, it was demonstrated that VacA activates PERK, an endoplasmic reticulum (ER) stress sensor protein, and stimulates phosphorylation of eIF2α, followed by induction of C/EBP homologous protein (CHOP), a key protein of ER stress-induced apoptosis [99], suggesting that ER stress is also involved in VacA-induced apoptosis in AZ-521 and dendritic cells. More recently, Radin et al. have reported that connexin 43 (Cx43), a major component of the gap junction, in AZ-521 cells contributes to VacA-induced cell death [100].

On the other hand, in eosinophils, VacA induced apoptosis via p38 mitogen-activated protein kinase (MAPK) activation, leading to Bax translocation and cytochrome *c* release, although VacA promotes inhibitor of apoptosis protein (c-IAP)-2 expression at an early step in the process [94].

4.6.3 Immunosuppressive Activity of VacA

VacA might contribute to chronic infection of *H. pylori* in the stomach by preventing protective immunity [51, 94, 101]. Molinari et al. have reported that VacA inhibits the Ii-dependent process of antigen presentation through MHC class II [102]. Other groups also indicated that VacA could not only interfere with T cell activation by preventing NFAT activity, resulting in suppression of IL-2 expression, but also activate MAPK kinase (MKK) 3/6 and Vav/Rac1, leading to massive actin reorganization in T cells (Fig. 4.3e) [103, 104]. In addition, Kim et al. reported that VacA treatment inhibits LPS-induced dendritic cell maturation, suggesting that VacA negatively regulates dendritic cell maturation through the restoration of E2F1 [105].

ROS also plays important roles in host defense [106]. In macrophages, VacA inhibits ROS production by interfering with expression of integrin-linked kinase (ILK), followed by suppression of endothelial nitric oxygen synthase (eNOS), resulting in increased survival of VacA-positive *H. pylori* [107].

4.6.4 Cell Signal Transduction by VacA

VacA has diverse roles in signal transduction including proliferation, apoptosis, and host defense [18, 108]. Nakayama et al. reported that in AZ-521 cells, VacA activates p38 MAPK/activating transcription factor 2 (ATF-2) signaling pathway, which requires interaction with an unknown GPI-anchored membrane protein in lipid rafts [72]. Furthermore, VacA induced phosphoinositide-3-kinase activation, leading to phosphorylation of Akt and glycogen synthase kinase-3β (GSK3β). This signal transduction pathway interfered with β-catenin translocation from cytoplasm to nucleus by inhibiting β-catenin release from GSK3β/β-catenin complex (Fig. 4.3d) [109]. Another study showed that VacA treatment activates prostaglandin E2 production through induction of cyclooxygenase 2 (COX-2) expression via the p38 MAPK/ATF-2 pathway, leading to stimulation of cis-acting replication element (CRE) site in the COX-2 promoter [110].

In T cells and monocytic cell line U937, VacA enhanced NF-κB activation [111, 112]. In U937 cells, VacA induced IL-8 production via activation of p38 MAPK through intracellular Ca^{2+} release, leading to stimulation of IL-8 promoter activation by binding of transcription factors, ATF-2, CRE-binding protein (CREB), and NF-κB [112]. These results indicate that *H. pylori* VacA is a pleiotropic toxin, able to affect various signaling transduction pathways and disrupt cell homeostasis.

4.7 Conclusion

VacA has pleiotropic actions in vitro and is considered to be one of the most important virulence factors of *H. pylori*. In addition to the findings related to VacA toxicity, it is interesting to note that the biological activities of VacA and the prevalence of some types of gastroduodenal disease caused by *H. pylori* infection are strongly associated with specific *vacA* gene alleles, suggesting that the *vacA* gene allele is a putative genetic biomarker of *H. pylori*-induced gastric diseases. Although many studies have demonstrated more clearly the importance of VacA in *H. pylori* infection, current investigations are focused on understanding the mechanisms by which VacA causes the pathogenesis of gastroduodenal diseases. Therefore, research on VacA may lead to a better understanding of its role in the development of gastric disorders and modulation of host immune defense in *H. pylori* infection. Thus, further studies may provide new insights into the role of VacA as a factor that promotes gastric diseases.

Acknowledgments MN was supported by Takeda Science Foundation. TH and KY were supported by the Cooperative Research Grant of the Institute of Tropical Medicine, Nagasaki University, 2012 and 2014. JM was supported by the Intramural Research Program, National Institutes of Health, and National Heart, Lung, and Blood Institute. KY was also supported by Grants-in-Aid for Scientific Research from the Ministry of Education, Culture, Sports, Science and

Technology of Japan and Improvement of Research Environment for Young Researchers from the Japan Science and Technology Agency.

References

1. Telford JL, Ghiara P, Dell'Orco M, Comanducci M, Burroni D, Bugnoli M, et al. Gene structure of the *Helicobacter pylori* cytotoxin and evidence of its key role in gastric disease. J Exp Med. 1994;179:1653–8.
2. Atherton JC, Cao P, Peek Jr RM, Tummuru MK, Blaser MJ, Cover TL. Mosaicism in vacuolating cytotoxin alleles of *Helicobacter pylori*. Association of specific *vacA* types with cytotoxin production and peptic ulceration. J Biol Chem. 1995;270:17771–7.
3. Tee W, Lambert JR, Dwyer B. Cytotoxin production by *Helicobacter pylori* from patients with upper gastrointestinal tract diseases. J Clin Microbiol. 1995;33:1203–5.
4. Memon AA, Hussein NR, Miendje Deyi VY, Burette A, Atherton JC. Vacuolating cytotoxin genotypes are strong markers of gastric cancer and duodenal ulcer-associated *Helicobacter pylori* strains: a matched case–control study. J Clin Microbiol. 2014;52:2984–9.
5. Leunk RD, Johnson PT, David BC, Kraft WG, Morgan DR. Cytotoxic activity in broth-culture filtrates of *Campylobacter pylori*. J Med Microbiol. 1988;26:93–9.
6. Cover TL, Blaser MJ. Purification and characterization of the vacuolating toxin from *Helicobacter pylori*. J Biol Chem. 1992;267:10570–5.
7. Cover TL, Tummuru MK, Cao P, Thompson SA, Blaser MJ. Divergence of genetic sequences for the vacuolating cytotoxin among *Helicobacter pylori* strains. J Biol Chem. 1994;269:10566–73.
8. de Bernard M, Papini E, de Filippis V, Gottardi E, Telford J, Manetti R, et al. Low pH activates the vacuolating toxin of *Helicobacter pylori*, which becomes acid and pepsin resistant. J Biol Chem. 1995;270:23937–40.
9. Yahiro K, Niidome T, Kimura M, Hatakeyama T, Aoyagi H, Kurazono H, et al. Activation of *Helicobacter pylori* VacA toxin by alkaline or acid conditions increases its binding to a 250-kDa receptor protein-tyrosine phosphatase beta. J Biol Chem. 1999;274:36693–9.
10. Cover TL, Vaughn SG, Cao P, Blaser MJ. Potentiation of *Helicobacter pylori* vacuolating toxin activity by nicotine and other weak bases. J Infect Dis. 1992;166:1073–8.
11. Rassow J. *Helicobacter pylori* vacuolating toxin A and apoptosis. Cell Commun Signal. 2011;9:26.
12. Papini E, Satin B, Bucc C, de Bernard M, Telford JL, Manetti R, et al. The small GTP binding protein rab7 is essential for cellular vacuolation induced by *Helicobacter pylori* cytotoxin. EMBO J. 1997;16:15–24.
13. Genisset C, Puhar A, Calore F, de Bernard M, Dell'Antone P, Montecucco C. The concerted action of the *Helicobacter pylori* cytotoxin VacA and of the v-ATPase proton pump induces swelling of isolated endosomes. Cell Microbiol. 2007;9:1481–90.
14. Galmiche A, Rassow J, Doye A, Cagnol S, Chambard JC, Contamin S, et al. The N-terminal 34 kDa fragment of *Helicobacter pylori* vacuolating cytotoxin targets mitochondria and induces cytochrome *c* release. EMBO J. 2000;19:6361–70.
15. Kimura M, Goto S, Wada A, Yahiro K, Niidome T, Hatakeyama T, et al. Vacuolating cytotoxin purified from *Helicobacter pylori* causes mitochondrial damage in human gastric cells. Microb Pathog. 1999;26:45–52.
16. Yamasaki E, Wada A, Kumatori A, Nakagawa I, Funao J, Nakayama M, et al. *Helicobacter pylori* vacuolating cytotoxin induces activation of the proapoptotic proteins Bax and Bak, leading to cytochrome *c* release and cell death, independent of vacuolation. J Biol Chem. 2006;281:11250–9.

17. Cover TL, Blanke SR. *Helicobacter pylori* VacA, a paradigm for toxin multifunctionality. Nat Rev Microbiol. 2005;3:320–32.
18. Isomoto H, Moss J, Hirayama T. Pleiotropic actions of *Helicobacter pylori* vacuolating cytotoxin, VacA. Tohoku J Exp Med. 2010;220:3–14.
19. Yahiro K, Satoh M, Nakano M, Hisatsune J, Isomoto H, Sap J, et al. Low-density lipoprotein receptor-related protein-1 (LRP1) mediates autophagy and apoptosis caused by *Helicobacter pylori* VacA. J Biol Chem. 2012;287:31104–15.
20. Kim IJ, Blanke SR. Remodeling the host environment: modulation of the gastric epithelium by the *Helicobacter pylori* vacuolating toxin (VacA). Front Cell Infect Microbiol. 2012;2:37.
21. Greenfield LK, Jones NL. Modulation of autophagy by *Helicobacter pylori* and its role in gastric carcinogenesis. Trends Microbiol. 2013;21:602–12.
22. Oldani A, Cormont M, Hofman V, Chiozzi V, Oregioni O, Canonici A, et al. *Helicobacter pylori* counteracts the apoptotic action of its VacA toxin by injecting the CagA protein into gastric epithelial cells. PLoS Pathog. 2009;5, e1000603.
23. Akada JK, Aoki H, Torigoe Y, Kitagawa T, Kurazono H, Hoshida H, et al. *Helicobacter pylori* CagA inhibits endocytosis of cytotoxin VacA in host cells. Dis Model Mech. 2010;3:605–17.
24. Tsugawa H, Suzuki H, Saya H, Hatakeyama M, Hirayama T, Hirata K, et al. Reactive oxygen species-induced autophagic degradation of *Helicobacter pylori* CagA is specifically suppressed in cancer stem-like cells. Cell Host Microbe. 2012;12:764–77.
25. Fischer W, Buhrdorf R, Gerland E, Haas R. Outer membrane targeting of passenger proteins by the vacuolating cytotoxin autotransporter of *Helicobacter pylori*. Infect Immun. 2001;69:6769–75.
26. Adrian M, Cover TL, Dubochet J, Heuser JE. Multiple oligomeric states of the *Helicobacter pylori* vacuolating toxin demonstrated by cryo-electron microscopy. J Mol Biol. 2002;318:121–33.
27. Cover TL, Hanson PI, Heuser JE. Acid-induced dissociation of VacA, the *Helicobacter pylori* vacuolating cytotoxin, reveals its pattern of assembly. J Cell Biol. 1997;138:759–69.
28. El-Bez C, Adrian M, Dubochet J, Cover TL. High resolution structural analysis of *Helicobacter pylori* VacA toxin oligomers by cryo-negative staining electron microscopy. J Struct Biol. 2005;151:215–28.
29. Chambers MG, Pyburn TM, González-Rivera C, Collier SE, Eli I, Yip CK, et al. Structural analysis of the oligomeric states of *Helicobacter pylori* VacA toxin. J Mol Biol. 2013;425:524–35.
30. Iwamoto H, Czajkowsky DM, Cover TL, Szabo G, Shao Z. VacA from *Helicobacter pylori*: a hexameric chloride channel. FEBS Lett. 1999;450:101–4.
31. Torres VJ, McClain MS, Cover TL. Interactions between p-33 and p-55 domains of the *Helicobacter pylori* vacuolating cytotoxin (VacA). J Biol Chem. 2004;279:2324–31.
32. Torres VJ, McClain MS, Cover TL. Mapping of a domain required for protein-protein interactions and inhibitory activity of a *Helicobacter pylori* dominant-negative VacA mutant protein. Infect Immun. 2006;74:2093–101.
33. Vinion-Dubiel AD, McClain MS, Czajkowsky DM, Iwamoto H, Ye D, Cao P, et al. A dominant negative mutant of *Helicobacter pylori* vacuolating toxin (VacA) inhibits VacA-induced cell vacuolation. J Biol Chem. 1999;274:37736–42.
34. McClain MS, Iwamoto H, Cao P, Vinion-Dubiel AD, Li Y, Szabo G, et al. Essential role of a GXXXG motif for membrane channel formation by *Helicobacter pylori* vacuolating toxin. J Biol Chem. 2003;278:12101–8.
35. Boquet P, Ricci V. Intoxication strategy of *Helicobacter pylori* VacA toxin. Trends Microbiol. 2012;20:165–74.
36. Torres VJ, Ivie SE, McClain MS, Cover TL. Functional properties of the p33 and p55 domains of the *Helicobacter pylori* vacuolating cytotoxin. J Biol Chem. 2005;280:21107–14.

37. Ivie SE, McClain MS, Torres VJ, Algood HM, Lacy DB, Yang R, et al. *Helicobacter pylori* VacA subdomain required for intracellular toxin activity and assembly of functional oligomeric complexes. Infect Immun. 2008;76:2843–51.
38. Bridge DR, Merrell DS. Polymorphism in the *Helicobacter pylori* CagA and VacA toxins and disease. Gut Microbes. 2013;4:101–17.
39. Miehlke S, Kirsch C, Agha-Amiri K, Günther T, Lehn N, Malfertheiner P, et al. The *Helicobacter pylori vacA* s1, m1 genotype and *cagA* is associated with gastric carcinoma in Germany. Int J Cancer. 2000;87:322–7.
40. Yamaoka Y, Kodama T, Kita M, Imanishi J, Kashima K, Graham DY. Relationship of *vacA* genotypes of *Helicobacter pylori* to *cagA* status, cytotoxin production, and clinical outcome. Helicobacter. 1998;3:241–53.
41. van Doorn LJ, Figueiredo C, Sanna R, Pena S, Midolo P, Ng EK, et al. Expanding allelic diversity of *Helicobacter pylori vacA*. J Clin Microbiol. 1998;36:2597–603.
42. Letley DP, Atherton JC. Natural diversity in the N terminus of the mature vacuolating cytotoxin of *Helicobacter pylori* determines cytotoxin activity. J Bacteriol. 2000;182:3278–80.
43. McClain MS, Cao P, Iwamoto H, Vinion-Dubiel AD, Szabo G, Shao Z, et al. A 12-amino-acid segment, present in type s2 but not type s1 *Helicobacter pylori* VacA proteins, abolishes cytotoxin activity and alters membrane channel formation. J Bacteriol. 2001;183:6499–508.
44. Letley DP, Rhead JL, Twells RJ, Dove B, Atherton JC. Determinants of non-toxicity in the gastric pathogen *Helicobacter pylori*. J Biol Chem. 2003;278:26734–41.
45. Ji X, Fernandez T, Burroni D, Pagliaccia C, Atherton JC, Reyrat J, et al. Cell specificity of *Helicobacter pylori* cytotoxin is determined by a short region in the polymorphic midregion. Infect Immun. 2000;68:3754–57.
46. Skibinski DA, Genisset C, Barone S, Telford JL. The cell-specific phenotype of the polymorphic *vacA* midregion is independent of the appearance of the cell surface receptor protein tyrosine phosphatase beta. Infect Immun. 2006;74:49–55.
47. Pagliaccia C, de Bernard M, Lupetti P, Ji X, Burroni D, Cover TL, et al. The m2 form of the *Helicobacter pylori* cytotoxin has cell type-specific vacuolating activity. Proc Natl Acad Sci USA. 1998;95:10212–17.
48. De Guzman BB, Hisatsune J, Nakayama M, Yahiro K, Wada A, Yamasaki E, et al. Cytotoxicity and recognition of receptor-like protein tyrosine phosphatases, RPTPalpha and RPTPbeta, by *Helicobacter pylori* m2VacA. Cell Microbiol. 2005;7:1285–93.
49. Bindayna KM, Al Mahmeed A. *vacA* genotypes in *Helicobacter pylori* strains isolated from patients with and without duodenal ulcer in Bahrain. Indian J Gastroenterol. 2009;28:175–9.
50. Rhead JL, Letley DP, Mohammadi M, Hussein N, Mohagheghi MA, Eshagh Hosseini M, et al. A new *Helicobacter pylori* vacuolating cytotoxin determinant, the intermediate region, is associated with gastric cancer. Gastroenterology. 2007;133:926–36.
51. González-Rivera C, Algood HM, Radin JN, McClain MS, Cover TL. The intermediate region of *Helicobacter pylori* VacA is a determinant of toxin potency in a Jurkat T cell assay. Infect Immun. 2012;80:2578–88.
52. Ogiwara H, Sugimoto M, Ohno T, Vilaichone RK, Mahachai V, Graham DY, et al. Role of deletion located between the intermediate and middle regions of the *Helicobacter pylori vacA* gene in cases of gastroduodenal diseases. J Clin Microbiol. 2009;47:3493–500.
53. Kidd M, Lastovica AJ, Atherton JC, Louw JA. Heterogeneity in the *Helicobacter pylori vacA* and *cagA* genes: association with gastroduodenal disease in South Africa? Gut. 1999;45:499–502.
54. Atherton JC, Peek Jr RM, Tham KT, Cover TL, Blaser MJ. Clinical and pathological importance of heterogeneity in *vacA*, the vacuolating cytotoxin gene of *Helicobacter pylori*. Gastroenterology. 1997;112:92–9.
55. Shimoyama T, Yoshimura T, Mikami T, Fukuda S, Crabtree JE, Munakata A. Evaluation of *Helicobacter pylori vacA* genotype in Japanese patients with gastric cancer. J Clin Pathol. 1998;51:299–301.

56. Yamaoka Y, Kodama T, Gutierrez O, Kim JG, Kashima K, Graham DY. Relationship between *Helicobacter pylori iceA*, *cagA*, and *vacA* status and clinical outcome: studies in four different countries. J Clin Microbiol. 1999;37:2274–9.
57. Basso D, Zambon CF, Letley DP, Stranges A, Marchet A, Rhead JL, et al. Clinical relevance of *Helicobacter pylori cagA* and *vacA* gene polymorphisms. Gastroenterology. 2008;135:91–9.
58. Ogiwara H, Graham DY, Yamaoka Y. *vacA* i-region subtyping. Gastroenterology. 2008;134:1267.
59. Yahiro K, Niidome T, Hatakeyama T, Aoyagi H, Kurazono H, Padilla PI, et al. *Helicobacter pylori* vacuolating cytotoxin binds to the 140-kDa protein in human gastric cancer cell lines, AZ-521 and AGS. Biochem Biophys Res Comm. 1997;238:629–32.
60. Seto K, Hayashi-Kuwabara Y, Yoneta T, Suda H, Tamaki H. Vacuolation induced by cytotoxin from *Helicobacter pylori* is mediated by the EGF receptor in HeLa cells. FEBS Lett. 1998;431:347–50.
61. Yahiro K, Wada A, Nakayama M, Kimura T, Ogushi K, Niidome T, et al. Protein-tyrosine phosphatase alpha, RPTP alpha, is a *Helicobacter pylori* VacA receptor. J Biol Chem. 2003;278:19183–9.
62. de Bernard M, Moschioni M, Papini E, Telford JL, Rappuoli R, Montecucco C. TPA and butyrate increase cell sensitivity to the vacuolating toxin of *Helicobacter pylori*. FEBS Lett. 1998;436:218–22.
63. Padilla PI, Wada A, Yahiro K, Kimura M, Niidome T, Aoyagi H, et al. Morphologic differentiation of HL-60 cells is associated with appearance of RPTPbeta and induction of *Helicobacter pylori* VacA sensitivity. J Bio Chem. 2000;275:15200–6.
64. McClain MS, Schraw W, Ricci V, Boquet P, Cover TL. Acid activation of *Helicobacter pylori* vacuolating cytotoxin (VacA) results in toxin internalization by eukaryotic cells. Mol Microbiol. 2000;37:433–42.
65. Fujikawa A, Shirasaka D, Yamamoto S, Ota H, Yahiro K, Fukada M, et al. Mice deficient in protein tyrosine phosphatase receptor type Z are resistant to gastric ulcer induction by VacA of *Helicobacter pylori*. Nat Genet. 2003;33:375–81.
66. Utt M, Danielsson B, Wadstrom T. *Helicobacter pylori* vacuolating cytotoxin binding to a putative cell surface receptor, heparan sulfate, studied by surface plasmon resonance. FEMS Immunol Med Microbiol. 2001;30:109–13.
67. Hennig EE, Godlewski MM, Butruk E, Ostrowski J. *Helicobacter pylori* VacA cytotoxin interacts with fibronectin and alters HeLa cell adhesion and cytoskeletal organization in vitro. FEMS Immunol Med Microbiol. 2005;44:143–50.
68. Gupta VR, Patel HK, Kostolansky SS, Ballivian RA, Eichberg J, Blanke SR. Sphingomyelin functions as a novel receptor for *Helicobacter pylori* VacA. PLoS Pathog. 2008;4, e1000073.
69. Sewald X, Gebert-Vogl B, Prassl S, Barwig I, Weiss E, Fabbri M, et al. Integrin subunit CD18 is the T-lymphocyte receptor for the *Helicobacter pylori* vacuolating cytotoxin. Cell Host Microbe. 2008;3:20–9.
70. Sewald X, Jimenez-Soto L, Haas R. PKC-dependent endocytosis of the *Helicobacter pylori* vacuolating cytotoxin in primary T lymphocytes. Cell Microbiol. 2011;13:482–96.
71. Schraw W, Li Y, McClain MS, van der Goot FG, Cover TL. Association of *Helicobacter pylori* vacuolating toxin (VacA) with lipid rafts. J Biol Chem. 2002;277:34642–50.
72. Nakayama M, Hisatsune J, Yamasaki E, Nishi Y, Wada A, Kurazono H, et al. Clustering of *Helicobacter pylori* VacA in lipid rafts, mediated by its receptor, receptor-like protein tyrosine phosphatase beta, is required for intoxication in AZ-521 cells. Infect Immun. 2006;74:6571–80.
73. Gauthier NC, Monzo P, Kaddai V, Doye A, Ricci V, Boquet P. *Helicobacter pylori* VacA cytotoxin: a probe for a clathrin-independent and Cdc42-dependent pinocytic pathway routed to late endosomes. Mol Biol Cell. 2005;16:4852–66.

74. Gupta VR, Wilson BA, Blanke SR. Sphingomyelin is important for the cellular entry and intracellular localization of *Helicobacter pylori* VacA. Cell Microbiol. 2010;12:1517–33.
75. Gauthier NC, Monzo P, Gonzalez T, Doye A, Oldani A, Gounon P, et al. Early endosomes associated with dynamic F-actin structures are required for late trafficking of *H. pylori* VacA toxin. J Cell Biol. 2007;177:343–54.
76. Li Y, Wandinger-Ness A, Goldenring JR, Cover TL. Clustering and redistribution of late endocytic compartments in response to *Helicobacter pylori* vacuolating toxin. Mol Biol Cell. 2004;15:1946–59.
77. Molinari M, Galli C, Norais N, Telford JL, Rappuoli R, Luzio JP, et al. Vacuoles induced by *Helicobacter pylori* toxin contain both late endosomal and lysosomal markers. J Biol Chem. 1997;272:25339–44.
78. Papini E, de Bernard M, Milia E, Bugnoli M, Zerial M, Rappuoli R, et al. Cellular vacuoles induced by *Helicobacter pylori* originate from late endosomal compartments. Proc Natl Acad Sci USA. 1994;91:9720–4.
79. Ikonomov OC, Sbrissa D, Yoshimori T, Cover TL, Shisheva A. PIKfyve Kinase and SKD1 AAA ATPase define distinct endocytic compartments. Only PIKfyve expression inhibits the cell-vacoulating activity of *Helicobacter pylori* VacA toxin. J Biol Chem. 2002;277:46785–90.
80. Suzuki J, Ohnsihi H, Shibata H, Wada A, Hirayama T, Iiri T, et al. Dynamin is involved in human epithelial cell vacuolation caused by the *Helicobacter pylori*-produced cytotoxin VacA. J Clin Invest. 2001;107:363–70.
81. Hotchin NA, Cover TL, Akhtar N. Cell vacuolation induced by the VacA cytotoxin of *Helicobacter pylori* is regulated by the Rac1 GTPase. J Biol Chem. 2000;275:14009–12.
82. Suzuki J, Ohnishi H, Wada A, Hirayama T, Ohno H, Ueda N, et al. Involvement of syntaxin 7 in human gastric epithelial cell vacuolation induced by the *Helicobacter pylori*-produced cytotoxin VacA. J Biol Chem. 2003;278:25585–90.
83. Papini E, Bugnoli M, de Bernard M, Figura N, Rappuoli R, Montecucco C. Bafilomycin A1 inhibits *Helicobacter pylori*-induced vacuolization of HeLa cells. Mol Microbiol. 1993;7:323–7.
84. Papini E, Gottardi E, Satin B, de Bernard M, Massari P, Telford J, et al. The vacuolar ATPase proton pump is present on intracellular vacuoles induced by *Helicobacter pylori*. J Med Microbiol. 1996;45:84–9.
85. Terebiznik MR, Raju D, Vazquez CL, Torbricki K, Kulkarni R, Blanke SR, et al. Effect of *Helicobacter pylori*'s vacuolating cytotoxin on the autophagy pathway in gastric epithelial cells. Autophagy. 2009;5:370–9.
86. Raju D, Jones NL. Methods to monitor autophagy in *H. pylori* vacuolating cytotoxin A (VacA)-treated cells. Autophagy. 2010;6:138–43.
87. Raju D, Hussey S, Ang M, Terebiznik MR, Sibony M, Galindo-Mata E, et al. Vacuolating cytotoxin and variants in Atg16L1 that disrupt autophagy promote *Helicobacter pylori* infection in humans. Gastroenterology. 2012;142:1160–71.
88. Tang B, Li N, Gu J, Zhuang Y, Li Q, Wang HG, et al. Compromised autophagy by MIR30B benefits the intracellular survival of *Helicobacter pylori*. Autophagy. 2012;8:1045–57.
89. Kuck D, Kolmerer B, Iking-Konert C, Krammer PH, Stremmel W, Rudi J. Vacuolating cytotoxin of *Helicobacter pylori* induces apoptosis in the human gastric epithelial cell line AGS. Infect Immun. 2001;69:5080–7.
90. Galmiche A, Rassow J. Targeting of *Helicobacter pylori* VacA to mitochondria. Gut microbes. 2010;1:392–5.
91. Domanska G, Motz C, Meinecke M, Harsman A, Papatheodorou P, Reljic B, et al. *Helicobacter pylori* VacA toxin/subunit p34: targeting of an anion channel to the inner mitochondrial membrane. PLoS Pathog. 2010:6:e1000878.

92. Willhite DC, Blanke SR. *Helicobacter pylori* vacuolating cytotoxin enters cells, localizes to the mitochondria, and induces mitochondrial membrane permeability changes correlated to toxin channel activity. Cell Microbiol. 2004;6:143–54.
93. Foo JH, Culvenor JG, Ferrero RL, Kwok T, Lithgow T, Gabriel K. Both the p33 and p55 subunits of the *Helicobacter pylori* VacA toxin are targeted to mammalian mitochondria. J Mol Biol. 2010;401:792–8.
94. Kim JM, Kim JS, Lee JY, Sim YS, Kim YJ, Oh YK, et al. Dual effects of *Helicobacter pylori* vacuolating cytotoxin on human eosinophil apoptosis in early and late periods of stimulation. Eur J Immunol. 2010;40:1651–62.
95. Lan CH, Sheng JQ, Fang DC, Meng QZ, Fan LL, Huang ZR. Involvement of VDAC1 and Bcl-2 family of proteins in VacA-induced cytochrome c release and apoptosis of gastric epithelial carcinoma cells. J Dig Dis. 2010;11:43–9.
96. Calore F, Genisset C, Casellato A, Rossato M, Codolo G, Esposti MD, et al. Endosome-mitochondria juxtaposition during apoptosis induced by *H. pylori* VacA. Cell Death Differ. 2010;17:1707–16.
97. Matsumoto A, Isomoto H, Nakayama M, Hisatsune J, Nishi Y, Nakashima Y, et al. *Helicobacter pylori* VacA reduces the cellular expression of STAT3 and pro-survival Bcl-2 family proteins, Bcl-2 and Bcl-XL, leading to apoptosis in gastric epithelial cells. Dig Dis Sci. 2011;56:999–1006.
98. Jain P, Luo ZQ, Blanke SR. *Helicobacter pylori* vacuolating cytotoxin A (VacA) engages the mitochondrial fission machinery to induce host cell death. Proc Natl Acad Sci USA. 2011;108:16032–7.
99. Akazawa Y, Isomoto H, Matsushima K, Kanda T, Minami H, Yamaghchi N, et al. Endoplasmic reticulum stress contributes to *Helicobacter pylori* VacA-induced apoptosis. PLoS One. 2013:8:e82322.
100. Radin JN, Gonzalez-Rivera C, Frick Cheng AE, Sheng J, Gaddy JA, Rubin DH, et al. Role of connexin 43 in *Helicobacter pylori* VacA-induced cell death. Infect Immun. 2014;82:423–32.
101. Schmees C, Gerhard M, Treptau T, Voland P, Schwendy S, Rad R, et al. VacA-associated inhibition of T-cell function: reviewed and reconsidered. Helicobacter. 2006;11:144–6.
102. Molinari M, Salio M, Galli C, Norais N, Rappuoli R, Lanzavecchia A, et al. Selective inhibition of Ii-dependent antigen presentation by *Helicobacter pylori* toxin VacA. J Exp Med. 1998;187:135–40.
103. Gebert B, Fischer W, Weiss E, Hoffmann R, Haas R. *Helicobacter pylori* vacuolating cytotoxin inhibits T lymphocyte activation. Science. 2003;301:1099–102.
104. Boncristiano M, Paccani SR, Barone S, Ulivieri C, Patrussi L, Ilver D, et al. The *Helicobacter pylori* vacuolating toxin inhibits T cell activation by two independent mechanisms. J Exp Med. 2003;198:1887–97.
105. Kim JM, Kim JS, Yoo DY, Ko SH, Kim N, Kim H, et al. Stimulation of dendritic cells with *Helicobacter pylori* vacuolating cytotoxin negatively regulates their maturation via the restoration of E2F1. Clin Exp Immunol. 2011;166:34–45.
106. Geiszt M, Leto TL. The Nox family of NAD(P)H oxidases: host defense and beyond. J Biol Chem. 2004;2798:51715–8.
107. Yuan J, Li P, Tao J, Shi X, Hu B, Chen H, et al. *H. pylori* escape host immunoreaction through inhibiting ILK expression by VacA. Cell Mol Immunol. 2009;6:191–7.
108. Backert S, Tegtmeyer N. The versatility of the *Helicobacter pylori* vacuolating cytotoxin vacA in signal transduction and molecular crosstalk. Toxins. 2010;2:69–92.
109. Nakayama M, Hisatsune J, Yamasaki E, Isomoto H, Kurazono H, Hatakeyama M, et al. *Helicobacter pylori* VacA-induced inhibition of GSK3 through the PI3K/Akt signaling pathway. J Bio Chem. 2009;284:1612–9.
110. Hisatsune J, Yamasaki E, Nakayama M, Shirasaka D, Kurazono H, Katagata Y, et al. *Helicobacter pylori* VacA enhances prostaglandin E2 production through induction of cyclooxygenase 2 expression via a p38 mitogen-activated protein kinase/activating transcription factor 2 cascade in AZ-521 cells. Infect Immun. 2007;75:4472–81.

111. Takeshima E, Tomimori K, Takamatsu R, Ishikawa C, Kinjo F, Hirayama T, et al. *Helicobacter pylori* VacA activates NF-kappaB in T cells via the classical but not alternative pathway. Helicobacter. 2009;14:271–9.
112. Hisatsune J, Nakayama M, Isomoto H, Kurazono H, Mukaida N, Mukhopadhyay AK, et al. Molecular characterization of *Helicobacter pylori* VacA induction of IL-8 in U937 cells reveals a prominent role for p38MAPK in activating transcription factor-2, cAMP response element binding protein, and NF-kappaB activation. J Immunol. 2008;180:5017–27.

Chapter 5
Autophagy

Hitoshi Tsugawa and Hidekazu Suzuki

Abstract Autophagy is a eukaryotic, nonspecific degradation mechanism and serves as part of the innate immune system of host cells. In *Helicobacter pylori*-infected host cells, autophagy is activated by the bacterial vacuolating cytotoxin A (VacA) via the following pathway: VacA binds to low-density lipoprotein receptor-related protein-1, induces intracellular glutathione deficiency, and enhances activation of protein kinase B (Akt), which in turn induces Mdm2-mediated p53 degradation and then activates autophagy. Translocated bacterial proteins, VacA and CagA, are degraded by autophagy. However, autophagy is not activated in the CD44v9-expressing gastric cancer stem-like cells, leading to the specific accumulation of intracellular CagA. Therefore, the presence of CD44v9-expressing cells in *H. pylori*-infected patients is associated with the risk of developing gastric cancer. Autophagy in the host gastric epithelial cells plays an important role in *H. pylori* infectious disease via degradation of VacA and CagA.

Keywords VacA • CagA • Cancer stem cells • CD44

5.1 Induction of Autophagy by VacA in Host Epithelial Cells Infected with *Helicobacter pylori*

Autophagy is a nonspecific degradation mechanism seen in eukaryotic cells. Targets for degradation by autophagy include intracellular components such as mitochondria, where targets are surrounded by the cellular double membrane to form transient autophagosomes and are degraded by fusion with lysosomes. Autophagy is an important cellular recycling mechanism to overcome temporary starvation, in addition to removing intracellular foreign materials. In particular, intracellular parasitic bacteria such as group A streptococci, *Shigella* spp., *Salmonella* spp.,

H. Tsugawa (✉)
Department of Biochemistry and Integrative Medical Biology, School of Medicine, Keio University, 35 Shinanomachi, Shinjuku-ku, Tokyo 160-8582, Japan
e-mail: h.tsugawa@a6.keio.jp

H. Suzuki
Medical Education Center, Keio University School of Medicine, 35 Shinanomachi, Shinjuku-ku, Tokyo, 160-8582 Tokyo, Japan

H. Suzuki et al. (eds.), *Helicobacter pylori*, DOI 10.1007/978-4-431-55705-0_5

and *Listeria monocytogenes* are degraded by autophagy. In fact, autophagy serves as part of the innate immune system of host cells. In recent years, it has been revealed that autophagy is activated in *Helicobacter pylori*-infected cells, interestingly, by an exotoxin known as vacuolating cytotoxin A (VacA), which is produced by the bacterium and is involved in ulceration. VacA is biologically versatile and shows apoptosis-inducing activity via mitochondrial pathway [1]. In a recent study, Terebiznik et al. indicated that VacA induces autophagy [2, 3]. VacA is composed of p55 fragments involved in receptor recognition and an N-terminal (p33) fragment involved in vacuolization. Further, the p55 fragment can be divided into m1 type (m1VacA) and m2 type (m2VacA) due to the difference in primary structure. Receptor protein-tyrosine phosphatase (RPTP) α and β have been identified as receptors for VacA [4, 5]. Additionally, Yahiro et al. recently indicated that low-density lipoprotein receptor-related protein-1(LRP1) is a receptor for VacA and this binding is important for induction of autophagy [6]. Our study demonstrated that although m1VacA bound to LRP1, m2VacA did not [7]. Thus, m2VacA probably does not induce autophagy.

Interestingly, m1VacA causes a reduction in intracellular glutathione (GSH) levels in the host epithelial cells, resulting in the accumulation of intracellular reactive oxygen species (ROS) [7]. It is well known that accumulation of intracellular ROS activates autophagy. In fact, autophagy induced by m1VacA is repressed by treatment with antioxidants such as N-acetylcysteine [7]. In addition, m1VacA also induces p53 downregulation during autophagy. Tasdemir et al. reported that p53 inactivation by chemical inhibition or knockdown induces autophagy via inhibition of mTOR [8]. Our study showed that p53 downregulation by m1VacA is necessary for the induction of autophagy [7]. In fact, the autophagic pathway via m1VacA is as follows: m1VacA induces GSH deficiency by binding to LRP1 and enhances activation of protein kinase B (Akt), which in turn induces Mdm2-mediated p53 degradation, activating autophagy [7] (Fig. 5.1).

Recent studies have revealed that p62 binds to LC3 on the autophagosomal membrane to target ubiquitinated aggregates for selective degradation [9]. Yahiro

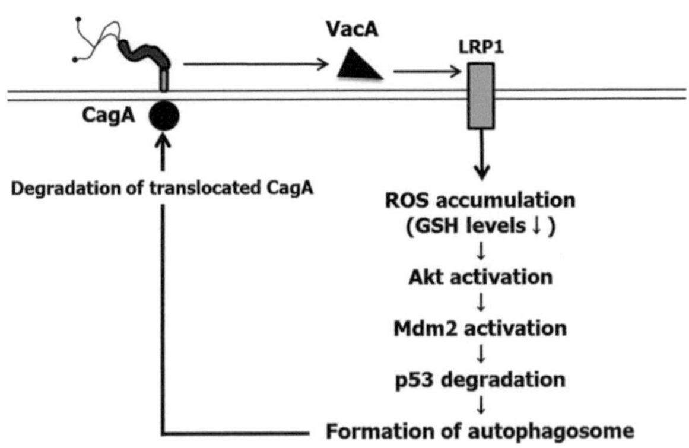

Fig. 5.1 Autophagy leading signals by VacA. VacA induced GSH deficiency via binding to LRP1 and then enhances activation of Akt. Activation of Akt induces Mdm2-mediated p53 degradation and then activates autophagy

et al. demonstrated that LC3-II is colocalized with p62 on the autophagosomal puncta induced by VacA [6]. Therefore, autophagy activated by VacA is considered selective and involves targeting by ubiquitination. Further studies are needed to clarify the selective mechanisms of VacA-activated autophagy.

5.2 Significance of Autophagy in Host Epithelial Cells Infected with *H. pylori*

H. pylori is a significant risk factor for the development of peptic ulcers and gastric cancer; therefore, in this chapter, we focus on the role of autophagy in *H. pylori* infection.

VacA binds to host cell-surface receptors and is internalized via endocytosis. An in vitro study demonstrated that VacA induces vacuolization and apoptosis in host epithelial cells [1]. These activities have been implicated to cause induced gastric mucosal injury in gastric ulcers during *H. pylori* infection [10]. Recently, it has been reported that intracellular VacA is degraded by autophagy activated by VacA itself [2, 3]. Therefore, decreased activation of autophagy by genetic polymorphisms of autophagy-related genes *ATG16L1* contributes to increase in VacA-mediated toxicity [2, 3].

The CagA effector protein is delivered into *H. pylori*-attached host epithelial cells via type IV secretion system (TFSS). Translocated CagA binds to and dysregulates SHP-2 tyrosine phosphatase and specifically interacts with the polarity-regulating kinase partitioning-defective 1 (Par1b)/microtubule affinity-regulating kinase 2 (MARK2) to disrupt tight junctions and cause a loss of epithelial apical-basolateral cell polarity. Additionally, it has been reported that *CagA*-transgenic mice developed gastrointestinal carcinomas. These observations indicate that CagA effector protein is a bacterial oncoprotein, involved in gastric carcinogenesis. However, translocated CagA is degraded by autophagy induced by VacA, and, as a result, CagA does not persist in host gastric epithelial cells [7].

Thus, it is thought that autophagy induced by VacA contributes to a reduction in gastric mucosal injury and gastric cancer risk associated with *H. pylori* infection, via degradation of VacA and CagA. On the other hand, characteristic alterations in the host cell, involved in the repression of autophagy, induce the accumulation of intracellular VacA and CagA. Thus, it is presumed that inhibition of autophagy may lead to increased gastric cancer risk via specific accumulation of CagA. Therefore, autophagy response in host gastric epithelial cells is considered to play an important role in the development of gastric carcinogenesis.

5.3 Suppression of Autophagy and Accumulation of CagA in CD44v9–Expressing Cancer Stem-Like Cells

CD44 is a cell-surface marker associated with cancer stem cells in various tumors. Gastric cancer stem-like cells express a variant isoform of CD44 (CD44v9). A recent study reported that the recurrence rate of early gastric cancer (EGC) is significantly higher in CD44v9-positive individuals than in CD44v9-negative individuals [11]. CD44v9-expressing gastric cancer cells suppress ROS accumulation via control of intracellular GSH levels by stabilizing xCT, a cystine transporter [12]. Since induction of autophagy by VacA requires the reduction of GSH, we hypothesize that autophagy is not activated by VacA in CD44v9-expressing cells. In fact, intracellular GSH levels in CD44v9-expressing cells are not decreased by VacA [7]. Further, increase of Akt, Mdm2 phosphorylation, and p53 degradation are also not observed in CD44v9-expressing cells; hence, autophagy is not induced by VacA [7]. In addition, translocated CagA accumulates in CD44v9-expressing cells (Fig. 5.2), which is reversed with sulfasalazine, a potent xCT inhibitor and a well-known anti-inflammatory drug for rheumatic arthritis and inflammatory bowel disease. Moreover, autophagy is activated in CD44v9-expressing cells by treatment with sulfasalazine [7]. These observations reveal that specific accumulation of intracellular CagA in CD44v9-expressing cancer stem-like cells is caused by the repression of autophagy. Therefore, the presence of CD44v9-expressing cells in *H. pylori*-infected patients is associated with the risk of developing gastric cancer.

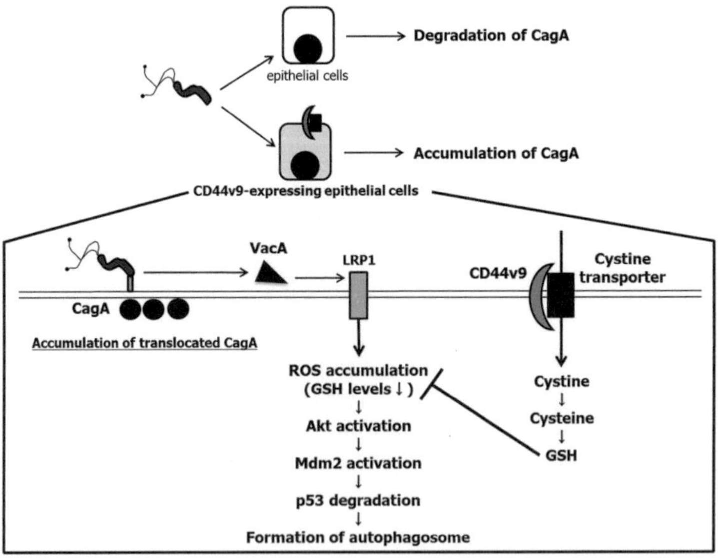

Fig. 5.2 Accumulation of intracellular CagA in the CD44v9-expressing cells through the repression of autophagy

References

1. Yamasaki E, Wada A, Kumatori A, et al. *Helicobacter pylori* vacuolating cytotoxin induces activation of the proapoptotic proteins Bax and Bak, leading to cytochrome c release and cell death, independent of vacuolation. J Biol Chem. 2006;281:11250–9.
2. Terebiznik MR, Raju D, Vazquez CL, et al. Effect of *Helicobacter pylori*'s vacuolating cytotoxin on the autophagy pathway in gastric epithelial cells. Autophagy. 2009;5:370–9.
3. Raju D, Hussey S, Ang M, et al. Vacuolating cytotoxin and variants in Atg16L1 that disrupt autophagy promote *Helicobacter pylori* infection in humans. Gastroenterology. 2012;142:1160–71.
4. Yahiro K, Niidome T, Kimura M, et al. Activation of *Helicobacter pylori* VacA toxin by alkaline or acid conditions increases its binding to a 250-kDa receptor protein-tyrosine phosphatase beta. J Biol Chem. 1999;274:36693–9.
5. Yahiro K, Wada A, Nakayama M, et al. Protein-tyrosine phosphatase alpha, RPTP alpha, is a *Helicobacter pylori* VacA receptor. J Biol Chem. 2003;278:19183–9.
6. Yahiro K, Satoh M, Nakano M, et al. Low-density Lipoprotein Receptor-related Protein-1 (LRP1) mediates autophagy and apoptosis caused by *Helicobacter pylori* VacA. J Biol Chem. 2012;287:31104–15.
7. Tsugawa H, Suzuki H, Saya H, et al. Reactive oxygen species-induced autophagic degradation of *Helicobacter pylori* CagA is specifically suppressed in cancer stem-like cells. Cell Host Microbe. 2012;12:764–77.
8. Tasdemir E, Maiuri MC, Galluzzi L, et al. Regulation of autophagy by cytoplasmic p53. Nat Cell Biol. 2008;10:676–87.
9. Kirkin V, McEwan DG, Novak I, Dikic I. A role for ubiquitin in selective autophagy. Mol Cell. 2009;34:259–69.
10. Fujikawa A, Shirasaka D, Yamamoto S, et al. Mice deficient in protein tyrosine phosphatase receptor type Z are resistant to gastric ulcer induction by VacA of *Helicobacter pylori*. Nat Genet. 2003;33:375–81.
11. Hirata K, Suzuki H, Imaeda H, et al. CD44 variant 9 expression in primary early gastric cancer as a predictive marker for recurrence. Br J Cancer. 2013;109:379–86.
12. Ishimoto T, Nagano O, Yae T, et al. CD44 variant regulates redox status in cancer cells by stabilizing the xCT subunit of system xc(−) and thereby promotes tumor growth. Cancer Cell. 2011;19:387–400.

Chapter 6
Ghrelin and Gut Hormone

Yoon Jin Choi and Nayoung Kim

Abstract One of the important gastric physiologies is the endocrine function. Among peptide hormones produced in the stomach, gastrin and somatostatin which are produced from G and D cell, respectively, are closely related with acid secretion, while ghrelin and leptin are known to be involved in gut motility as well as the regulation of appetite and body weight. Since *Helicobacter pylori* (*H. pylori*) is the major etiologic agent of chronic active gastritis, *H. pylori* infection can alter gastric hormone production. While the effect of *H. pylori* infection and its eradication on G and D cells and subsequent acid secretion has been relatively well investigated, its effect on ghrelin and leptin levels has been inconclusive. This discrepancy originates from different measurement methodology or existence of several types in one hormone. In addition, the different stage of *H. pylori* infection or site of the infection may cause a different result on ghrelin levels in the stomach and blood similar to the case of acid secretion. *H. pylori* infection seems to raise gastric leptin production, but its effect on circulating level is not much. Understanding the effect of *H. pylori* infection and eradication on gastric hormonal changes might provide a critical clinical implication for the management of gastrointestinal diseases as well as obesity or eating disorder.

Keywords Ghrelin • Leptin • *Helicobacter pylori*

Y.J. Choi
Department of Internal Medicine, Seoul National University Bundang Hospital, 82 Gumi-ro, 173 beon-gil, Bundang-gu, Seongnam, Gyeonggi-do 463-707, South Korea

N. Kim (✉)
Department of Internal Medicine, Seoul National University College of Medicine, 101, Daehak-ro, Jongno-gu, Seoul, South Korea

Department of Internal Medicine, Seoul National University Bundang Hospital, 82 Gumi-ro, 173 beon-gil, Bundang-gu, Seongnam, Gyeonggi-do 463-707, South Korea
e-mail: nayoungkim49@snu.ac.kr

H. Suzuki et al. (eds.), *Helicobacter pylori*, DOI 10.1007/978-4-431-55705-0_6

6.1 Introduction

Although gut is not the classical endocrine tissues such as the pituitary, thyroid, and adrenal glands, it produces an array of peptide hormones which are important for both enteric and non-enteric physiology [1]. That is, enteroendocrine cells are distributed throughout the gastrointestinal tract, but a significant proportion of endocrine hormones which are involved in appetite regulation or intestinal motility are produced and secreted by the gastric mucosa.

Traditionally, gastrin and somatostatin have been well investigated related with gastric acid secretion. Both endocrine and paracrine mediators also exert control over the maintenance of gastric secretory functions. Gastrin is released from the G cells of the antral mucosa and travels through the bloodstream to the corpus where gastrin stimulates enterochromaffin-like (ECL) cells to secrete histamine which, in turn, stimulates the parietal cells to secrete acid. In contrast, somatostatin (SST) and prostaglandins inhibit acid secretion [2].

Other two hormones, ghrelin and leptin which are secreted from the stomach, play critical roles in controlling appetite and satiety. Ghrelin is a 28-amino acid peptide hormone, primarily produced in, and secreted from, the gastric mucosa [3]. Meanwhile, leptin is a hormone mainly produced by adipose tissues with modulatory effects on feeding behavior and weight control [4]. However, recently stomach has been also identified as an important source of leptin. Both ghrelin and leptin act in an autocrine/paracrine manner and can travel through blood and act in the central nerve system (CNS).

Helicobacter pylori (*H. pylori*) is a predominant etiologic factor of gastric inflammation leading to atrophy of gastric glands. Its infection of the human gastric mucosa alters the normal gastric physiology. Especially, cytokines deregulate secretion of gastric hormones including gastrin, SST, ghrelin, and leptin during the inflammation. Finally, altered gut–brain interactions by this deregulation of these hormones underlie gastrointestinal (GI) symptom generation. Furthermore, the change of these hormones can induce GI disorder. This chapter aims to review the data on hormones which are produced in stomach, focusing on ghrelin and the effect of *H. pylori* infection and its eradication on local and circulating levels of these hormones.

6.2 Ghrelin

6.2.1 Production of Ghrelin

6.2.1.1 Source of Ghrelin

The stomach is considered as the major source of circulating ghrelin, since severe reduction in blood ghrelin levels is observed in patients that undergone gastrectomy

[5]. Ghrelin is produced in oxyntic cells that are prominent in the corpus of stomach. The ghrelin-producing cells are located from the base to the neck of the glands. It is also secreted from the small intestine and the colon [6]. In addition, ghrelin is also expressed in the hypothalamus [7], the pituitary [8], and several tissues in the periphery [9].

6.2.1.2 Process of Ghrelin Production

Although the major active product of the ghrelin gene is the 28-amino acid peptide acylated at Ser^3 with C8:0, recent studies have revealed that the ghrelin gene can generate various molecules besides ghrelin, which include des-acyl ghrelin [10].

The human ghrelin gene, located on the short arm of chromosome 3 (3p25–26), is composed of five exons and four introns [11]. There are two different transcriptional initiation sites in the ghrelin gene, resulting in two distinct mRNA transcripts: transcript-A and transcript-B (Fig. 6.1). The main ghrelin mRNA transcript in human codes for a 117-amino acid long peptide: preproghrelin (1–117). The signal peptide sequence, preproghrelin (1–23) of preproghrelin (1–117), is cleaved to form proghrelin (1–94). A series of posttranslational steps, including the process of protease cleavage and acyl modification of the ghrelin precursor peptide, results in the production of mature ghrelin peptides (acyl and des-acyl ghrelin) or other ghrelin gene-associated peptides (C-ghrelin and obestatin).

The ghrelin peptide is acylated by the enzyme ghrelin O-acyltransferase (GOAT) [12], which is expressed predominantly in the stomach, gut, and pancreas but also at other sites [13]. This acyl modification of ghrelin is easily cleaved during sample extraction. Thus, acyl ghrelin should be isolated from blood specimens by adding ethylenediaminetetraacetic acid (EDTA) with aprotinin or p-hydroxymercuribenzoic acid, separating the plasma by centrifugation and immediate acidification before freezing at −80 °C to ensure the stability of acyl ghrelin during storage. The nonacylated form of ghrelin, without octanoic acid modification at Ser3 residue, des-acyl ghrelin is also present at significant level in both the stomach and blood [10, 14]. Des-acyl ghrelin is the most abundant ghrelin-related molecule in the body, comprising 80–90 % of the total circulating ghrelin, and has a longer half-life.

Besides these peptides, several other ghrelin gene-derived peptides with or without acyl modification (des-Gln 14-ghrelin, ΔEx3-C-ghrelin, etc.) are predicted to be produced from the splicing variants of the ghrelin gene transcripts [11].

6.2.2 *Regulation of Ghrelin*

Many factors which affect serum ghrelin levels were listed in Table 6.1 [15].

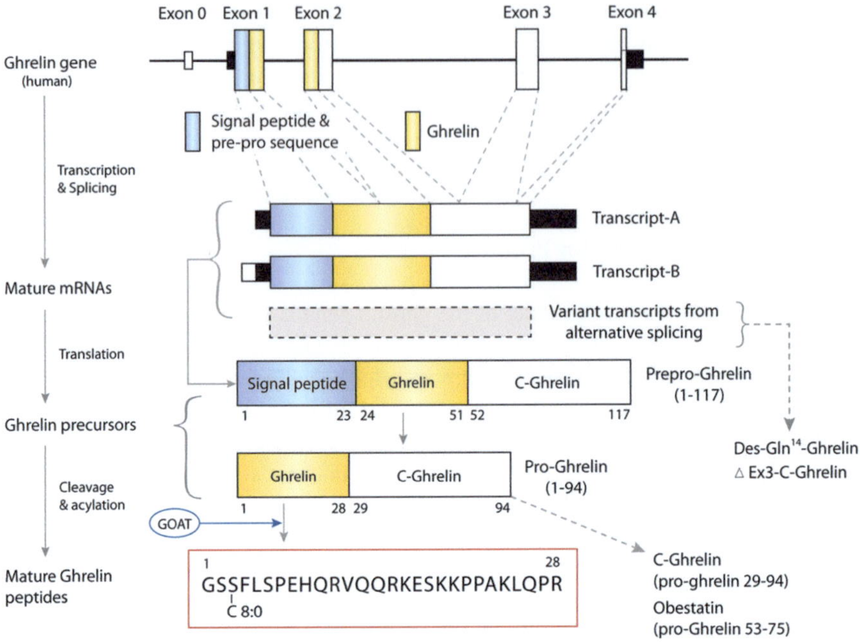

Fig. 6.1 Structure of the human ghrelin gene and processing steps from the ghrelin gene to acyl ghrelin, des-acyl ghrelin, or other ghrelin-associated peptides (Adapted from Nishi et al. [11]). The human ghrelin gene is composed of four exons. The major mRNA transcript (transcript-A) of the ghrelin gene is translated into a 117-amino acid ghrelin precursor: preproghrelin (1–117). The signal peptide sequence, preproghrelin (1–23) of preproghrelin (1–117), is cleaved to form proghrelin (1–94). A series of posttranslational steps, including the process of protease cleavage and acyl modification of the ghrelin precursor peptide, results in the production of mature ghrelin peptides (acyl and des-acyl ghrelin) or other ghrelin gene-associated peptides (C-ghrelin and obestatin). Besides these peptides, several other ghrelin gene-derived peptides with or without acyl modification (des-Gln14-ghrelin, ΔEx3-C-ghrelin, etc.) are predicted to be produced from the splicing variants of the ghrelin gene transcripts. *GOAT* ghrelin-O-acyltransferase

Table 6.1 Variable situations which affect blood ghrelin level (Adapted from Yin et al. [15])

Group	Elevated ghrelin level	Decreased ghrelin level
Nutrients	Fatty acids[a], amino acids[a]	Glucose, fatty acids[a]
Hormones	Glucagon[a], IGF, estrogen[a]	Insulin, growth hormone, somatostatin, leptin[a], estrogen[a]
Autonomic nervous system	Vagus nerve activation'	Sympathetic nerve activation
Physiological status	Fasting, lean, youth	Feeding[a], obesity, aging[a]
Pathological status	Prader–Willi syndrome, anorexia nervosa, cachexia	Metabolic syndrome, diabetes mellitus

[a]Data is controversial

6.2.2.1 Posttranslational Modification of Ghrelin Precursor Protein

As above mentioned, GOAT is a member of the family membrane-bound O-acyltransferases which shows highly specific expression in the gastric mucosa. This discovery manifests its significance in the regulation of ghrelin secretion, because the amount and activity of GOAT likely affect the level of acyl ghrelin. An in vitro study has demonstrated that GOAT activity could be inhibited potently by an octanoylated ghrelin pentapeptide and other end products [16], suggesting the existence of a negative feedback regulation on the production of acyl ghrelin. Plasma esterases have been reported to des-acylate acyl ghrelin, whereas plasma proteases account for the degradation of circulating ghrelin [17]. The circulating level of ghrelin is determined by the balance among its secretion rate, degradation rate, and clearance rate.

6.2.2.2 Nutrients Regulating Ghrelin Expression and Secretion

Glucose markedly inhibits ghrelin secretion (Table 6.1). That is, oral infusion of glucose can decrease the plasma concentration of total ghrelin 30 min after ingestion in humans [18] and in rats [19]. Ingestion of crude fiber has the similar effect with glucose [19]. Insulin-induced hypoglycemia upregulates ghrelin mRNA expression [20] and serum acyl ghrelin level [21] in the stomach. With regard to fatty acid, it appears that the effects of fatty acids and triglycerides on ghrelin secretion are dependent on the length of their chain. Generally, lipid ingestion leads to a smaller decline in ghrelin relative to the administration of glucose or amino acids [22]. This observation may explain the weight gain effect of high-fat diet. Oral ingestion of a physiological dose of essential amino acids leads to a continuous rise in serum ghrelin level in humans [23].

6.2.2.3 Hormones Regulating Ghrelin Expression and Secretion

Insulin

In rats, gastric artery perfusion of insulin inhibits ghrelin release from isolated stomach tissue significantly (Table 6.1) [24]. In humans, infusion of insulin significantly decreases plasma ghrelin level [25]. Fasting plasma acyl ghrelin level is negatively related to insulin concentration [21]. This inhibitory effect of insulin may underlie the suppression of glucose on ghrelin and the inverse relationship between body weight and ghrelin level.

Glucagon

Glucagon may contribute to the preprandial surge of ghrelin. Plasma acyl ghrelin concentration rises transiently, while des-acyl ghrelin increases persistently after administration of glucagon in rats [26].

Growth Hormone/Insulin-Like Growth Factor-1 (IGF-1)

Growth hormone exerts a negative feedback action on ghrelin production and secretion. Administration of growth hormone in cultured rat gastric tissue time dependently inhibits total ghrelin secretion [27]. Contrary to the growth hormone, administration of recombinant human IGF-1 in severely undernutritioned patients elevates plasma total ghrelin concentration [28].

Somatostatin

Somatostatin (SST) probably inhibits ghrelin synthesis directly (Table 6.1). Plasma acyl and total ghrelin levels fall after the infusion of somatostatin or octreotide, somatostatin analog [29]. Since ghrelin increases the level of somatostatin in plasma [30], the inhibitory effect of somatostatin on ghrelin may be considered as a negative feedback modulation.

Leptin

It is generally agreed that leptin inhibits ghrelin synthesis (Table 6.1). Leptin is mainly synthesized and secreted by adipose tissue. Leptin concentration in obese is significantly higher than normal, whereas ghrelin is lower [31]. Leptin correlates with ghrelin in a complex pattern, which depends on the body weight (normal or obesity) and insulin sensitivity or insulin concentration. As shown by recent studies, ghrelin mRNA increases in the stomach during fasting, whereas leptin and leptin mRNA decrease [32]. Leptin dose-dependently inhibits ghrelin transcription in vitro [32] and decreases ghrelin release from isolated rat stomach [24]. Central leptin gene therapy decreases plasma leptin level and increases ghrelin level significantly in the mouse fed with high-fat diet [33], indicating that leptin exerts its inhibition on ghrelin secretion only in peripheral tissues. Thus, peripheral, especially gastric, leptin probably represses ghrelin expression through its receptor in gastric mucosa cells.

Estrogen

Many studies report that estrogen upregulates ghrelin level. Administration of estrogen elevates plasma total ghrelin concentration in female patients with anorexia nervosa [28]. Ghrelin mRNA level rises significantly after estrogen administration in cultured stomach cells [34]. However, there also exist discrepant results. Estrogen replacement therapy in postmenopausal women induces serum total and acyl ghrelin secretion only to an insignificant extent or even decreases serum total ghrelin level [15]. Variable methods (i.e., per oral or transdermal), duration for estrogen administration, and physiological status could attribute to these contradictable results.

6.2.2.4 Autonomic Nervous System Regulating Ghrelin Expression and Secretion

Autonomic nervous system, especially the parasympathetic nerve, plays an important role in the regulation of ghrelin (Table 6.1). Excitation of the vagus nerve can stimulate ghrelin secretion. In rats and humans, ghrelin level rises after administration of muscarinic agonists and falls after administration of muscarinic antagonists [35].

6.2.2.5 Physiological Status

Ghrelin level is negatively correlated with body mass index (BMI) in humans. Fasting acyl ghrelin [36] and total ghrelin [37] are significantly lower in the aged population than in the youth (Table 6.1). However, this age-dependent decline of ghrelin is not observed in the obese population [38]. Many studies report an elevated serum ghrelin level in female subjects relative to male ones. Serum total ghrelin level is about threefold higher in women during the late follicular stage of the cycle than in men [39].

6.2.2.6 *H. pylori* Infection

H. pylori-induced chronic gastritis is characterized by chronic inflammatory changes in the gastric mucosa leading to extensive mucosal atrophy and eventual epithelial metaplasia. It is speculated that destruction of oxyntic mucosa could result in a negative correlation with local or circulating ghrelin level. This will be further discussed in Sect. 6.2.4.

6.2.3 *Role of Ghrelin*

6.2.3.1 The Mechanisms of Action of Acyl Ghrelin

Acyl ghrelin has been identified as the endogenous ligand of the growth hormone secretagogue receptor (GHS-R) [3]. It crosses the blood–brain barrier in both directions using a saturable transport system that requires the presence of the unique octanoyl residue of the ghrelin molecule [40]. GHS-Rs are also widely expressed in the central nervous system. They are found in the pituitary, brainstem, and hypothalamus, whereas peripheral receptor expression has been described in the myocardium, GI tract, adipose tissue, liver, kidney, placenta, and the T cells [41]. Acyl ghrelin acts in GHS-R on vagal afferent nerve fibers in the stomach [42], which transmit this signal to the nucleus of the solitary tract (NTS) (Fig. 6.2). From the NTS, the information is projected to the arcuate nucleus (ARC) of the hypothalamus, where neuropeptide Y (NPY) neurons are activated (Fig. 6.2). The NPYY2 and/or Y4 receptor in the CNS may be involved in upper GI motility because Y2 and Y4 receptor agonists can induce phase III-like contractions in the duodenum when given to animals in the fed state [43]. From the ARC, the signal is finally transmitted to the dorsal motor nucleus of the vagus nerve (DVC, dorsal vagal complex) and via vagal efferent fibers, and fasted motor activity is induced in the gut (Fig. 6.2) [44].

6.2.3.2 The Representable Roles of Acyl Ghrelin

Acyl Ghrelin Is a Signal for Hunger

Acyl ghrelin is considered as a short regulator of food intake in both animals and humans [45, 46]. In rats, acute and chronic administration of ghrelin enhances food intake and weight gain [47, 48]. Peripheral administration of ghrelin produced a 28 % increase of food intake in normal-weight healthy volunteers [45]. The subjects who received exogenous ghrelin reported an increase in appetite and showed a higher caloric intake than after placebo [49].

Acyl Ghrelin Influences Gut Motility

As previously mentioned, in humans, ghrelin stimulates gastric motility [50] and acid secretion [51]. This fasted motor activity of the GI tract has been considered to play a role of a mechanical cleansing of the stomach and the intestine in preparation for the next meal. In healthy volunteers, the peripheral administration of ghrelin induces the occurrence of phase III of the migrating motor complex after about 20 min. Moreover, it induces a premature phase III originating in the stomach about 14 min after its injection [52]. A positive correlation was reported between preprandial ghrelin concentration and gastric emptying time. The duration of gastric emptying is considered as an important factor for the duration of satiety [53, 54].

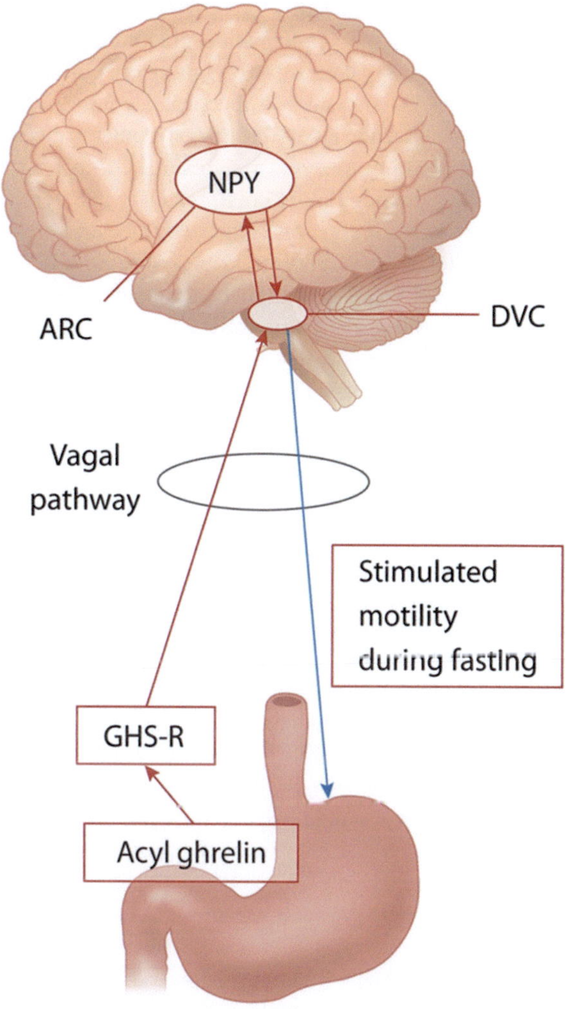

Fig. 6.2 The effects of acyl ghrelin via the brain–gut axis (Adapted from Fujimiya et al. [44]). *ARC* arcuate nucleus, *DVC* dorsal vagal complex, *GHS-R* growth hormone secretagogue receptors, *NPY* neuropeptide Y

Ghrelin and GI Disease

Preprandial ghrelin levels have significantly decreased in patients with functional dyspepsia (FD) with delayed gastric emptying in Lee's report [55]. Moreover, low preprandial ghrelin levels were observed in patients with dysmotility-like FD [55]. Shindo et al. [56] also reported that the maximum gastric emptying time, Tmax, was significantly prolonged with significant lower acyl ghrelin levels in postprandial distress syndrome (PDS) patients not in epigastric pain syndrome (EPS) patients. Lower acyl ghrelin levels were also found in nonerosive reflux disease patients [56]. This correlation between acyl ghrelin levels and Tmax in the PDS patients indicates acyl ghrelin's role in gastric emptying of PDS patients

[56]. Interestingly, El-Salhy et al. [57] demonstrated that ghrelin-producing cells were suppressed in constipation-predominant IBS patients, while diarrhea-predominant IBS patients have significantly higher ghrelin-positive cells in the oxyntic mucosa compared with normal controls. Since there was no difference either in the plasma levels or gastric contents between IBS groups and control subjects, it could be speculated that some compensatory mechanism may exist and disruption of this regulation could result in IBS symptoms.

Furthermore, ghrelin was suggested to be an important biomarker for activity in IBD patients. Serum ghrelin levels were found to be higher in patients with ulcerative colitis and also higher in patients with ileal Crohn's disease compared with colonic disease patients [58]. Ghrelin is also significantly elevated in active IBD patients and positively correlated with serum inflammatory markers such as tumor necrosis factor-α (TNF-α), C-reactive protein (CRP), erythrocyte sedimentation rate (ESR), and sedimentation fibrinogen [59].

Acyl Ghrelin Is Associated with GI Malignancy

Significant reduction of ghrelin mRNA and peptide expression was found in esophagogastric adenocarcinomas compared to adjacent nonneoplastic gastric mucosa. This finding suggests that ghrelin production was suppressed due to damage to normal ghrelin-secreting mucosa from adenocarcinoma [60]. An et al. [61] also reported that lower ghrelin levels were present in differentiated tumor tissue than in undifferentiated tissue. These findings suggested that the development of cancer may lead to an inability to produce ghrelin, which is also influenced by the state of differentiation. In colorectal cancer studies, Waseem et al. [62] showed that colorectal cancer cells excessively secrete ghrelin in vitro to promote proliferation. Malignant colorectal tissue samples also showed enhanced stage-dependent expression of ghrelin. However, expression of ghrelin and its functional receptor (GHS-R1a) were suppressed in advanced state, poorly differentiated tumors. GHS-R1a expression was lost in malignant colorectal cells, while GHS-R1b expression was enhanced [62].

Glucose Control

The ghrelin system, using both the acylated and des-acylated molecules, is actively involved in the acute and the long-term control of glucose metabolism and insulin concentrations [63]. It has been demonstrated that glucose output by primary hepatocytes is time- and dose-dependently stimulated by acyl ghrelin and inhibited by des-acyl ghrelin. Apparently, the two forms of peptides must be considered as separate hormones able to modify each other's actions on glucose handling [41].

6.2.3.3 The Function of Des-acyl Ghrelin

Des-acyl ghrelin was first identified as the inactive form of ghrelin unable to bind to GHS-R. Some authors suggested that des-acyl ghrelin might have an anorexigenic activity that is contrary to the orexigenic activity of acylated ghrelin [64]. Conversely, a recent study showed that both ghrelin and des-acyl ghrelin function as orexigenic peptides in the hypothalamus [65]. Des-acyl ghrelin was later proposed to have nonendocrine functions including cardioprotective, antiproliferative, and adipogenic activities and antagonizing octanoyl-ghrelin-induced effects on insulin secretion and blood glucose levels in humans. Furthermore, it has been reported that des-acyl ghrelin is able to antagonize acyl ghrelin-induced glucose output. These actions might be mediated by a different receptor than GHS-R1a, which is not expressed in the hepatocytes. In humans, the acute administration of acyl ghrelin induced a rapid rise of glucose and insulin levels [66]. In contrast, des-acyl ghrelin prevented the acyl ghrelin-induced rise of insulin and glucose when coadministered [66]. Further studies examining the physiological interaction of des-acyl ghrelin and its targets will help to highlight the roles of ghrelin-related peptides in the regulation of feeding and energy homeostasis.

6.2.4 The Effect of H. pylori *Infection on Ghrelin*

Since ghrelin is predominantly produced in the stomach, many researchers evaluated the effect of *H. pylori* infection on ghrelin expression in the gastric mucosa and blood. Theoretically, *H. pylori*-related gastritis may progress to atrophy with loss of oxyntic glands leading to a negative effect on ghrelin production in the stomach. Among the studies of this theme, Isomoto et al. [67] most well demonstrated the stepwise changes of ghrelin mRNA, peptide, and its blood level according to *H. pylori* infection status. The gastric ghrelin mRNA expression level of *H. pylori*-positive patients (1.64 ± 1.27 in arbitrary units) was significantly lower than in *H. pylori*-negative subjects (4.87 ± 4.1, $P < 0.0001$). A similar trend was noted for ghrelin peptide contents (31.2 ± 27.5 vs 81.2 ± 64.1 ng/mg protein, respectively, $P < 0.0001$). Although there was no significant difference in the number of ghrelin-immunoreactive cells/mm^2 in terms of *H. pylori* status, plasma ghrelin concentrations in *H. pylori*-infected patients ($144.6 \pm 7.8.8$ fmol/ml) were significantly lower than in uninfected subjects (196.1 ± 97.2, $P < 0.05$) and increased following the cure of the infection. In the study, plasma ghrelin levels correlated positively with the expression levels of ghrelin mRNA and peptide products. There was a significant stepwise decrease in gastric ghrelin mRNA expression, peptide contents, and density of ghrelin-immunoreactive cells with progression of histological severity of glandular atrophy in the corpus. However, there have been other different results and they will be discussed in the following Sects. (6.2.4.1, 6.2.4.2 and 6.2.5).

6.2.4.1 Ghrelin in Gastric Mucosa in Regard to *H. pylori* Infection

Comparisons of ghrelin levels in specimens from the stomach between *H. pylori*-infected and *H. pylori*-noninfected subjects have showed rather variable results (Table 6.2) [67–79]. First of all, the reports were heterogeneous in terms of diseases, specimen, and assay methods. Direct assays of ghrelin levels in gastric fluid samples [67, 69, 70, 73] and radioimmunoassay in gastric mucosa [67, 73] from infected and noninfected subjects were highly variable (Table 6.2). However, ghrelin mRNA expression looks like a more reliable method for assessment of gastric ghrelin. In most studies, mRNA expression was lower in *H. pylori*-positive subjects compared with *H. pylori*-negative subjects (Table 6.2) [67, 72, 74, 76]. Quantitative assessment of ghrelin-immunoreactive cells also showed relatively consistent results, fewer ghrelin-producing cells in *H. pylori*-positive subjects [67, 76, 78, 80, 81], indicating that the number of ghrelin-producing cells might decrease proportionally to the severity of *H. pylori* gastritis [67, 80, 81].

6.2.4.2 Ghrelin in Blood in Regard to *H. pylori* Infection

The circulating level of ghrelin is determined by the balance among its secretion rate, degradation rate, and clearance rate. While plasma esterases des-acylate acyl ghrelin, plasma proteases degrade circulating ghrelin. Therefore, the effect of *H. pylori* on circulating level of ghrelin seems like not simple. Consequently, studies which have compared blood ghrelin levels between *H. pylori*-infected and *H. pylori*-noninfected subjects showed inconsistent results (Table 6.3). Recently, Nweneka and Prentice [98] concluded that circulating ghrelin was significantly lower in *H. pylori*-positive subjects in a meta-analysis. Nevertheless, *H. pylori*-associated gastric ulcer was reported to be associated with high plasma ghrelin, but atrophic gastritis in *H. pylori*-positive subjects [99], or even in *H. pylori*-negative subjects [100], showed the lowest plasma ghrelin levels. These results suggest that *H. pylori* infection may have different effects on circulating levels of ghrelin according to different stage or disease of the infection.

Most researchers presume that total ghrelin levels were considered as a good surrogate marker not only for des-acyl but also acylated ghrelin. Majority of these studies in which only total ghrelin levels (not acylated ghrelin) were measured determined that *H. pylori* infection decreases plasma ghrelin levels. Few studies have evaluated acyl ghrelin for this theme, and the findings are again contradictory. In a Japanese data [96], the acylated ghrelin/total ghrelin ratio as well as plasma acyl ghrelin levels were reduced. However, Italian study [97] showed a significant increase in acyl ghrelin and the ratio of acylated ghrelin/total ghrelin for which the authors speculated a compensatory increase in the acylation process in response to a loss of total ghrelin secretion.

These controversial results might be also related with the uncertain optimum method of plasma ghrelin measurement. Several methods with different sensitivity

Table 6.2 Summary of studies that compared gastric levels of ghrelin between *H. pylori*-positive and *H. pylori*-negative subjects (including eradication results[a])

Author	Subjects	Sex	Age	No	Specimen	Method	Results in HP (+) subjects
Gokcel A et al. 2003, Turkey [79]	Healthy	W	A	39	NA	Commercial EIA	→Total ghrelin (Ghrl)
Tatsuguchi A et al[a] 2004, Japan [78]	Healthy + PU/CG	B	A	50	Body	IHC	↓Ghrl-expressing cell/ ↑ Ghrl-expressing cell after cure (12 weeks)
Isomoto H et al. 2005, Japan [77]	NUD	B	A	61	Fundus	RT-PCR	↓ Ghrl mRNA, associated with virulence strain (type 1)
Isomoto H et al[a] 2005, Japan [67]	Dyspepsia	B	A	81	Body	RT-PCR RIA/ICH	↓ Ghrl mRNA and peptides (atrophy and chronic inflammation), → Ghrl-expressing cell /→mRNA, Ghrl, and Ghrl-expressing cell after cure (4 weeks)
Osawa et al. 2005, Japan [76]	Healthy	B	A	160	Mid-body	RT-PCR/ ICH	↓Ghrl mRNA and Ghrl-expressing cell more so as atrophy increases
Osawa et al[a] 2006, Japan [75]	Healthy +PU/ CG	B	A	134 HP (+)	Mid-body	RT-PCR/ ICH	↑Ghrl mRNA and Ghrl-expressing cells after eradication (12 weeks)
Salles N et al. 2006, France [74]	Hospitalized in geriatric	B	A (>75 years)	62	Body	RT-PCR	↓Ghrl mRNA more so as atrophy increases
Jun DW et al. 2007, Korea [72]	CG	B	A	63	Body	RT-PCR	↓Ghrl mRNA in HP (+), but $P = 0.07$
Choe YH et al. 2007, Korea [73]	Not stated	B	A	41	Antrum/ body/ fundus	ELISA	→Ghrl in HP(+)/ after cure (4 weeks)
Jang EJ et al[a] 2008, Korea [71]	PU	B	A	22	Antrum and fundus	RT-PCR	↑Ghrl mRNA after cure (duration not stated),
Roper J et al. 2008, USA [70]	Healthy	M	A	256	Antrum and fundus	EIA	→Ghrl in juice/mucosa (both antrum and body)

(continued)

Table 6.2 (continued)

Author	Subjects	Sex	Age	No	Specimen	Method	Results in HP (+) subjects
Stec-Michalska K et al. 2009, Poland [69]	FD	B	A	88	Antrum and body	ELISA/RT-PCR	↑Ghrl mRNA and Ghrl in juice
Lee et al[a] 2010, Korea [68]	Healthy	B	A	9	Fundus	RT-PCR	↑Ghrl mRNA after cure (5 weeks)

[a]Studies including results after eradication

A adults, *B* both, *C* child, *W* women, *M* men, *Ghrl* ghrelin, *H* healthy subjects, HP *H. pylori*, *No* sample size, *NA* not available, *CG* chronic gastritis, *DM* diabetes mellitus, *PU* peptic ulcer, *Sick* patients who visited clinics, but the specific condition is not stated, *EIA* enzyme immunoassay, *ELISA* enzyme-linked immunosorbent assay, *ICH* immunohistochemistry, *RIA* radioimmunoassay, *RT-PCR* real-time polymerase chain reaction

↑ increased (or higher), ↓ decreased (or lower), → no significant change (or difference)

Table 6.3 Summary of studies that compared circulating levels of ghrelin between *H. pylori*-positive and *H. pylori*-negative subjects[a]

Author	Subjects	Sex	Age	No	Method	Specimen	Ghrl type	Results in HP (+)
Gokcel et al. 2003, Turkey [79]	Not stated	W	A	39	Commercial EIA	Plasma	Total	→
Isomoto et al. 2004, Japan [82]	Sick	B	A	68	Commercial RIA	Plasma	Total	↓
Isomoto et al. 2005, Japan [67]	Dyspepsia	B	A	81	In-house RIA	Plasma	Total	↓
Isomoto et al. 2005, Japan [83]	Sick	B	A	89	In-house RIA	Plasma	Total	↓
Osawa et al. 2005, Japan [76]	Healthy (health checkup)	M	A	160	In-house RIA	Plasma	Total	↓
Shiotani et al. 2005, Japan [84]	Healthy	B	A	132	Commercial ELISA	Serum	Total	↓
Konturek et al. 2006, Poland [85]	Healthy	B	B	180	Human RIA	Serum	Total	↓
Plonka et al. 2006, Poland [86]	Healthy	B	B	538	Commercial RIA	Serum	Total	↓
Plonka et al. 2006, Poland [87]	Healthy	B	C	287	Commercial RIA	Serum	Total	↓
Salles et al. 2006, France [74]	Hospitalized in geriatric	B	A	62	Commercial RIA	Plasma	Total	↓
Alonso et al. 2007,Spain [88]	Type I DM	B	A	15	Commercial RIA	Plasma	Total	↓
An et al. 2007, Korea [61]	Gastric cancer	B	A	41	Commercial ELISA	Plasma	Total	→
Cindoruk et al. 2007, Turkey [89]	Sick (normal 24pH without atrophy)	B	A	50	RIA	Plasma	Total	→
D'Onghia et al. 2007, Italy [90]	Colon cancer	B	A	29	RIA	Serum	Total	↓
de Martel 2007, USA [91]	Healthy + sick	B	A	110	Commercial ELISA	Serum	Total	→
Jun et al. 2007, Korea [72]	Chronic gastritis	B	A	63	Commercial RIA	Plasma	Total	→
Pacifico et al. 2008, Italy [92]	Healthy + GI symptom	B	C	85	Commercial RIA	Serum	Total	→
Roper et al. 2008, USA [70]	Healthy	M	A	256	Commercial EIA	Serum	Total	→
Shak et al. 2008, USA [93]	BMI >35Kg/m^2	B	A	24	Commercial EIA	Plasma	Total and acyl	→

(continued)

Table 6.3 (continued)

Author	Subjects	Sex	Age	No	Method	Specimen	Ghrl type	Results in HP (+)
Chuang et al[a] 2009, Taiwan [94]	PU + FD	M	A	145	Commercial RIA	Plasma	Total	↓
Chuang et al[a] 2009, Taiwan [94]	PU + FD	W	A	196	Commercial RIA	Plasma	Total	→
Gao et al. 2009, China [95]	Healthy (health checkup)	B	A	100	Commercial RIA	Plasma	Total	↓
Kawashima et al. 2009, Japan [96]	Healthy + sick (PU + CG)	B	A	220	Commercial EIA	Plasma	Acyl	↓
Campana et al. 2009, Italy [97]	Healthy + CG (HP (−))	B	A	50	RIA	Plasma	Acyl	↑ in atrophy

[a]Articles written in English and assay method was stated

A adults, *B* both, *C* child, *W* women, *M* men, *Ghrl* ghrelin, *HP H. pylori*, *No* sample size, *CG* chronic gastritis, *DM* diabetes mellitus, *FD* functional dyspepsia, *PU* peptic ulcer, *Sick* patients who visited clinics, but the specific condition is not stated, *EIA* enzyme immunoassay, *ELISA* enzyme-linked immunosorbent assay, *ICH* immunohistochemistry, *RIA* radioimmunoassay, *RT-PCR* real-time polymerase chain reaction

↑ higher, ↓ lower, → no significant difference

Table 6.4 Current methods for measuring ghrelin degradation and levels in vitro (Adapted from Satou M et al. [101])

Method	Principle	Merit/demerit	Sensitivity	Efficiency	Cost
HPLC	Chromatography	Simple/time consumption	+	+	++
RIA	Immunoreaction	High sensitivity/hazardous	+++	++	+++
ELISA	Immunoreaction	Easy handling/high costs	++	++	+++
Western blot	Immunoreaction	Easy handling/not accurately quantitative	++	++	+
MS	Ionization	Fast/not accurately quantitative	++	+++	+++

HPLC high-performance liquid chromatography, *RIA* radioimmunoassay, *ELISA* enzyme-linked immunosorbent assay, *MS* mass spectrometry

such as radioimmunoassay or enzyme immunoassay/enzyme-linked immunosorbent assay have been used (Table 6.4) [101]. However, more basically, there are numerous factors in addition to atrophy that could affect plasma ghrelin concentration, including gastrin, IGF-1, obesity, insulin resistance, hyperinsulinemia, cholesterol, and urine excretion (Table 6.1), and *H. pylori* infection may not be the only major determinant that affects plasma ghrelin concentration.

Taken together, to conclude the issue of the formation of different types of ghrelin as well as the true effect of *H. pylori* infection on blood ghrelin levels, further well-designed research is needed.

6.2.5 *The Effect of Eradication of* H. pylori *on Ghrelin Regarding FD Symptom*

6.2.5.1 The Effect of Eradication of *H. pylori* on Ghrelin in the Gastric mRNA or Blood Samples

Comparisons on the gastric ghrelin parameters before and after eradication have not been much performed (Table 6.2). Three studies demonstrated that *H. pylori* eradication increases ghrelin mRNA (including one randomized controlled study in which a control group did not receive eradication) [68, 71, 75]. Osawa et al. [75] demonstrated median preproghrelin mRNA expression was increased nearly fourfold, 12 weeks after *H. pylori* eradication in the study with 134 subjects with successful eradication. Two Korean studies showed an increased ghrelin mRNA after cure of the H. pylori infection [68, 71]. The number of ghrelin-immunoreactive cells also increased in 50 *H. pylori*-eradicated patients although atrophy and intestinal metaplasia were not significantly changed [78]. However, Choe et al. [73] failed to demonstrate a change in ghrelin expression 4weeks after the eradication (Table 6.2).

The effect of *H. pylori* eradication on circulating level of ghrelin was more evaluated than on the gastric levels, since the systemic level of ghrelin is an attractive issue to other fields than gastroenterology. Unfortunately, the results were found to be still inconsistent (Table 6.5). At first, Nwokolo et al. [102] reported a rise in circulating plasma ghrelin levels following eradication of *H. pylori* in their study of 12 healthy subjects. Further studies reported elevated levels of blood ghrelin after the cure of *H. pylori* [71, 96, 103]. However, subsequent studies found no change [67, 68, 73, 82, 83, 89] or even decrease [75, 92] in circulating ghrelin levels. Nweneka and Prentice [98] performed a meta-analysis that showed that *H. pylori* eradication does not have an effect on circulating ghrelin levels. Nevertheless, pre-eradication elevation of ghrelin may be a predictor of a fall in plasma levels post-eradication [76, 98].

6.2.5.2 The Effect of Eradication of *H. pylori* on Ghrelin Regarding Functional Dyspepsia Symptom

Possible mechanisms by which *H. pylori* may elicit dyspeptic symptoms include alterations of gastric motility and endocrine and acid secretory abnormalities. Since ghrelin regulates acid secretion, appetite, and gastrointestinal motility, whether *H. pylori* infection or eradication could affect ghrelin levels might be closely related with the role of *H. pylori* in developing FD. Even though there have been some studies which evaluated the association between plasma ghrelin level and FD, there was no study which analyzed the effect of *H. pylori* infection status or its eradication on the change of both ghrelin levels and dyspeptic symptoms, so far (Table 6.6).

Two studies reported that fasting total ghrelin levels were significantly lower in patients with dysmotility-like FD than healthy volunteers [55, 104]. In addition, Shindo et al. [56] also showed significantly lower plasma acyl ghrelin levels in patients with PDS compared with healthy volunteers. On the contrary, Nishizawa et al. [105] reported that plasma acyl ghrelin levels were significantly higher in each dysmotility or ulcer-like dyspepsia group not in the nonspecific type compared with the control group, but *H. pylori* infection status was not considered in his study [105]. Two other studies with only women showed no difference between control and FD groups [106, 107]. It suggests the importance of classification of exact subgroup in the research regarding the relationship between ghrelin and FD. Acyl ghrelin may be closely associated with PDS than EPS.

In our study, plasma acyl ghrelin was lower in PDS than in control but not in EPS type [108]. However, plasma des-acyl ghrelin and gastric mRNA expression of preproghrelin did not show any significant difference [108]. Decreased expression of gastric preproghrelin mRNA in the subjects with atrophy was also shown. Importantly, plasma acyl ghrelin level and preproghrelin mRNA expression in the corpus were elevated 1 year after the eradication of *H. pylori* [108]. Moreover, the symptom improvement correlated with the upregulation of plasma acyl ghrelin but not with gastric ghrelin mRNA [108]. These results suggest that plasma acyl ghrelin could be one of the causal factors of dyspepsia, especially PDS, and anti-*H. pylori*

Table 6.5 Summary of studies that compared circulating levels of ghrelin before and after *H. pylori* eradication [a]

Author	Subjects	Sex	Age	No	Method	Specimen	Ghrl type	FU (weeks)	Results after eradication
Nwokolo et al. 2003, UK [102]	Healthy	B	A	10	Commercial RIA	Plasma	Total	6	↑
Isomoto et al. 2004, Japan [82]	Sick	B	A	9	Commercial RIA	Plasma	Total	4	→
Isomoto et al. 2005, Japan [67]	Dyspepsia	B	A	43	In-house RIA	Plasma	Total	4	→
Isomoto et al. 2005, Japan [83]	Sick	B	A	10	In-house RIA	Plasma	Total	4	→
Osawa et al. 2006, Japan [75]	Healthy +PU/CG	M	A	134	In-house RIA	Plasma	Total	12	↓
Choe et al. 2007, Korea [73]	Sick	B	A	8	Commercial ELISA	Plasma	Total	4	→
Cindoruk et al. 2007, Turkey [89]	Sick (normal 24pH without atrophy)	B	A	23	RIA	Plasma	Total	12	→
Czesnikiewicz-Guzik et al. 2007, Poland [103]	Sick	W	A	49	Commercial RIA	Plasma	Total	4	↑
Jang et al. 2008, Korea [71]	PU	B	A	15	Commercial RIA	Plasma	Total	NA	↑
Pacifico et al. 2008, Italy [92]	Healthy + sick (GI symptom)	B	C	22	RIA	Serum	Total	52	↓
Kawashima et al. 2009, Japan [96]	Both (PU + CG)	B	A	49	Commercial EIA	Plasma	Acyl	23	↑
Lee et al. 2010, Korea [68]	Healthy	B	A	9	ELISA	Plasma	Total	5	→

[a] Articles written in English and assay method was stated

[b] From the same article

A adults, *B* both, *C* child, *W* women, *M* men, *Ghrl* ghrelin, *HP H. pylori*, *NA* not available, *No* sample size, *CG* chronic gastritis, *FU* follow-up, *GI* gastrointestinal, *PU* peptic ulcer, *Sick* patients who visited clinics, but the specific condition is not stated, *EIA* enzyme immunoassay, *ELISA* enzyme-linked immunosorbent assay, *ICH* immunohistochemistry, *RIA* radioimmunoassay, *RT-PCR* real-time polymerase chain reaction

↑ increased, ↓ decreased, → no significant change

Table 6.6 Summary of studies that analyzed the relationship between plasma ghrelin levels and functional dyspepsia

Author	C	Patients (n)	Con (n)	Acyl Ghrl	Des-acyl Ghrl	Total Ghrl	Correlation
Shinomiya et al. 2005, Japan [106][a]	II	Dysmotility-like (14) Ulcer-like (4)	18	No difference	No difference	(−)	Acyl Ghrl positively correlated with symptom score
Nishizawa et al. 2006, Japan [105][b]	II	Dysmotility-like (16) Ulcer-like (12) Nonspecific type (19)	17	↑in FD ↑in dysmotility-like ↑in ulcer-like	(−)	↑in FD ↑ in dysmotility-like	Ghrl (acyl & total) positively correlated with indigestion score
Takamori et al. 2007, Japan [104]	II	Dysmotility-like (16)	19	No difference	↓ in FD	↓ in FD	Plasma Ghrl (acyl, des-acyl, total) did not correlate with T_{max}
Lee et al. 2009, Korea [55]	II	Dysmotility-like (42)	14	(−)	(−)	↓ in FD	Pre- or postprandial Ghrl levels did not significantly correlated with gastric half-emptying time
Shindo et al. 2009, Japan [56]	III	PDS (76), EPS (36)	20	↓ in PDS	No difference	(−)	Acyl Ghrl negatively correlated with T_{max}
Kim et al. 2012, Korea [107][a, c]	III	PDS (13), EPS (9)	12	No difference	(−)	(−)	Fasting acyl Ghrl negatively correlated with epigastric pain score. Postprandial/fasting Ghrl ratio positively correlated with score of early satiety

[a]Only women were enrolled
[b]*H. pylori* infection status was not mentioned
[c]Patients are older than control
C criteria, *Con* control subjects, *EPS* epigastric pain syndrome, *FD* functional dyspepsia, *Ghrl* ghrelin, *PDS* postprandial syndrome, T_{max} maximum gastric-emptying time, *II* ROME II, *III* ROME III criteria, (−) not applicable
↑ higher, ↓ lower, → no significant difference

therapy may be considered with a priority as a *H. pylori*-associated functional dyspepsia.

In spite of strong positive correlations among total, acyl, and des-acyl ghrelin, only a weak positive correlation between ghrelin mRNA and plasma acyl ghrelin was demonstrated in our study. As a result, some subjects with lower expression levels of ghrelin mRNA or reduced plasma total ghrelin did not show dyspeptic symptom. Moreover, at 1-year follow-up after *H. pylori* eradication, some subjects with increased ghrelin mRNA did not show upregulation of plasma acyl ghrelin in our study. This strongly suggests that there might not be a certain correlation between gastric mRNA and blood ghrelin (total or acyl) levels. Previously, Osawa et al. [75] also reported no correlation among the changes of plasma ghrelin or gastric preproghrelin mRNA or the number of ghrelin-positive cells after the *H. pylori* cure. Furthermore, after *H. pylori* cure plasma ghrelin concentration was more strongly influenced by body weight change than by the increase in gastric preproghrelin mRNA or the number of ghrelin-producing cells [75]. According to our study, the upregulation of plasma acyl ghrelin does not seem to be only epiphenomenon following the resolution of *H. pylori* infection, since there were some correlations between symptom improvement and the increase in plasma acyl ghrelin even in *H. pylori*-negative patients. Even though there was no significant difference in prevalence of *H. pylori* positivity between the control group and FD group, reduced plasma acyl ghrelin levels may attribute to dyspeptic symptoms in some subset of *H. pylori*-infected FD patients. Taken together, further studies are necessary regarding the dynamics of gastric ghrelin production and regulation of its circulating levels. In addition, it is very important to identify the true effect of *H. pylori* eradication on the gastric and circulating ghrelin levels as well as on the symptom changes in patients with FD.

6.3 Leptin

6.3.1 Production of Leptin

Leptin was discovered in 1994 as a hormone produced by adipose tissue with a modulatory effect on feeding behavior and weight control [4]. Leptin is a product of the obese (Ob) gene, which is located on chromosome 7 in humans and acts through its receptor Ob-R. Interestingly, the stomach has been identified as an important source of leptin [4], and leptin-producing cells were found to be localized in the lower half of the fundic glands, a site similar to that of the pepsinogen-secreting chief cells. The soluble isoform of its receptor (Ob-R) is secreted by chief cells in the gastric mucosa [109]. Moreover, intestines do express membrane-bound leptin receptors on their brush border [4, 109, 110] (Fig. 6.3). Collectively, gastric exocrine and endocrine secretions of leptin constitute a gastroenteric axis to coordinate its roles in the GI tract [110].

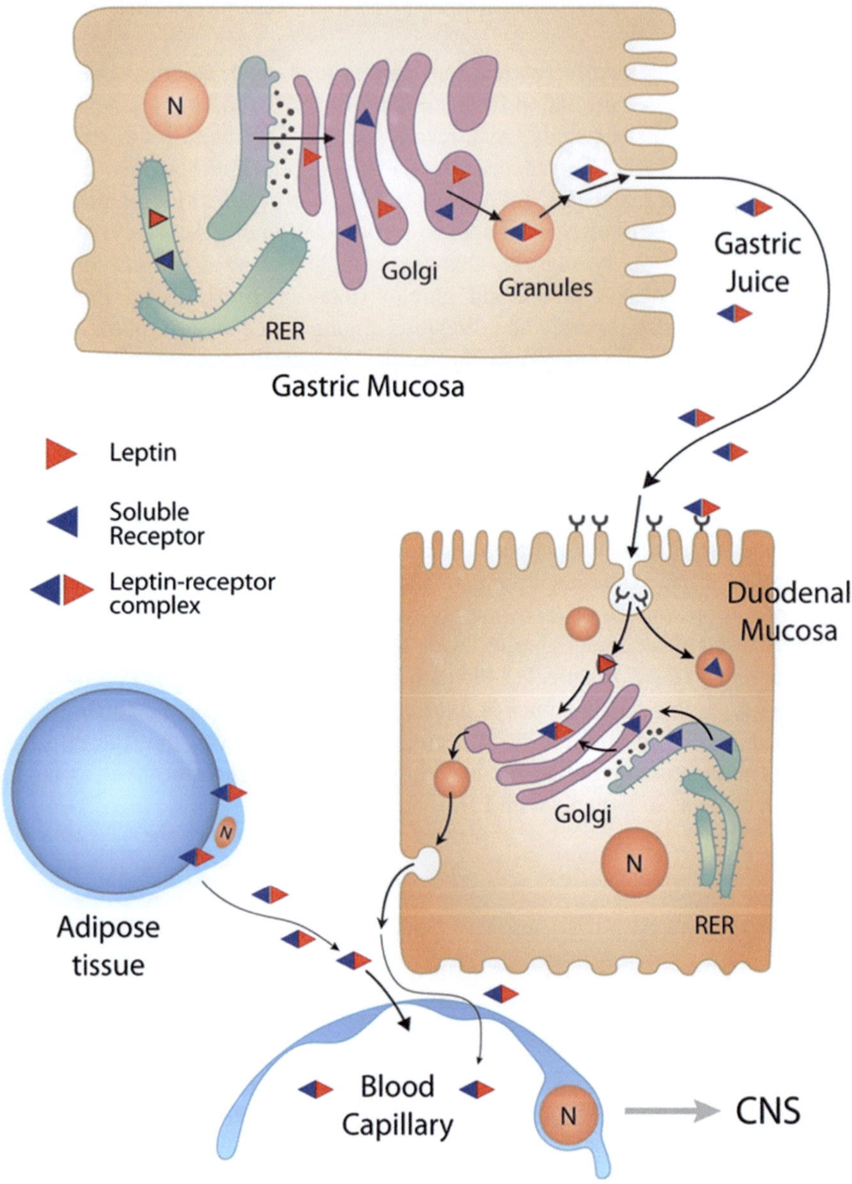

Fig. 6.3 Schematic drawing that illustrates the secretion of leptin by the adipocyte and the gastric chief cell (Adapted from Cammisotto et al. [110]). Both types of cell secrete the leptin receptor. Both the leptin and the leptin receptor are synthesized in the rough endoplasmic reticulum, transferred to the Golgi apparatus, and packaged into either small vesicles (adipocytes) or secretory granules (gastric cells). At the level of the trans-cisternae of the Golgi and in the secretory granules, leptin binds its soluble receptor to form the leptin–leptin receptor complex. This complex is discharged by both cell types through an exocytotic event. The adipose tissue secretes toward the blood circulation, while the gastric cells secrete in an exocrine fashion into the

6.3.2 Regulation and Role of Gastric Leptin

6.3.2.1 Regulation of Gastric Leptin

Gastric leptin is sensitive to the nutritional status of the body. Fasting for 48 h induced a decrease in gastric leptin expression and content in rats [32]. Food intake quickly depletes gastric leptin, while sustained feeding stimulates the leptin gene expression and leptin synthesis [32, 111]. These observations are similar in rodents and humans [111].

Leptin mRNA expression in rat gastric mucosa is upregulated by sucrose-rich but not by fat-rich diets [112]. Fasted rats refed with a carbohydrate-rich diet have their gastric leptin synthesis increased [46]. Intravenous infusions of cholecystokinin (CCK), pentagastrin, or secretin trigger leptin release into the gastric juice [109]. Insulin also stimulates gastric leptin secretion, but this effect is dependent on the integrity of the vagal nerve [113]. Finally, leptin secretion seems to downregulate its own production in the stomach, as Zucker fa/fa rats with no functional leptin receptor are characterized by an upregulation of gastric leptin mRNA and gastric leptin content [114]. Interestingly, leptin infusion in rats leads to an increase in histamine gastric tissue content, which may suggest a counter-regulatory process [115].

6.3.2.2 Role of Gastric Leptin

Leptin derived from the stomach can be distinguished from adipocyte leptin through its rapid increased secretion following a meal and through its exocrine secretion (i.e., mainly in the gastric lumen). While the gastric mucosa secretes leptin within minutes after the beginning of food intake, it takes several hours for adipocytes to release significant amounts of leptin. Differences in time frame of secretion between adipose tissue and gastric mucosa may reflect different roles. Gastric leptin remains stable in gastric juice even at pH 2. Moss and Calam [116] also suggests that it could act locally in the stomach to affect gastric functions.

It has been considered that gastric leptin may exert paracrine effects within the gastric mucosa or stimulating vagal afferents to signal the central nervous system [109]. Gastric leptin controls food intake and satiety sensations by acting on the stomach itself. It potentiates the effect of CCK by slowing gastric emptying and promoting gastric distension [117], not directly affecting the CNS [118]. It also

Fig. 6.3 (continued) gastric juice. Leptin in the gastric juice is vehiculated to the duodenal lumen. In the duodenum leptin–leptin receptor complex is internalized and separated. The leptin is channeled toward the trans-Golgi cisternae where it binds a newly synthesized soluble leptin receptor. The complex reaches the blood circulation. *RER* rough endoplasmic reticulum, *CNS* central nervous system, *N* nucleus

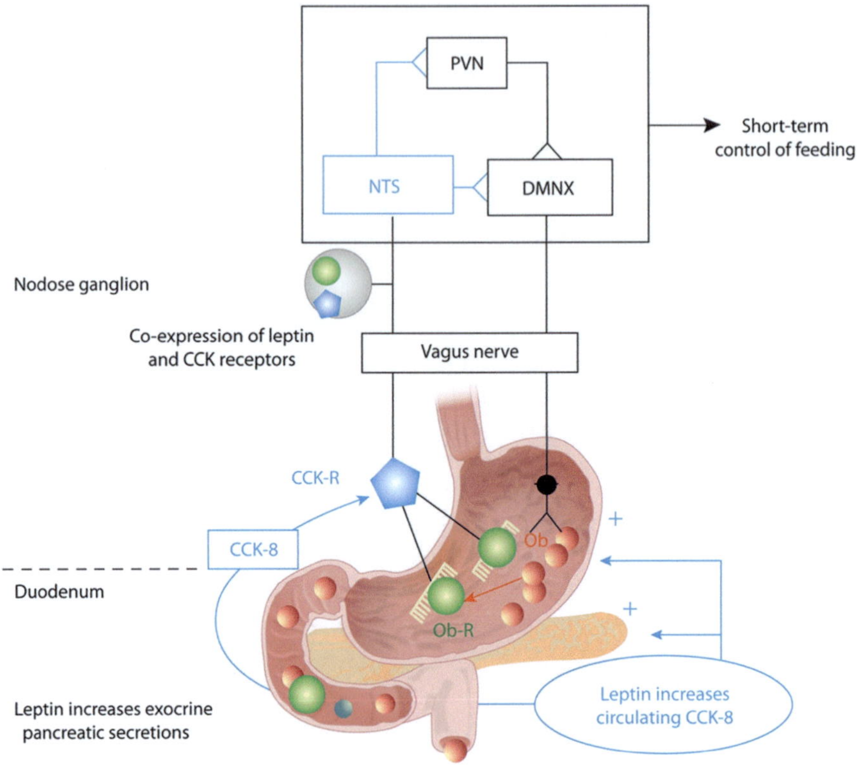

Fig. 6.4 Model of the action of gastric leptin (Adapted from Guilmeau et al. [111]). Leptin (secreted by the stomach) and CCK can be considered as short-term gastrointestinal signals in the control of feeding. These signals locally activate their receptors on vagal terminals to generate signals that are processed in the NTS and the paraventricular nucleus (PVN) of the hypothalamus. The stomach-derived leptin secreted in the lumen enters the intestine in an active form. *CCK-R* cholecystokinin receptor, *DMNX* dorsomotor nucleus of the vagus nerve, *Ob* leptin, *Ob-R* leptin receptors

stimulates the production of glucagon-like peptides 1 and 2 (GLP1 and GLP2) that inhibit gastric emptying [119, 120].

Leptin receptor isoforms have been found in the rat nodose ganglion, which contains the cell bodies of vagal afferent neurons, and in the vagus nerve proper [121, 122]. Signals arising from the upper gastrointestinal tract are conveyed by the viscero-sensitive vagal afferent neurons to the nucleus of the solitary tract (NTS), then to the hypothalamus where they participate in the process of meal-induced termination of food intake [123] (Fig. 6.4). CCK secreted from duodenal endocrine I cells typically functions as one of these short-term satiety signals via activation of the CCK-1 receptor [124].

On the other hand, exocrine luminal leptin has been shown to act directly on intestinal cells through their specific receptors present on enterocyte microvilli. It regulates the transport of nutrients and enhances di- and tripeptide uptake by increasing the number of PepT1 transporters on microvilli [110].

In short, while adipose leptin acts on the long term mainly through its interactions with the central nervous system, gastric leptin acts locally directly on the gastric mucosa for regulating food intake. Gastric leptin influences both the intestinal track and the central nervous system (Fig. 6.4).

6.3.3 Leptin and H. pylori

Changes in gastric and serum leptin levels in *H. pylori*-infected patients have been the matter of several investigations. *H. pylori* infection significantly increased gastric leptin expression, and the cure of the infection significantly reduced this expression with a concomitant increase in BMI [125]. *H. pylori*-infected gastric mucosa was found to be capable of releasing larger amounts of leptin mRNA than that without *H. pylori* infection [4, 72]. Although the majority of studies have shown an increase of gastric leptin mRNA in *H. pylori*-infected subjects (Table 6.7), still there is a discrepancy [70, 74].

Leptin-secreting endocrine cells are present in the gastric mucosa, but they are few in number and scattered in the gastric mucosa close to blood capillaries [111]. These cells do not reach gastric lumen but secrete leptin to blood circulation. Since adipocytes are predominant source of leptin, whether *H. pylori* infection could affect circulating levels of leptin could be more complex (Table 6.8). It has been reported that rapid increase in the concentration of plasma leptin in response to CCK was involved in the mobilization of a gastric leptin store [126]. Similar to gastric leptin mRNA, higher serum leptin levels were reported in the subjects with *H. pylori* infection compared with *H. pylori*-negative subjects [70, 85, 127]. However, many studies reported no significant difference in plasma leptin according to *H. pylori* infection status [4, 72, 74, 84, 94]. Similarly, serum leptin level did not change significantly after curing *H. pylori* infection [102, 125].

6.3.4 Leptin in Comparison with Ghrelin in the GI Tract

Effects of ghrelin on appetite and observed changes in ghrelin levels in response to fasting or eating are opposite to those of leptin [128]. Ghrelin induces adiposity in adipose tissues, increases appetite, and initiates eating behavior [129]. Circadian changes in the level of circulating ghrelin and leptin are reciprocal [130]. Leptin and ghrelin also have contradictory effects on intestinal inflammation. For example, ghrelin has ameliorated inflammation in the GI tract in a mouse model of colitis [131]. Ghrelin's effect on GI tract motility is also generally opposite to that of leptin. Ghrelin stimulates gastric acid secretion and motility and accelerates gastric emptying and small intestinal transit [132]. In contrast, leptin has decreased the expression and secretion of ghrelin from gastric mucosa [24]. Interestingly, leptin

Table 6.7 Summary of studies that compared gastric levels of leptin between *H. pylori*-positive and *H. pylori*-negative subjects (including *H. pylori* eradication)[a]

Authors	Subjects	Sex	Age	No	Specimen	Method	Results in HP (+) subjects
Breidert M et al. 1999, USA [4]	Dyspepsia	B	A	39	Antrum and body	Commercial ELISA	↑Leptin mRNA in HP (+) in only body
Azuma T et al. 2001, Japan [125]	CG	B	A	201 HP (+)	Fundus	RT-PCR	↑Leptin mRNA in HP (+) /↓Leptin mRNA after cure (12 weeks)
Salles N et al. 2006, France [74]	Hospitalized in geriatric	B	A (>75 years)	62	Body	RT-PCR	↓Leptin mRNA more so as atrophy increases
Jun DW et al. 2007, Korea [72]	CG	B	A	63	Body	RT-PCR	↑Leptin mRNA in HP (+)
Roper J et al. 2008, USA [70]	Healthy	M	A	256	Antrum and fundus	Commercial ELISA	→ Leptin in mucosa

[a]Articles written in English and assay method was stated

A adults, *B* both, *C* child, *W* women, *M* men, *Ghrl* ghrelin, *H* healthy subjects, *HP H. pylori*, *No* sample size, *CG* chronic gastritis, *PU* peptic ulcer, *Sick* patients who visited clinics, but the specific condition is not stated, *EIA* enzyme immunoassay, *ELISA* enzyme-linked immunosorbent assay, *ICH* immunohistochemistry, *RIA* radioimmunoassay, *RT-PCR* real-time polymerase chain reaction

↑ increased (or higher), ↓ decreased (or lower), → no significant change (or difference)

cells are adjacent to ghrelin cells in the gastric mucosa, surrounding ghrelin cells in the lower half of the stomach, possibly providing a paracrine regulation of ghrelin secretion [128]. Interestingly, ghrelin shows some effects on the gut that are similar to those of leptin. For example, ghrelin decreases the rate of apoptosis in the intestinal cells [133], ameliorates ischemic–reperfusion injury, and decreases the permeability of the intestine in case of shock [129]. Further studies are required to elucidate the nature of the interaction between ghrelin and leptin in health and disease, clearly.

Table 6.8 Summary of studies that compared circulating levels of leptin between *H. pylori*-positive and *H. pylori*-negative subjects (including *H. pylori* eradication)

Authors	Subjects	Sex	Age	No	Specimen	Method	Results after eradication
Breidert M et al. 1999, USA [4]	Dyspepsia	B	A	39	Plasma	Commercial ELISA	→
Azuma T et al. 2001, Japan [125]	CG	B	A	201 HP (+)	Serum	Commercial ELISA	→after cure (12 weeks)
Nwokolo et al. 2003, UK [102]	Healthy	B	A	10	Plasma	Commercial RIA	→after cure (6 weeks)
LanKarani KB et al. 2004, Iran [127]	Dyspepsia + healthy	B	A	66	Serum	Commercial ELISA	↑
Shiotani et al. 2005, Japan [84]	Healthy	B	A	132	Serum	Commercial ELISA	→
Konturek et al. 2006, Poland [85]	Healthy	B	B	180	Serum	Human RIA	↑
Plonka et al. 2006, Poland [87]	Healthy	B	C	287	Serum	Commercial RIA	↓
Salles N et al. 2006, France [74]	Hospitalized in geriatric	B	A (>75 years)	62	Plasma	Commercial RIA	→
Jun DW et al. 2007, Korea [72]	CG	B	A	63	Plasma	Commercial RIA	→
Roper J et al. 2008, USA [70]	Healthy	M	A	256	Serum	Commercial ELISA	↑
Pacifico et al. 2008, Italy [92]	Healthy + sick (GI symptom)	B	C	85	Serum	Commercial RIA	↓in HP (+)/ ↑ after cure
Chuang et al.[a] 2009, Taiwan [94]	PU + FD	M	A	145	Plasma	Commercial RIA	→
Chuang et al[a] 2009, Taiwan [94]	PU + FD	W	A	196	Plasma	Commercial RIA	→

A adults, *B* both, *C* child, *W* women, *M* men, *Ghrl* ghrelin, *H* healthy subjects *HP H. pylori*; *No* sample size, *CG* chronic gastritis, *FD* functional dyspepsia, *GI* gastrointestinal, *PU* peptic ulcer, *Sick* patients who visited clinics, but the specific condition is not stated, *EIA* enzyme immunoassay, *ELISA* enzyme-linked immunosorbent assay, *ICH* immunohistochemistry, *RIA* radioimmunoassay, *RT-PCR* real-time polymerase chain reaction

↑ increased (or higher), ↓ decreased (or lower), → no significant change (or difference)

6.4 Gastrin and Somatostatin Related with Acid Secretion

One of the important functions in gastric physiology is to regulate and sustain acid secretion to sterilize ingested nutrients in which several hormones are involved. There are three regulatory molecules that stimulate acid secretion (acetylcholine, histamine, and gastrin) and one regulatory molecule that inhibits acid secretion (somatostatin). Acetylcholine is a neurotransmitter that is released by enteric neurons, while histamine is a paracrine that is released from enterochromaffin-like (ECL) cells. Gastrin is a hormone that is released by G cells, and somatostatin (SST) is produced from D cells in the stomach (Fig. 6.5) [134]. Acetylcholine and histamine directly stimulate parietal cells to increase acid secretion. Gastrin stimulates acid secretion by stimulating histamine release from ECL cells. Gastrin also has a direct effect on parietal cells, which stimulates parietal cells' proliferation. In oxyntic glands of the gastric body and fundus, SST-releasing cells are anatomically and functionally coupled to parietal and ECL cells. When the pH of the stomach

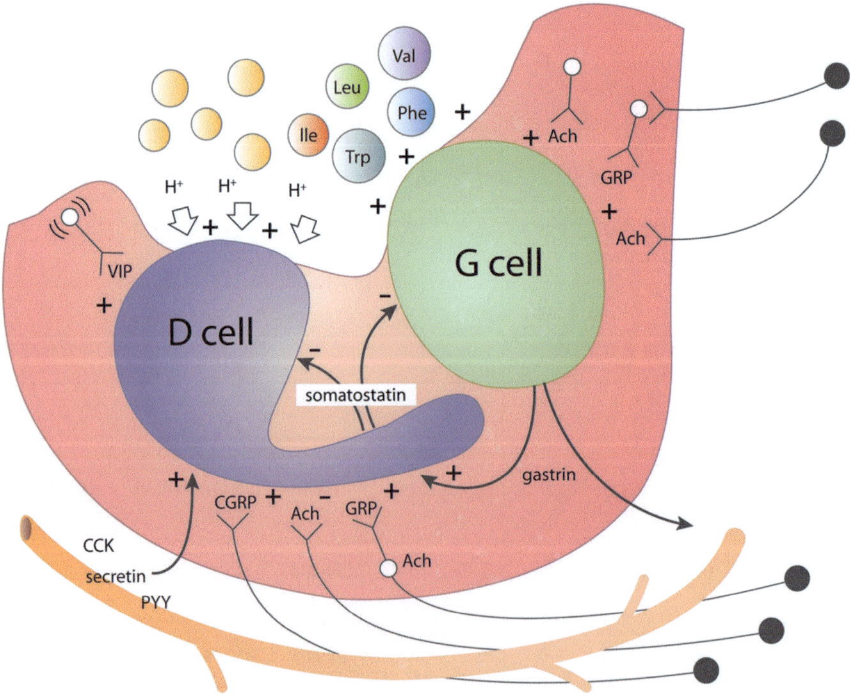

Fig. 6.5 Gastrin release is stimulated by extramural cholinergic and intramural cholinergic and non-cholinergic factors and inhibited by somatostatin (Adapted from Kaneko et al. [134]). Somatostatin release is stimulated by luminal (acid), paracrine, and hormonal factors (gastrin), and by intramural non-cholinergic factors, and is inhibited by extramural cholinergic factors. *CCK* cholecystokinin, *PYY* peptide YY, *CGRP* calcitonin gene-related peptide, *VIP* vasoactive intestinal peptide, *GRP* gastrin-releasing peptide, *Ach* acetylcholine

gets too low, somatostatin secretion is stimulated. Somatostatin inhibits acid secretion by direct effects on parietal cells and also by inhibiting release of the positive regulators, histamine and gastrin. Inversely, endogenous gastrin release ceased when the pH of the perfusate dropped below 2.5. Gastrin release is suppressed primarily by direct contact of acid with the antrum [135]. Taken together, acid secretion and physiology of G and D cells are closely related to each other, and the effect of *H. pylori* on these cells needs to be understood in association with the acid secretory system. This part focuses on the two major acid regulatory hormones, gastrin and somatostatin, and acid secretion related with *H. pylori* infection.

6.4.1 Gastrin and H. pylori *Infection*

Levi et al. [136] reported that both basal/stimulated acid secretion and basal/meal-stimulated plasma gastrin levels were significantly higher in *H. pylori*-positive duodenal ulcer (DU) patients (n = 25) compared with *H. pylori*-negative DU patients (n = 6), in which the phenomenon has been named as "the gastrin link." Their data clearly demonstrated that *H. pylori* infection-induced hypergastrinemia was followed by an increase in acid secretion in DU patients. Eradication of *H. pylori* reduced the output of gastrin in the stomach [137].

6.4.2 Somatostatin and H. pylori *Infection*

SST is an originally discovered in sheep hypothalamus, which showed an inhibitory effect on growth hormone release. In mammals, the GI tract and pancreas contain the largest amounts of somatostatin. Most of the gastrointestinal SST immunoreactivity is confined to the mucosal layer [138], where epithelial endocrine cells, D cells, are localized.

In the antrum, the D cells have apical membranes that are exposed to the lumen ("open cells"). In the corpus, the D cells are of the "closed" type; they are not exposed to the luminal surface of the mucosa [139]. SST plays an important role in the regulation of gastric acid secretion by inhibiting gastrin release. In the antral mucosa, the open D cell releases SST in response to increased acidity in the gastric lumen. Because the apical surface of D cells opens onto the gastric lumen, changes in pH may be sensed directly through chemoreceptors on the apical membranes.

The effect of *H. pylori* infection on gastric levels of SST has been relatively well investigated compared with ghrelin and leptin. SST release cannot be exactly assessed by measuring plasma levels, because SST is released from many organs and destroyed locally [139]. Instead, SST levels in both gastric mucosa and juice reflect the local SST regulation [140]. Antral SST concentrations were decreased in *H. pylori*-infected patients, but not the corpus [141] which has been demonstrated by measuring the SST concentration in the biopsy specimens from both sites

[142, 143]. Moss et al. [144] demonstrated that eradication of *H. pylori* from patients with duodenal ulcer caused an approximately twofold increase in SST mRNA in antral, but not in corpus biopsies. They also showed an increase in D cell numbers after *H. pylori* eradication in subjects with active duodenal ulcer [144].

6.4.3 *Acid Secretion and* H. pylori *Infection*

The effect of *H. pylori* on acid secretion depends on the stage of infection or predominant location where the infection occurs. That is, once *H. pylori* infection is established, transient hypochlorhydria occurs. Chronic infection in the antrum causes upregulation of gastrin and subsequent elevated acid secretion, while chronic infection in the corpus leads to impaired acid secretion by direct suppression of H^+-K^+-ATPase or involvement of cytokines such as IL-1β or TNF-α. Therefore, the direct inhibition of H^+-K^+-ATPase, indirect inhibition through cytokines, and loss of parietal cells by ongoing inflammation were three mechanisms for the low acid secretion.

The recent progress of *H. pylori*-dependent molecular mechanism of acid secretion is stated in the next sections.

6.4.3.1 Mechanism for Hypochlorhydria by Acute *H. pylori* Infection

In the acute phase of *H. pylori* infection, transient hypochlorhydria occurs [145], while low acidic status continues from several weeks to months. Low acid secretion by acute *H. pylori* infection was demonstrated in the absence of loss of parietal cells [146], impaired permeability of gastric mucosa [146], and glandular atrophy [147]. Moreover, IL-1β which is produced by neutrophils and inhibits acid secretion was not involved in this stage. Instead, this low acid secretion in the acute stage is likely to result from direct contact with parietal cells by *H. pylori* or its product [148, 149]. Studies about ultrastructure of stomach reported that *H. pylori* was observed adjacent to parietal cells and even detected in secretory canaliculi of these cells [150, 151]. In human, *H. pylori* infection also suppressed acid secretion via histamine, acetylcholine, and cAMP [152, 153], and this inhibition was resolved soon after the eradication of *H. pylori* [154]. This transient inhibition of acid secretion in the acute phase facilitates the successful settlement of *H. pylori* in the stomach.

6.4.3.2 The Effect of *H. pylori* Infection on H^+-K^+-ATPase

H^+-K^+-ATPase has α-subunit (HKα) and β-subunit. *H. pylori*-infected gastric mucosa or gastric epithelial cell lines showed the inhibition of HKα promoter activity in the endogenous or transfected H^+-K^+-ATPase [145]. Specifically,

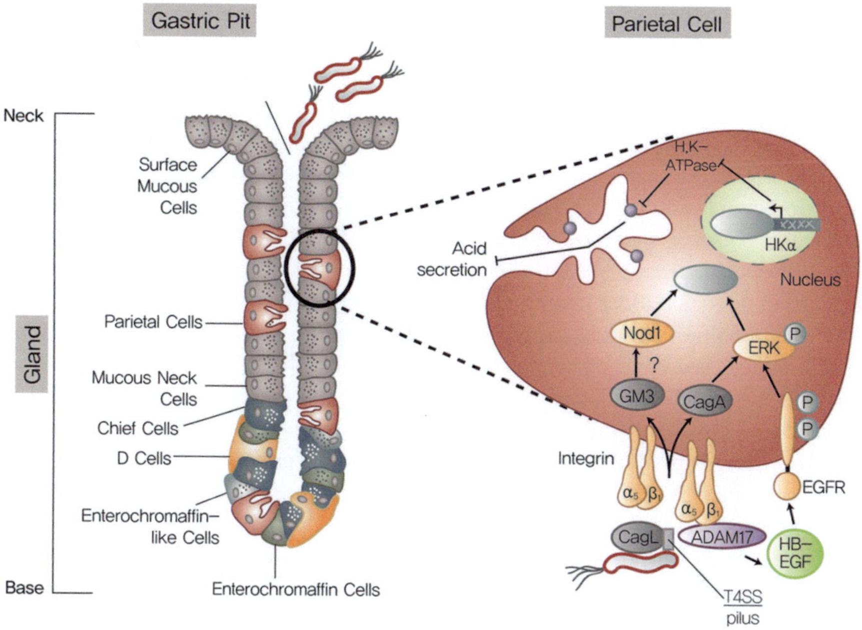

Fig. 6.6 Schematic illustration of a corporal gland in the human stomach focusing on the acid-secreting region (Adapted from Smolka et al. [145]). The location of various cell types in a gland of the human corpus was indicated in different colors. A gastric parietal cell is enlarged to the right, showing *Helicobacter pylori* interacting with integrins through CagL, injecting CagA and possibly the bacterial peptidoglycan-derived glycosylated tripeptide GM-3, leading to the activation of diverse host signaling pathways. The consequent mobilization of nuclear factor-κB (NF-κB) p50 homodimers to the nucleus results in the repression of gastric H, K-adenosine triphosphatase (H, K-ATPase) α subunit transcription and the inhibition of acid secretion as indicated *ADAM 17*, a disintegrin and a metalloprotease 17, *EGFR* EGF receptor, *HB-EGF* heparin-binding epidermal growth factor, *T4SS* type IV secretion system

H. pylori inserts its protein to gastric cells through the type IV secretion system (T4SS), which look like cylindrical channel, and this inserted protein upregulated NF-κB. Interestingly, the site which NF-κB combined in the promoter of H^+-K^+-ATPase was identified, and this fusion of NF-κB p50 homodimer resulted in repression of the transcription of HKα [145].

CagA protein encoded by *cag* pathogenicity island (cag PAI); CagL, CagE, and Cag M, which consist of the T4SS; and lytic transglycosylase are mechanistically involved in NF-κB activation and repression of HKα transcription (Fig. 6.6). CagL, a T4SS pilus component, binds to the integrin α5 β 1 to mediate translocation of virulence factors into the host cell and initiate signaling. During acute *H. pylori* infection, CagL dissociates ADAM 17 (a disintegrin and a metalloprotease 17) from the integrin α5 β 1 complex and stimulates ADAM17-dependent release of heparin-binding epidermal growth factor (HB-EGF), EGF receptor (EGFR) stimulation, ERK1/2 kinase activation, and NF-κB-mediated repression of HKα [145].

6.4.3.3 The Effect of Acid Secretion of IL-1β and TNF-α

As noted above, the inflammation via multiple cytokines seems to be a key mechanism for the changes in the endocrine system by *H. pylori* infection. It has been explained that the destroyed acid homeostasis influences the distribution of *H. pylori* in the stomach itself and finally the degree of gastritis. That is, *H. pylori* infection could diminish the number or function of D cells in the antrum, while it elevates the gastrin secretion of G cells via IL-8, IL-1β, or TNF-α [155, 156]. This disorganized endocrine system dumps excessive amount of acid into duodenum leading to duodenal ulcer. On the other hand, upregulation of IL-1β and TNF-α by *H. pylori* infection strongly suppresses acid secretion [157], and at the same time, IL-1β reduces secretion of histamine from ECL cells [158]. This makes *H. pylori* thrive in the corpus and destroy parietal cells and finally aggravating impaired acid secretion. *H. pylori* infection, simultaneously, enhances the gastrin release. Consequently, the gastrin level in blood was further accelerated due to reduced SST secretion, since IL-1β and TNF-α inhibit SST release during the Th1 immune response [159]. This chronic low acidic milieu accompanying atrophy and elevated gastrin has been considered to be a good condition for developing gastric cancer.

6.4.3.4 The Changes of Acid Secretion After *H. pylori* Eradication

Progression to certain disease by *H .pylori* infection is known to be determined by the degree of acid secretion when the organism invades the stomach. In the case of subjects with high acid secretion, *H. pylori* escapes the corpus and settles in the antrum leading to antrum-dominant gastritis with excessive gastrin release. By contrast, when the organism comes into the subjects with low acid secretion, it migrates into the corpus with adequate acidity. This subsequently causes atrophy of parietal cells and aggravates hypochlorhydria. These diverse situations when the *H. pylori* infection occurs inevitably draw different effects on acid secretion after *H. pylori* eradication. Basically, the changes of acid secretion after *H. pylori* eradication depend on the degree of inflammation in the corpus. More specifically, in the setting of antrum-dominant gastritis with intact corporal glands in spite of the infection, the eradication reduces gastrin and subsequently acid secretion, as well. On the contrary, in the setting of severe corporal inflammation, the inhibition of parietal cells by *H. pylori* disappears, leading to the elevated acid secretion [159].

Several studies reported immediate increase in acid secretion after administration of anti-*H. pylori* agents [154, 159]. These results came from the termination of direct contact of parietal cells with *H. pylori* or its products [148, 149] and reduction of IL-1β or TNF-α which represses the acid secretion. Osawa et al. [160] reported that mRNA of H^+-K^+-ATPase increased 3 months after *H. pylori* eradication even in severe atrophy without changes in the number of parietal cells. The changes of genes encoding H_2 receptors, muscarinic M_3 receptor, and anion exchanger 2, which are involved in acid secretion export in the basal and apical sides of parietal cells, were not observed [160], while the increase in H^+-K^+-ATPase and reduction in the concentration of IL-1β were demonstrated. This result suggests that upregulation of H^+-K^+-ATPase and reduction in the concentration of IL-1β are attributable to the short-term increase in acid secretion after *H. pylori* eradication.

This increase of acid secretion has continued until 5 years after *H. pylori* eradication [161]. Nine of 23 subjects showed the restoration of normal acid secretion at 7-month follow-up after the eradication, and its steady and significant increase has been observed until 2 years after the cure [161]. However, more than 2 years after the eradication, an additional increase of acid secretory function was not found, and most subjects showed lower optimal levels of acid secretion compared with healthy control subjects [161]. This suggests numeric restoration of the parietal cells might have only minor effect on increasing acid secretion.

6.5 Mechanisms Underlying *H. pylori*-Induced Hormone Change in the Stomach

H. pylori may control the production of gastric hormones directly or indirectly. In this section, possible mechanisms of *H. pylori*-induced hormonal change will be discussed.

6.5.1 Urease and Ammonia

Levi et al. [136] originally proposed that the alkaline condition generated locally by *H. pylori* urease increased gastrin release. Measurements of the pH in the gastric mucus layer have shown that *H. pylori* infection causes a more alkaline milieu, although the difference is only 0.3–0.8 of a pH point [162]. Long-term exposure of the antral mucosa to elevated levels of ammonia in the gastric juice induced G-cell hyperfunction in rats [163]. An inverse correlation between gastric juice ammonia levels and antral SST concentrations was also observed in humans [141]. In addition, 4-week-long oral treatment with 0.01 % ammonia, which was clinically estimated as the concentration in the gastric juice in patients with *H. pylori* infection, decreased the release of SST and the number of D cells in the rat stomach [164]. However, acute infection of *H. pylori* can lead to different results in D-cell secretion. Previously, increasing intragastric urea did not elevate gastrin in infected persons [165]. Inhibition of urease by acetohydroxamic acid or bismuth plus antibiotics did not decrease gastrin release in a short-term experiment [166].

6.5.2 Lipopolysaccharide

After a 24-h incubation with *H. pylori*-derived culture broth, reduced ghrelin expression in gastric biopsies from *H. pylori*-negative subjects has been reported [167]. Piotrowski et al. [168] demonstrated that the binding of SST to its receptor on gastric mucosal cell membranes was inhibited by LPS from *H. pylori*, suggesting that *H. pylori*, through its LPS, is capable of interfering with SST regulatory effects on gastric mucosal G-cell function. It has been demonstrated that LPS suppressed fasted plasma ghrelin through production of IL-1 and prostaglandin and that exogenous ghrelin can normalize LPS-induced-altered digestive functions [169]. The expression and activity of GOAT under *H. pylori* infection, chronic gastritis, and gastric atrophy is an important issue. Two hours after LPS administration, significantly greater decrease of acyl ghrelin than des-acyl ghrelin has been reported [170]. The rapid decrease in plasma GOAT levels and slightly increased gastric GOAT protein levels at 2 h after the injection suggest that the inhibition of gastric GOAT release and an important role of circulating GOAT in the formation of acyl ghrelin during *H. pylori* infection are still to be investigated.

6.5.3 Somatostatin Receptors

SST acts through a family of homologous receptors, SST receptors (SSTRs). SSTR-2, which is the most extensively investigated of the five SSTRs, has been thought to be involved in gastric secretion. SSTR-2-positive cells were co-localized in 85 % of G cells and one-third of D cells [171]. The percentage of SSTR-2-labeling cells among D cells was significantly increased in rat antrum after 2–4-week treatments with 0.01 % ammonia solution. Ammonia may cause a decrease in the inhibitory potency of SST on G-cell function not only through a decrease in D-cell number but also through the additional inhibition of SST release, via an SSTR-2 in the residual D cells, in a paracrine manner, resulting in an increase in serum gastrin levels in rats [171]. Retallack et al. [172] speculated that a decrease in SSTR-2 mRNA on G cells may be responsible for the *H. pylori* infection-induced increase in gastrin secretion from the G-cell-rich fraction of *H. pylori*-infected human antral cell preparations.

6.5.4 Inflammatory Mediators

After the successful establishment of *H. pylori* infection, inflammatory responses were followed in the host mucosa. Since most of gastric hormone-producing cells are localized from base to neck of glands, where *H. pylori* rarely approaches, *H. pylori* may not directly affect release of these hormones, but it could influence the production of the peptide via inflammatory mediators.

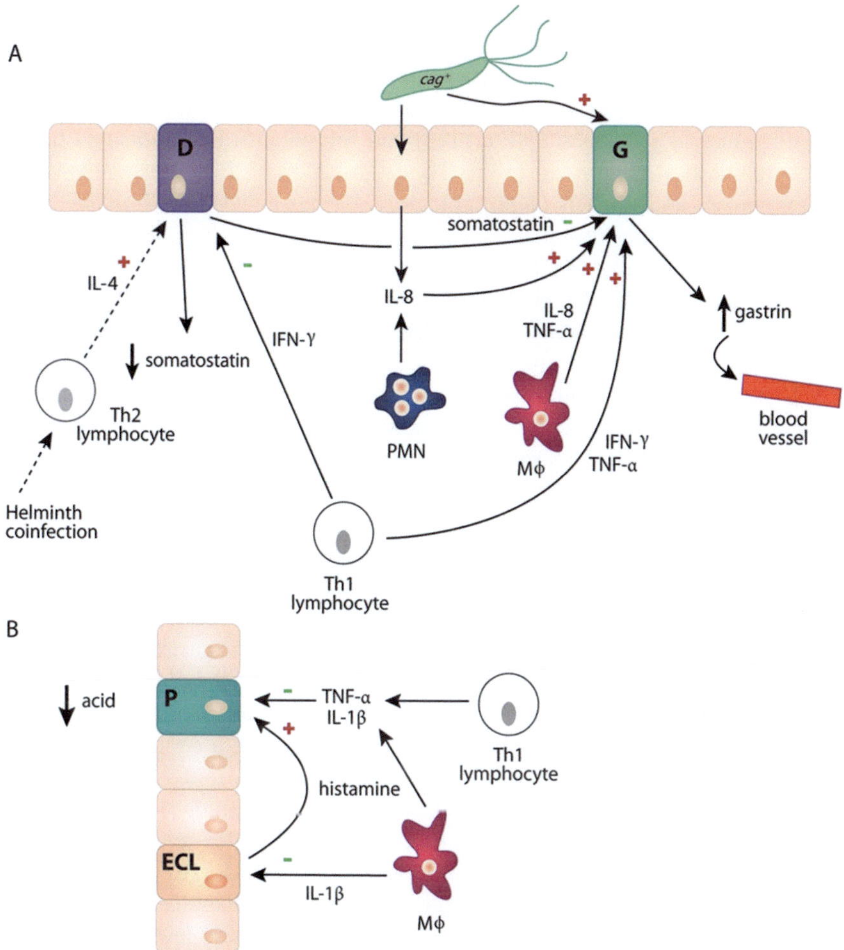

Fig. 6.7 Cytokine-mediated alterations in gastrin and somatostatin secretion (**a**) and enterochromaffin-like (ECL) and parietal (P) cell function (**b**) in *H. pylori* infection (Adapted from Peek et al. [159]). (**a**) IFN-γ from Th1 helper cells, IL-8, and TNF-α stimulate gastrin secretion from G cells. In contrast, IFN-γ inhibits somatostatin release from D cells, decreasing its inhibitory effect on gastrin. IL-4 stimulates somatostatin secretion, which may be a potential mechanism by which helminth coinfection, and stimulation of Th2 responses, can protect against the development of corpus atrophy. (**b**) IL-1β suppresses ECL cell histamine release, and IL-1β and TNF-α inhibit acid secretion from parietal cells *PMN*, polymorphonuclear cell

It has been reported that a series of the cytokines are increased in *H. pylori* gastritis: interleukin (IL)-1, IL-6, and IL-8, TNF-α, interferon-γ, macrophage inflammatory protein-1α, and platelet-activating factor [173, 174]. Cytokines have a key role in the disruption of acid homeostasis through affecting G or D cells (Fig. 6.7a) [159]. T helper (Th)1 cytokines such as interferon (IFN)-γ [175, 176] and the pro-inflammatory cytokines IL-1β, TNF-α, and IL-8

[177, 178] stimulate gastrin secretion from cultured G cells. IL-8 increases gastrin secretion from isolated canine G cells [176, 177]. TNF-α directly affects G cells in dogs and humans to increase gastric secretion [179]. The secretion of somatostatin from D cells, which negatively regulates gastrin, is inhibited by the pro-inflammatory cytokines TNF-α [180] and IFN-γ [175] but stimulated by the Th2 cytokine IL-4 (Fig. 6.7a) [175]. The polarization of T-cell responses in the gastric mucosa thus impacts on physiological changes induced by *H. pylori* infection. In addition, the pro-inflammatory cytokines IL-1β and TNF-α are potent inhibitors of acid secretion by parietal cells [157], and IL-1β also decreases histamine release from ECL cells (Fig. 6.7b) [181].

By contrast, eradication of *H. pylori* has been reported to induce a decrease in antral IL-8 levels and an improvement in histological inflammatory findings [182]. Antral SST concentrations were significantly increased after eradication therapy [183]. A negative correlation has been demonstrated between antral SST concentrations and IL-8 secretion in organ cultures of mucosal biopsies [184]. These findings suggest that certain mutual interactions between IL-8 and SST might be present in *H. pylori* infection in humans. In addition, a close correlation between an increase in gastric SST levels and the normalization of neutrophil infiltration indicated peptide inflammation interactions in *H. pylori*-induced gastritis [183].

Even though the relationship between these cytokines and gastric production of ghrelin or leptin has been far less evaluated, it is assumed that cytokines have been involved in the reduced production of gastric ghrelin and enhanced leptin synthesis. *H. pylori*-induced gastritis is thought to be associated with a strong activation of both Th1 and Th17 cells [167]. It is possible that factors released by *H. pylori* stimulate macrophages and dendritic cells which, in turn, produce factors driving Th1 and Th17 responses. The possible mechanism of *H. pylori*-related impairment of ghrelin synthesis in the stomach was illustrated in Fig. 6.8 [167].

It has also been reported that multiple cytokines and inflammation raise circulating leptin levels. In experimental animals, blood leptin levels are acutely increased by inflammatory stimuli, such as endotoxin, LPS, and turpentine, and by the administration of pro-inflammatory cytokines such as TNF-α and IL-1 [185, 186]. In rats, elevated leptin levels in blood are present during infection with the nematode *Nippostrongylus brasiliensis* and in the course of intestinal inflammation [187, 188]. Leptin itself can induce a shift of T cells in ob/ob mice to a predominantly Th1 response by increasing interferon-γ and IL-2 and decreasing IL-4 cytokine production in vivo [189]. In colonic mucosa of patients with ulcerative colitis, elevated levels of leptin and leptin receptors were reported [190]. Against this background, it is likely that *H. pylori* infection could raise gastric leptin production. Since most of the present evidences reported relationships between inflammation and only systemic circulating leptin levels, whether gastric leptin has a similar pro-inflammatory effect as circulating leptin or this condition is involved in other GI disorder needs further evaluation.

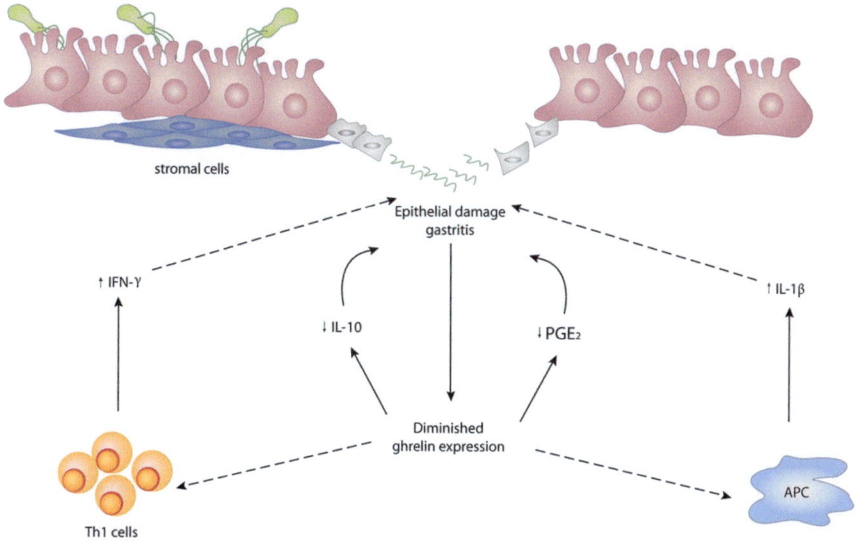

Fig. 6.8 Downregulation of ghrelin expression in the stomach during *Helicobacter pylori* infection (Adapted from Paoluzi et al. [167]). Epithelial damage and gastritis induced by *Helicobacter pylori* determine a diminished expression of ghrelin which, in turn, sustains the ongoing T helper (Th) 1 cells' response. Downregulation of ghrelin is also followed by a reduced release of prostaglandin E_2 (PGE_2) and interleukin (IL)-10 which, together with pro-inflammatory factors as IL-1b, contribute to the detrimental immune response and damage in the stomach *APC* antigen-presenting cell

6.5.5 Mucosal Atrophy and Bacterial Factors of H. pylori

H. pylori is a predominant etiologic factor of gastric inflammation and atrophy of gastric glands. It is assumed that atrophic gastritis may cause the loss of ghrelin-producing cells. This hypothesis has been supported by studies showing lower levels of gastric [75, 76] and circulating ghrelin [96] in patients with atrophic gastritis. It has been also reported that serum ghrelin level represents the most sensitive and specific noninvasive marker for selecting patients at high risk for atrophic body gastritis [191]. Probably, the extent of gastric atrophy and the duration of the infection may play a key role in modulation of ghrelin levels by *H. pylori* [80]. The atrophic changes especially in oxyntic glands were thought to be a resulting state from an array of inflammatory processes, and this leads to a negative effect of gastric hormone production. Inversely, it seems that recovery of ghrelin-producing capacity after *H. pylori* eradication might depend on the recovery from numeric or functional damage of the oxyntic glands.

Moreover, it is not surprising that certain strains may produce greater changes in endocrine function as these strains are more likely to cause clinical diseases. A major factor in highly virulent *H. pylori* isolates is the *cag* pathogenicity island

(PAI), a 40-kb DNA segment that encodes about 32 proteins. Some of them including accessory Cag proteins are forming components of a type IV secretion system (T4SS) [145]. For example, CagL, a structural component of the T4SS pilus in pathogenic *H. pylori*, is essential to pathogenesis because its deletion abolishes almost completely *H. pylori*'s ability to induce host cell secretion of the pro-inflammatory cytokine IL-8 [192].

Among patients who were infected with cytotoxin-associated gene A (*cagA*)-positive strain have higher plasma gastrin concentrations than those who with *cagA*-negative strain [193]. Those with the s1/m1 variant of the vacuolating cytotoxin (*vacA*) gene, which is more likely to cause ulcers, have less antral SST peptide than those with the less virulent type s2/m2 [194]. Kim et al. [195] demonstrated that hypergastrinemia, with a decrease in the number of antral D cells in *H. pylori*-associated gastritis, is relevant to the presence of CagA. These findings seem to support the concept that D-cell deficiency may be relevant in toxigenic *H. pylori*-associated chronic active gastritis [196].

Similarly, the effect of *H. pylori* on ghrelin production was different according to *H. pylori* virulence. Patients with type I strain *H. pylori*, expressing CagA and VacA, have lower circulating ghrelin levels than those with the less virulent type II strain [77].

6.5.6 Neuron

It has been recently reported that acute administration of *H. pylori* is capable of inhibiting acid secretion directly as well as indirectly by activating intramural CGRP sensory neurons coupled to stimulation of SST and inhibition of histamine secretion [197]. The reciprocal changes in SST and histamine secretion were due to release of CGRP from sensory neurons.

Since *H. pylori* is present in the upper regions of gastric mucosa, whereas SST and ECL cells are located from base to the neck of glands, it is speculated that activation of CGRP sensory neurons may be one of the explanations to how initial patchy superficial colonization of the stomach can induce acute hypochlorhydria.

6.5.7 Secondary Endocrine Changes to Gastric Acid

H. pylori-infected patients with hyposecretion of acid tend to have corpus gastritis that is believed to be related to reduction in acid secretion brought about by a specific *H. pylori* product or by inflammatory cytokines, including IL-1β and TNF-α, which inhibit parietal cells [157]. IL-1β also inhibits ECL cells [181]. Finally, *H. pylori* infection accelerates the development of corpus atrophy, which further diminishes acid secretion through the loss of parietal cells [198]. A recent study clearly demonstrated that *H. pylori* (*cagA*-positive) infection induced a

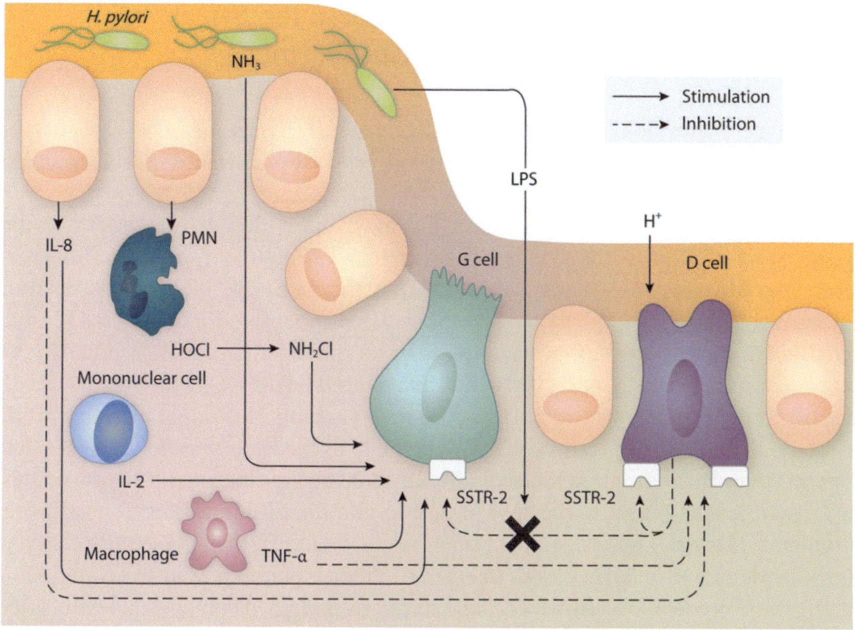

Fig. 6.9 Mechanisms speculated that are to underlie *Helicobacter pylori*-induced gut peptide change (Adapted from Kaneko et al. [134]). *PMN* polymorphonuclear leukocyte, *IL* interleukin, *LPS* lipopolysaccharide, *TNF-α* tumor necrosis factor-α, *SSTR-2* somatostatin receptor subtype 2

decrease in acid secretion and an increase in serum gastrin, with these phenomena returning to control levels after treatment with an IL-1 receptor antagonist in Mongolian gerbils [147]. Because gastrin is a physiological stimulant of acid secretion, a decrease in intragastric acidity induced by *H. pylori*-related corpus gastritis with atrophy might precede the gastrin release.

Interestingly, the acid-stimulating effect of ghrelin has been postulated by gastrin release [199] or vagal stimulation [200]. Indeed, Lee et al. [199] reported that gastrin was released in response to ghrelin. Whether a negative feedback to ghrelin production by elevated acid secretion might exist has not been evaluated. The ghrelin-producing cells, ECL cells, and D cells interspersed with each other in the oxyntic mucosa [201] suggests that there may be a functional link between them.

Finally, the possible mechanisms discussed in Sect. 6.4 as an underling *H. pylori*-induced gastric gastrin and somatostatin change are summarized in Fig. 6.9 [134].

6.6 Possible Reasons for Discrepancy in the Effect of *H. pylori* on the Levels of Gut Hormones

The change of physiology of G and D cells during *H. pylori* infection and after its eradication has been relatively well evaluated, while those of ghrelin and leptin are still under debate.

With regard to circulating ghrelin, the most often raised hypothesis for the inconsistent results is the compensatory release from other organs. However, since ghrelin is predominantly secreted from the stomach, its products derived from other tissues are unlikely to be sufficient to compensate for the altered plasma ghrelin dynamics. Another explanation may be due to differences in populations (age, race, geography, gender, BMI, diet, and overall health), extent of disease (e.g., whether atrophy is present or not), and *H. pylori* strain. This could partially result from the use of different immunoassay methods to measure ghrelin. The fact that ghrelin degrades to different kinds of ghrelin could be the fundamental cause on this discrepant results [101]. In the case of gastric ghrelin, different sites of gastritis or atrophy might make *H. pylori* infection draw different effects. As the same story, a different degree of recovery from damage or diverse duration after the completion of *H. pylori* eradication, probably, have different influences on ghrelin levels after cure of the infection. It has been reported that *H. pylori* infection alters gastric and blood level of ghrelin separately at different time points in the Mongolian gerbil model [202]. To confirm these hypotheses, serial measurements of blood ghrelin for long time after the eradication are needed.

On the other hand, gastric leptin levels according to *H. pylori* infection have been far less investigated compared with ghrelin. Although the exact mechanism is not clear, most studies have reported an increase in gastric leptin levels when subjects were infected by *H. pylori*. Nonetheless, as the primary contributor of circulating leptin is exclusively the adipose tissue, plasma leptin levels had strongly positive correlations with BMI, irrespective of *H. pylori* status.

6.7 Conclusion

Gut hormones such as ghrelin, leptin, gastrin, and SST could be regulated by neuronal, hormonal, and immune process under *H. pylori* infection. Understanding the changes of acid secretion related with *H. pylori* infection could give basic insights into identifying why the same *H. pylori* causes different outcomes among gastric ulcer, duodenal ulcer, gastric cancer, or asymptomatic histologic gastritis. In particular, ghrelin and leptin are involved not only in gastrointestinal physiology but in systemic energy regulation including adiposity, appetite, or circulation. Investigating these hormonal dynamics altered by *H. pylori* infection could provide an important clinical implication in the prevention and treatment of illnesses including obesity, functional dyspepsia, and GI cancer and give proper evidences of eradicating this microorganism.

References

1. Jeffery PL, McGuckin MA, Linden SK. Endocrine impact of *Helicobacter pylori*: focus on ghrelin and ghrelin o-acyltransferase. World J Gastroenterol. 2011;17:1249–60.
2. Ban S. In: Shepherd NA, Warren BF, Williams GT, Greenson JK, Lauwers GY, Novelli MR, editors. Stomach: Morson and Dawson's gastrointestinal pathology. 5th ed. Hoboken: Wiley-Blackwell; 2013. p. 89–103.
3. Kojima M, Hosoda H, Date Y, Nakazato M, Matsuo H, Kangawa K. Ghrelin is a growth-hormone-releasing acylated peptide from stomach. Nature. 1999;402:656–60.
4. Breidert M, Miehlke S, Glasow A, Orban Z, Stolte M, Ehninger G, et al. Leptin and its receptor in normal human gastric mucosa and in *Helicobacter* pylori-associated gastritis. Scand J Gastroenterol. 1999;34:954–61.
5. Ariyasu H, Takaya K, Tagami T, Ogawa Y, Hosoda K, Akamizu T, et al. Stomach is a major source of circulating ghrelin, and feeding state determines plasma ghrelin-like immunoreactivity levels in humans. J Clin Endocrinol Metab. 2001;86:4753–8.
6. Date Y, Kojima M, Hosoda H, Sawaguchi A, Mondal MS, Suganuma T, et al. Ghrelin, a novel growth hormone-releasing acylated peptide, is synthesized in a distinct endocrine cell type in the gastrointestinal tracts of rats and humans 1. Endocrinology. 2000;141:4255–61
7. Kamegai J, Tamura H, Shimizu T, Ishii S, Sugihara H, Wakabayashi I. Central effect of ghrelin, an endogenous growth hormone secretagogue, on hypothalamic peptide gene expression. Endocrinology. 2000;141:4797–800.
8. Korbonits M, Bustin SA, Kojima M, Jordan S, Adams EF, Lowe DG, et al. The expression of the growth hormone secretagogue receptor ligand ghrelin in normal and abnormal human pituitary and other neuroendocrine tumors 1. J Clin Endocrinol Metab. 2001;86:881–7.
9. Papotti M, Ghè C, Cassoni P, Catapano F, Deghenghi R, Ghigo E, et al. Growth hormone secretagogue binding sites in peripheral human tissues 1. J Clin Endocrinol Metab. 2000;85:3803–7.
10. Hosoda H, Kojima M, Matsuo H, Kangawa K. Ghrelin and des-acyl ghrelin: two major forms of rat ghrelin peptide in gastrointestinal tissue. Biochem Biophys Res Commun. 2000;279:909–13.
11. Nishi Y, Yoh J, Hiejima H, Kojima M. Structures and molecular forms of the ghrelin-family peptides. Peptides. 2011;32:2175–82.
12. Gutierrez JA, Solenberg PJ, Perkins DR, Willency JA, Knierman MD, Jin Z, et al. Ghrelin octanoylation mediated by an orphan lipid transferase. Proc Natl Acad Sci. 2008;105:6320–5.
13. Kang K, Schmahl J, Lee J-M, Garcia K, Patil K, Chen A, et al. Mouse ghrelin-O-acyltransferase (GOAT) plays a critical role in bile acid reabsorption. FASEB J. 2012;26:259–71.
14. Stengel A, Keire D, Goebel M, Evilevitch L, Wiggins B, Taché Y, et al. The RAPID method for blood processing yields new insight in plasma concentrations and molecular forms of circulating gut peptides. Endocrinology. 2009;150:5113–8.
15. Yin X, Li Y, Xu G, An W, Zhang W. Ghrelin fluctuation, what determines its production? Acta Biochim Biophys Sin. 2009;41:188–97.
16. Yang J, Zhao T-J, Goldstein JL, Brown MS. Inhibition of ghrelin O-acyltransferase (GOAT) by octanoylated pentapeptides. Proc Natl Acad Sci. 2008;105(31):10750–5.
17. Nishi Y, Hiejima H, Mifune H, Sato T, Kangawa K, Kojima M. Developmental changes in the pattern of ghrelin's acyl modification and the levels of acyl-modified ghrelins in murine stomach. Endocrinology. 2005;146:2709–15.

18. Baldelli R, Bellone S, Castellino N, Petri A, Rapa A, Vivenza D, et al. Oral glucose load inhibits circulating ghrelin levels to the same extent in normal and obese children. Clin Endocrinol. 2006;64:255–9.
19. Guo Z-F, Ren A-J, Zheng X, Qin Y-W, Cheng F, Zhang J, et al. Different responses of circulating ghrelin, obestatin levels to fasting, re-feeding and different food compositions, and their local expressions in rats. Peptides. 2008;29:1247–54.
20. Toshinai K, Mondal MS, Nakazato M, Date Y, Murakami N, Kojima M, et al. Upregulation of ghrelin expression in the stomach upon fasting, insulin-induced hypoglycemia, and leptin administration. Biochem Biophy Res Commun. 2001;281:1220–5.
21. Kim SW, Kim KW, Shin CS, Park dJ, Park KS, Cho BY. Acylated ghrelin secretion is acutely suppressed by oral glucose load or insulin-induced hypoglycemia independently of basal growth hormone secretion in humans. Horm Res. 2006;67:211–9.
22. Overduin J, Frayo RS, Grill HJ, Kaplan JM, Cummings DE. Role of the duodenum and macronutrient type in ghrelin regulation. Endocrinology. 2005;146:845–50.
23. Knerr I, Gröschl M, Rascher W, Rauh M. Endocrine effects of food intake: insulin, ghrelin, and leptin responses to a single bolus of essential amino acids in humans. Ann Nutr Metab. 2002;47:312–8.
24. Kamegai J, Tamura H, Shimizu T, Ishii S, Sugihara H, Oikawa S. Effects of insulin, leptin, and glucagon on ghrelin secretion from isolated perfused rat stomach. Regul Pept. 2004;119:77–81.
25. Flanagan DE, Evans ML, Monsod TP, Rife F, Heptulla RA, Tamborlane WV, et al. The influence of insulin on circulating ghrelin. Am J Physiol Endocrinol Metab. 2003;284:E313–E6.
26. Katayama T, Shimamoto S, Oda H, Nakahara K, Kangawa K, Murakami N. Glucagon receptor expression and glucagon stimulation of ghrelin secretion in rat stomach. Biochem Biophy Res Commun. 2007;357:865–70.
27. Seoane L, Al-Massadi O, Barreiro F, Dieguez C, Casanueva F. Growth hormone and somatostatin directly inhibit gastric ghrelin secretion. An in vitro organ culture system. J Endocrinol Investig. 2007;30:RC22–RC5.
28. Grinspoon S, Miller KK, Herzog DB, Grieco KA, Klibanski A. Effects of estrogen and recombinant human insulin-like growth factor-I on ghrelin secretion in severe undernutrition. J Clin Endocrinol Metab. 2004;89:3988–93.
29. Shimada M, Date Y, Mondal MS, Toshinai K, Shimbara T, Fukunaga K, et al. Somatostatin suppresses ghrelin secretion from the rat stomach. Biochem Biophys Res Commun. 2003;302:520–5.
30. Arosio M, Ronchi CL, Gebbia C, Cappiello V, Beck-Peccoz P, Peracchi M. Stimulatory effects of ghrelin on circulating somatostatin and pancreatic polypeptide levels. J Clin Endocrinol Metab. 2003;88:701–4.
31. Tschöp M, Weyer C, Tataranni PA, Devanarayan V, Ravussin E, Heiman ML. Circulating ghrelin levels are decreased in human obesity. Diabetes. 2001;50:707–9.
32. Zhao Z, Sakata I, Okubo Y, Koike K, Kangawa K, Sakai T. Gastric leptin, but not estrogen and somatostatin, contributes to the elevation of ghrelin mRNA expression level in fasted rats. J Endocrinol. 2008;196:529–38.
33. Dube MG, Beretta E, Dhillon H, Ueno N, Kalra PS, Kalra SP. Central leptin gene therapy blocks high-fat diet-induced weight gain, hyperleptinemia, and hyperinsulinemia increase in serum ghrelin levels. Diabetes. 2002;51:1729–36.
34. Sakata I, Tanaka T, Yamazaki M, Tanizaki T, Zheng Z, Sakai T. Gastric estrogen directly induces ghrelin expression and production in the rat stomach. J Endocrinol. 2006;190:749–57.
35. Broglio F, Gottero C, Van Koetsveld P, Prodam F, Destefanis S, Benso A, et al. Acetylcholine regulates ghrelin secretion in humans. J Clin Endocrinol Metab. 2004;89:2429–33.

36. Di Francesco V, Fantin F, Residori L, Bissoli L, Micciolo R, Zivelonghi A, et al. Effect of age on the dynamics of acylated ghrelin in fasting conditions and in response to a meal. J Am Geriatr Soc. 2008;56:1369–70.
37. Kozakowski J, Rabijewski M, Zgliczyński W. Ghrelin-growth hormone releasing and orexigenic hormone in men declines with age, insulin and with decrease in testosterone concentration. Neuro Endocrinol Lett. 2008;29:100–6.
38. Schutte AE, Huisman HW, Schutte R, van Rooyen JM, Malan L, Malan NT. Aging influences the level and functions of fasting plasma ghrelin levels: the POWIRS-Study. Regul Pept. 2007;139:65–71.
39. Barkan AL, Dimaraki EV, Jessup SK, Symons KV, Ermolenko M, Jaffe CA. Ghrelin secretion in humans is sexually dimorphic, suppressed by somatostatin, and not affected by the ambient growth hormone levels. J Clin Endocrinol Metab. 2003;88:2180–4.
40. Horvath TL, Diano S, Sotonyi P, Heiman M, Tschop M. Minireview: ghrelin and the regulation of energy balance – a hypothalamic perspective. Endocrinology. 2001;142:4163–9.
41. Perboni S, Inui A. Appetite and gastrointestinal motility: role of ghrelin-family peptides. Clin Nutr. 2010;29:227–34.
42. Patterson M. Ghrelin enhances gastric emptying in diabetic gastroparesis: a double blind, placebo controlled, crossover study. Gut. 2005;54:1693–8.
43. Fujimiya M, Asakawa A, Ataka K, Kato I, Inui A. Different effects of ghrelin, des-acyl ghrelin and obestatin on gastroduodenal motility in conscious rats. World J Gastroenterol. 2008;14:6318–26.
44. Fujimiya M, Asakawa A, Ataka K, Chen C-Y, Kato I, Inui A. Ghrelin, des-acyl ghrelin, and obestatin: regulatory roles on the gastrointestinal motility. Int J Pept. 2010. doi:10.1155/2010/305192.
45. Akamizu T, Takaya K, Irako T, Hosoda H, Teramukai S, Matsuyama A, et al. Pharmacokinetics, safety, and endocrine and appetite effects of ghrelin administration in young healthy subjects. Eur J Endocrinol. 2004;150:447–55.
46. Sánchez J, Oliver P, Palou A, Picó C. The inhibition of gastric ghrelin production by food intake in rats is dependent on the type of macronutrient. Endocrinology. 2004;145:5049–55.
47. Tschöp M, Smiley DL, Heiman ML. Ghrelin induces adiposity in rodents. Nature. 2000;407:908–13.
48. Nakazato M, Murakami N, Date Y, Kojima M, Matsuo H, Kangawa K, et al. A role for ghrelin in the central regulation of feeding. Nature. 2001;409:194–8.
49. Wren A, Seal L, Cohen M, Brynes A, Frost G, Murphy K, et al. Ghrelin enhances appetite and increases food intake in humans. J Clin Endocrinol Metab. 2001;86:5992–5.
50. Inui A, Asakawa A, Bowers CY, Mantovani G, Laviano A, Meguid MM, et al. Ghrelin, appetite, and gastric motility: the emerging role of the stomach as an endocrine organ. FASEB J. 2004;18:439–56.
51. Tack J, Depoortere I, Bisschops R, Verbeke K, Janssens J, Peeters T. Influence of ghrelin on gastric emptying and meal-related symptoms in idiopathic gastroparesis. Aliment Pharmacol Ther. 2005;22:847–53.
52. Peeters T. Central and peripheral mechanisms by which ghrelin regulates gut motility. J Physol Pharmacol. 2003;54:95–103.
53. Hosoda H, Kojima M, Kangawa K. Ghrelin and the regulation of food intake and energy balance. Mol Interv. 2002;2:494–503.
54. Fujino K, Inui A, Asakawa A, Kihara N, Fujimura M, Fujimiya M. Ghrelin induces fasted motor activity of the gastrointestinal tract in conscious fed rats. J Physiol. 2003;550:227–40.
55. Lee KJ, Cha DY, Cheon SJ, Yeo M, Cho SW. Plasma ghrelin levels and their relationship with gastric emptying in patients with dysmotility-like functional dyspepsia. Digestion. 2009;80:58–63.

56. Shindo T, Futagami S, Hiratsuka T, Horie A, Hamamoto T, Ueki N, et al. Comparison of gastric emptying and plasma ghrelin levels in patients with functional dyspepsia and non-erosive reflux disease. Digestion. 2009;79:65–72.
57. El-Salhy M, Lillebø E, Reinemo A, Salmelid L. Ghrelin in patients with irritable bowel syndrome. Int J Mol Med. 2009;23:703–7.
58. Karmiris K, Koutroubakis IE, Xidakis C, Polychronaki M, Voudouri T, Kouroumalis EA. Circulating levels of leptin, adiponectin, resistin, and ghrelin in inflammatory bowel disease. Inflamm Bowel Dis. 2006;12:100–5.
59. Ates Y, Degertekin B, Erdil A, Yaman H, Dagalp K. Serum ghrelin levels in inflammatory bowel disease with relation to disease activity and nutritional status. Dig Dis Sci. 2008;53:2215–21.
60. Mottershead M, Karteris E, Barclay JY, Suortamo S, Newbold M, Randeva H, et al. Immunohistochemical and quantitative mRNA assessment of ghrelin expression in gastric and oesophageal adenocarcinoma. J Clin Pathol. 2007;60:405–9.
61. An JY, Choi M-G, Noh JH, Sohn TS, Jin D-K, Kim S. Clinical significance of ghrelin concentration of plasma and tumor tissue in patients with gastric cancer. J Surg Res. 2007;143:344–9.
62. Waseem T, Ahmad F, Azam M, Qureshi MA. Role of ghrelin axis in colorectal cancer: a novel association. Peptides. 2008;29:1369–76.
63. Broglio F, Gottero C, Prodam F, Gauna C, Muccioli G, Papotti M, et al. Non-acylated ghrelin counteracts the metabolic but not the neuroendocrine response to acylated ghrelin in humans. J Clin Endocrinol Metab. 2004;89:3062–5.
64. Chen C-Y, Chao Y, Chang F-Y, Chien EJ, Lee S-D, Doong M-L. Intracisternal des-acyl ghrelin inhibits food intake and non-nutrient gastric emptying in conscious rats. Int J Mol Med. 2005;16:695–9.
65. Toshinai K, Yamaguchi H, Sun Y, Smith RG, Yamanaka A, Sakurai T, et al. Des-acyl ghrelin induces food intake by a mechanism independent of the growth hormone secretagogue receptor. Endocrinology. 2006;147:2306–14.
66. Lucidi P, Murdolo G, Di Loreto C, Parlanti N, De Cicco A, Fatone C, et al. Metabolic and endocrine effects of physiological increments in plasma ghrelin concentrations. Nutr Metab Cardiovasc Dis. 2005;15:410–7.
67. Isomoto H, Ueno H, Saenko VA, Mondal MS, Nishi Y, Kawano N, et al. Impact of *Helicobacter* pylori infection on gastric and plasma ghrelin dynamics in humans. Am J Gastroenterol. 2005;100:1711–20.
68. Lee ES, Yoon YS, Park C-Y, Kim H-S, Um TH, Baik HW, et al. Eradication of *Helicobacter* pylori increases ghrelin mRNA expression in the gastric mucosa. J Kor Med Sci. 2010;25:265–71.
69. Stec-Michalska K, Malicki S, Michalski B, Peczek L, Wisniewska-Jarosinska M, Nawrot B. Gastric ghrelin in relation to gender, stomach topography and *Helicobacter pylori* in dyspeptic patients. World J Gastroenterol. 2009;15:5409–17.
70. Roper J, Francois F, Shue PL, Mourad MS, Pei Z, Olivares de Perez AZ, et al. Leptin and ghrelin in relation to *Helicobacter* pylori status in adult males. J Clin Endocrinol Metab. 2008;93:2350–7.
71. Jang EJ, Park SW, Park JS, Park SJ, Hahm KB, Paik SY. The influence of the eradication of *Helicobacter* pylori on gastric ghrelin, appetite, and body mass index in patients with peptic ulcer disease. J Gastroenterol Hepatol. 2008;23(Supple2):S278–S85.
72. Jun DW, Lee OY, Lee YY, Choi HS, Kim TH, Yoon BC. Correlation between gastrointestinal symptoms and gastric leptin and ghrelin expression in patients with gastritis. Dig Dis Sci. 2007;52:2866–72.
73. Choe YH, Lee JH, Lee HJ, Paik KH, Jin DK, Song SY, et al. Ghrelin levels in gastric mucosa before and after Eradication of *Helicobacter* pylori. Gut Liver. 2007;1:132–7.

74. Salles N, Ménard A, Georges A, Salzmann M, De Ledinghen V, De Mascarel A, et al. Effects of *Helicobacter* pylori infection on gut appetite peptide (leptin, ghrelin) expression in elderly inpatients. J Gerontol A Biol Sci Med Sci. 2006;61:1144–50.
75. Osawa H, Kita H, Ohnishi H, Nakazato M, Date Y, Bowlus CL, et al. Changes in plasma ghrelin levels, gastric ghrelin production, and body weight after *Helicobacter* pylori cure. J Gastroenterol. 2006;41:954–61.
76. Osawa H, Nakazato M, Date Y, Kita H, Ohnishi H, Ueno H, et al. Impaired production of gastric ghrelin in chronic gastritis associated with *Helicobacter* pylori. J Clin Endocrinol Metab. 2005;90:10–6.
77. Isomoto H, Nishi Y, Ohnita K, Mizuta Y, Kohno S, Ueno H, et al. The relationship between plasma and gastric ghrelin levels and strain diversity in *Helicobacter* pylori virulence. Am J Gastroenterol. 2005;100:1425–7.
78. Tatsuguchi A, Miyake K, Gudis K, Futagami S, Tsukui T, Wada K, et al. Effect of *Helicobacter pylori* infection on ghrelin expression in human gastric mucosa. Am J Gastroenterol. 2004;99:2121–7.
79. Gokcel A, Gumurdulu Y, Kayaselcuk F, Serin E, Ozer B, Ozsahin AK, et al. *Helicobacter pylori* has no effect on plasma ghrelin levels. Eur J Endocrinol. 2003;148:423–6.
80. Liew P-L, Lee W-J, Lee Y-C, Chen W-Y. Gastric ghrelin expression associated with *Helicobacter pylori* infection and chronic gastritis in obese patients. Obes Surg. 2006;16:612–9.
81. Méndez-Sánchez N, Pichardo-Bahena R, Vásquez-Fernández F, Lezama-Mora JI, León-Canales AL, Barredo-Prieto B, et al. Effect of *Helicobacter pylori* infection on gastric ghrelin expression and body weight. Rev Gastroenterol Mex. 2007;72:359–64.
82. Isomoto H, Nakazato M, Ueno H, Date Y, Nishi Y, Mukae H, et al. Low plasma ghrelin levels in patients with *Helicobacter pylori*–associated gastritis. Am J Med. 2004;117:429–32.
83. Isomoto H, Nishi Y, Kohno S, Wen C-Y, Ueno H, Nakazato M. Impact of *Helicobacter pylori* infection on ghrelin and various neuroendocrine hormones in plasma. World J Gastroenterol. 2005;52:53.4–14.3.
84. Shiotani A, Miyanishi T, Uedo N, Iishi H. *Helicobacter pylori* infection is associated with reduced circulating ghrelin levels independent of body mass index. Helicobacter. 2005;10:373–8.
85. Konturek P, Czesnikiewicz-Guzik M, Bielanski W, Konturek S. Involvement of *Helicobacter pylori* infection in neuro-hormonal control of food intake. J Physiol Pharmacol. 2006;57:67–81.
86. Plonka M, Konturek P, Bielanski W, Pawlik T, Brzozowski T, Konturek S. Relationship between ghrelin and *Helicobacter pylori* infection in Polish adult shepherds and their children. Aliment Pharmacol Ther Symp Ser. 2006;2:160–8.
87. Plonka M, Bielanski W, Konturek S, Targosz A, Sliwowski Z, Dobrzanska M, et al. *Helicobacter pylori* infection and serum gastrin, ghrelin and leptin in children of Polish shepherds. Dig Liver Dis. 2006;38:91–7.
88. Alonso N, Granada ML, Salinas I, Reverter JL, Flores L, Ojanguren I, et al. Plasma ghrelin concentrations in type 1 diabetic patients with autoimmune atrophic gastritis. Eur J Endocrinol. 2007;157:763–9.
89. Cindoruk M, Yetkin I, Deger SM, Karakan T, Kan E, Unal S. Influence of *H. pylori* on plasma ghrelin in patients without atrophic gastritis. World J Gastroenterol. 2007;13:1595–8.
90. D'Onghia V, Leoncini R, Carli R, Santoro A, Giglioni S, Sorbellini F, et al. Circulating gastrin and ghrelin levels in patients with colorectal cancer: correlation with tumour stage, *Helicobacter pylori* infection and BMI. Biomed Pharmacother. 2007;61:137–41.
91. de Martel C, Haggerty TD, Corley DA, Vogelman JH, Orentreich N, Parsonnet J. Serum ghrelin levels and risk of subsequent adenocarcinoma of the esophagus. Am J Gastroenterol. 2007;102:1166–72.

92. Pacifico L, Anania C, Osborn JF, Ferrara E, Schiavo E, Bonamico M, et al. Long-term effects of *Helicobacter pylori* eradication on circulating ghrelin and leptin concentrations and body composition in prepubertal children. Eur J Endocrinol. 2008;158:323–32.
93. Shak JR, Roper J, Perez-Perez GI, Tseng C-h, Francois F, Gamagaris Z, et al. The effect of laparoscopic gastric banding surgery on plasma levels of appetite-control, insulinotropic, and digestive hormones. Obes Surg. 2008;18:1089–96.
94. Chuang CH, Sheu BS, Yang HB, Lee SC, Kao AW, Cheng HC, et al. Gender difference of circulating ghrelin and leptin concentrations in chronic *Helicobacter pylori* infection. Helicobacter. 2009;14:54–60.
95. Gao X-Y, Kuang H-Y, Liu X-M, Duan P, Yang Y, Ma Z-B. Circulating ghrelin/obestatin ratio in subjects with *Helicobacter pylori* infection. Nutrition. 2009;25:506–11.
96. Kawashima J, Ohno S, Sakurada T, Takabayashi H, Kudo M, Ro S, et al. Circulating acylated ghrelin level decreases in accordance with the extent of atrophic gastritis. J Gastroenterol. 2009;44:1046–54.
97. Campana D, Nori F, Pagotto U, De Iasio R, Morselli-Labate AM, Pasquali R, et al. Plasma acylated ghrelin levels are higher in patients with chronic atrophic gastritis. Clin Endocrinol. 2007;67:761–6.
98. Nweneka C, Prentice A. *Helicobacter pylori* infection and circulating ghrelin levels – a systematic review. BMC Gastroenterol. 2011;11:7.
99. Suzuki H, Masaoka T, Nomoto Y, Hosoda H, Mori M, Nishizawa T, et al. Increased levels of plasma ghrelin in peptic ulcer disease. Aliment Pharmacol Ther. 2006;24(s4):120–6.
100. Zub-Pokrowiecka A, Rembiasz K, Konturek SJ, Budzynski A, Konturek PC, Budzynski P. Ghrelin in diseases of the gastric mucosa associated with *Helicobacter pylori* infection. Med Sci Monit Basic Res. 2010;16:CR493–500.
101. Satou M, Nakamura Y, Ando H, Sugimoto H. Understanding the functional significance of ghrelin processing and degradation. Peptides. 2011;32:2183–90.
102. Nwokolo C, Freshwater D, O'Hare P, Randeva H. Plasma ghrelin following cure of *Helicobacter pylori*. Gut. 2003;52:637–40.
103. Czesnikiewicz-Guzik M, Loster B, Bielanski W, Guzik TJ, Konturek PC, Zapala J, et al. Implications of oral *Helicobacter pylori* for the outcome of its gastric eradication therapy. J Clin Gastroenterol. 2007;41:145–51.
104. Takamori K, Mizuta Y, Takeshima F, Akazawa Y, Isomoto H, Ohnita K, et al. Relation among plasma ghrelin level, gastric emptying, and psychologic condition in patients with functional dyspepsia. J Clin Gastroenterol. 2007;41:477–83.
105. Nishizawa T, Suzuki H, Nomoto Y, Masaoka T, Hosoda H, Mori M, et al. Enhanced plasma ghrelin levels in patients with functional dyspepsia. Aliment Pharmacol Ther. 2006;24:104–10.
106. Shinomiya T, Fukunaga M, Akamizu T, Irako T, Yokode M, Kangawa K, et al. Plasma acylated ghrelin levels correlate with subjective symptoms of functional dyspepsia in female patient. Scand J Gastroenterol. 2005;40:648–53.
107. Kim YS, Lee JS, Lee TH, Cho JY, Kim JO, Kim WJ, et al. Plasma levels of acylated ghrelin in patients with functional dyspepsia. World J Gastroenterol. 2012;18:2231–7.
108. Choi YJ, Kim N, Kim JY, Lee DH, Jung HC. Plasma ghrelin, leptin and serotoin levels in patients with functional dyspepsia. Gastroenterology. 2013;144(Sulppl 1):S928–9.
109. Sobhani I, Bado A, Vissuzaine C, Buyse M, Kermorgant S, Laigneau J, et al. Leptin secretion and leptin receptor in the human stomach. Gut. 2000;47:178–83.
110. Cammisotto P, Bendayan M. A review on gastric leptin: the exocrine secretion of a gastric hormone. Anat cell biolol. 2012;45:1–16.
111. Guilmeau S, Buyse M, Bado A. Gastric leptin: a new manager of gastrointestinal function. Curr Opin Pharmacol. 2004;4:561–6.
112. Cinti S, De Matteis R, Pico C, Ceresi E, Obrador A, Maffeis C, et al. Secretory granules of endocrine and chief cells of human stomach mucosa contain leptin. Int J Obes. 2000;24:789–93.

113. Lindqvist A, de la Cour CD, Stegmark A, Håkanson R, Erlanson-Albertsson C. Overeating of palatable food is associated with blunted leptin and ghrelin responses. Regul Pept. 2005;130:123–32.
114. Picó C, Sánchez J, Oliver P, Palou A. Leptin production by the stomach is up-regulated in obese (fa/fa) Zucker rats. Obes Res. 2002;10:932–8.
115. Erkasap N, Uzuner K, Serteser M, Köken T, Aydın Y. Gastroprotective effect of leptin on gastric mucosal injury induced by ischemia–reperfusion is related to gastric histamine content in rats. Peptides. 2003;24:1181–7.
116. Moss SF, Calam J. Acid secretion and sensitivity to gastrin in patients with duodenal ulcer: effect of eradication of *Helicobacter pylori*. Gut. 1993;34:888–92.
117. Moran TH, McHugh PR. Cholecystokinin suppresses food intake by inhibiting gastric emptying. Am J Physiol. 1982;242:R491–7.
118. Wang L, Barachina MD, Martınez V, Wei JY, Taché Y. Synergistic interaction between CCK and leptin to regulate food intake. Regul Pept. 2000;92:79–85.
119. Nauck MA, Niedereichholz U, Ettler R, Holst JJ, Ørskov C, Ritzel R, et al. Glucagon-like peptide 1 inhibition of gastric emptying outweighs its insulinotropic effects in healthy humans. Am J Physiol Endocrinol Metab. 1997;273:E981–E8.
120. Naveilhan P, Hassani H, Canals JM, Ekstrand AJ, Larefalk Å, Chhajlani V, et al. Normal feeding behavior, body weight and leptin response require the neuropeptide Y Y2 receptor. Nat Med. 1999;5:1188–93.
121. Andrews PL, Sanger GJ. Abdominal vagal afferent neurones: an important target for the treatment of gastrointestinal dysfunction. Curr Opin Pharmacol. 2002;2:650–6.
122. Burdyga G, Spiller D, Morris R, Lal S, Thompson D, Saeed S, et al. Expression of the leptin receptor in rat and human nodose ganglion neurones. Neuroscience. 2002;109:339–47.
123. Cammisotto P, Bendayan M. Leptin secretion by with adipose tissue and gastric mucosa. Histol Histopathol. 2007;22:199–210.
124. Weller A, Smith GP, Gibbs J. Endogenous cholecystokinin reduces feeding in young rats. Science. 1990;247:1589–91.
125. Azuma T, Suto H, Ito Y, Ohtani M, Dojo M, Kuriyama M, et al. Gastric leptin and *Helicobacter pylori* infection. Gut. 2001;49:324–9.
126. Bado A, Levasseur S, Attoub S, Kermorgant S, Laigneau J P, Bortoluzzi M-N, et al. The stomach is a source of leptin. Nature. 1998;394:790–3.
127. Lankarani KB, Moghadami M, Masoumpoor M, Geramizadeh B, Omrani GR. Serum leptin level in patients with functional dyspepsia. Dig Liver Dis. 2004;36:717–21.
128. Zhao Z, Sakai T. Characteristic features of ghrelin cells in the gastrointestinal tract and the regulation of stomach ghrelin expression and production. World J Gastroenterol. 2008;14:6306.
129. Yarandi SS, Hebbar G, Sauer CG, Cole CR, Ziegler TR. Diverse roles of leptin in the gastrointestinal tract: modulation of motility, absorption, growth, and inflammation. Nutrition. 2011;27:269–75.
130. Cummings DE, Purnell JQ, Frayo RS, Schmidova K, Wisse BE, Weigle DS. A preprandial rise in plasma ghrelin levels suggests a role in meal initiation in humans. Diabetes. 2001;50:1714–9.
131. Gonzalez–Rey E, Chorny A, Delgado M. Therapeutic action of ghrelin in a mouse model of colitis. Gastroenterology. 2006;130:1707–20.
132. Date Y, Nakazato M, Murakami N, Kojima M, Kangawa K, Matsukura S. Ghrelin acts in the central nervous system to stimulate gastric acid secretion. Biochem Biophys Res Commun. 2001;280:904–7.
133. Park JM, Kakimoto T, Kuroki T, Shiraishi R, Fujise T, Iwakiri R, et al. Suppression of intestinal mucosal apoptosis by ghrelin in fasting rats. Exp Biol Med. 2008;233:48–56.
134. Kaneko H, Konagaya T, Kusugami K. *Helicobacter pylori* and gut hormones. J Gastroenterol. 2002;37:77–86.

135. Itoh Z, Takeuchi S, Aizawa I, Honda R. The negative feedback mechanism of gastric acid secretion: significance of acid in the gastric juice in man and dog. Surgery. 1975;77:648–60.
136. Levi S, Beardshall K, Haddad G, Playford R, Ghosh P, Calam J. Campylobacter pylori and duodenal ulcers: the gastrin link. Lancet. 1989;1:1167–8.
137. Beardshall K, Moss S, Gill J, Levi S, Ghosh P, Playford R, et al. Suppression of *Helicobacter pylori* reduces gastrin releasing peptide stimulated gastrin release in duodenal ulcer patients. Gut. 1992;33:601–3.
138. Penman E, Wass J, Butler M, Penny E, Price J, Wu P, et al. Distribution and characterisation of immunoreactive somatostatin in human gastrointestinal tract. Regul Pept. 1983;7:53–65.
139. Walsh JH, Dockray GJ, Mitty RD. Gut peptides: biochemistry and physiology. Endocrinologist. 1994;4:487.
140. Arnold R, Lankisch P. Somatostatin and the gastrointestinal tract. Clin Gastroenterol. 1980;9:733–53.
141. Kaneko H, Nakada K, Mitsuma T, Uchida K, Furusawa A, Maeda Y, et al. *Helicobacter pylori* infection induces a decrease in immunoreactive-somatostatin concentrations of human stomach. Dig Dis Sci. 1992;37:409–16.
142. Zavros Y, Paterson A, Lambert J, Shulkes A. Expression of progastrin-derived peptides and somatostatin in fundus and antrum of nonulcer dyspepsia subjects with and without *Helicobacter pylori* infection. Dig Dis Sci. 2000;45:2058–64.
143. Tzaneva M, Julianov A. Chromogranin A-, somatostatin-and serotonin-containing endocrine cells in the corporal gastric mucosa of patients with *Helicobacter pylori* associated chronic gastritis. Endocr Regul. 1999;33:79–82.
144. Moss SF, Calam J, Legon S, Bishop A, Polak J. Effect of *Helicobacter pylori* on gastric somatostatin in duodenal ulcer disease. Lancet. 1992;340:930–2.
145. Smolka AJ, Backert S. How *Helicobacter pylori* infection controls gastric acid secretion. J Gastroenterol. 2012;47:609–18.
146. Murphy F, Read N, Taylor K, Trier J. Epidemic gastritis with hypochlorhydria. Gastroenterology. 1979;76:1449–57.
147. Takashima M, Furuta T, Hanai H, Sugimura H, Kaneko E. Effects of *Helicobacter pylori* infection on gastric acid secretion and serum gastrin levels in Mongolian gerbils. Gut. 2001;48:765–73.
148. Hoffman JS, King WW, Fox JG, Janik D, Cave DR. Rabbit and ferret parietal cell inhibition by *Helicobacter* species. Dig Dis Sci. 1995;40:147–52.
149. Cave D, Vargas M. Effect of a Campylobacter pylori protein on acid secretion by parietal cells. Lancet. 1989;334:187–9.
150. Tagkalidis PP, Royce SG, Macrae FA, Bhathal PS. Selective colonization by *Helicobacter pylori* of the deep gastric glands and intracellular canaliculi of parietal cells in the setting of chronic proton pump inhibitor use. Eur J Gastroenterol Hepatol. 2002;14:453–6.
151. Chen XG, Correa P, Offerhaus J, Rodriguez E, Janney F, Hoffmann E, et al. Ultrastructure of the gastric mucosa harboring Campylobacter-like organisms. Am J Clin Pathol. 1986;86:575–82.
152. Jablonowski HHK, Kramer N, Geis G, Opferkuch W, Strohmeyer G. Effect of *Helicobacter pylori* on dbc-AMP stimulated acid secretion by human parietal cells. Hepatogastroenterology. 1994;41:546–8.
153. Jablonowski H, Hengels K, Kraemer N, Geis G, Opferkuch W, Strohmeyer G. Effects of *Helicobacter pylori* on histamine and carbachol stimulated acid secretion by human parietal cells. Gut. 1994;35:755–7.
154. Furuta T, Baba S, Takashima M, Shirai N, Xiao F, Futami H, et al. H+/K+ –adenosine triphosphatase mRNA in gastric fundic gland mucosa in patients infected with *Helicobacter pylori*. Scand J Gastroenterol. 1999;34:384–90.
155. Wolfe M, Nompleggi D. Cytokine inhibition of gastric acid secretion – a little goes a long way. Gastroenterology. 1992;102:2177–8.

156. Wallace JL, Cucala M, Mugridge K, Parente L. Secretagogue-specific effects of interleukin-1 on gastric acid secretion. Am J Physiol Gastrointest Liver Physiol. 1991;261:G559–G64.
157. Beales I, Calam J. Interleukin 1β and tumour necrosis factor α inhibit acid secretion in cultured rabbit parietal cells by multiple pathways. Gut. 1998;42:227–34.
158. Amedei A, Munarp F, Bella CD, Niccolai E, Benagiano M, Bencini L, et al. *Helicobacter pylori* HP0175 promotes the production of IL-23, IL-6, IL-1β AND TGF-β. Eur J Inflamm. 2013;11:261–8.
159. Peek RM, Crabtree JE. *Helicobacter* infection and gastric neoplasia. J Pathol. 2006;208:233–48.
160. Osawa H, Kita H, Ohnishi H, Hoshino H, Mutoh H, Ishino Y, et al. *Helicobacter pylori* eradication induces marked increase in H+/K+ −adenosine triphosphatase expression without altering parietal cell number in human gastric mucosa. Gut. 2006;55:152–7.
161. Iijima K, Sekine H, Koike T, Imatani A, Ohara S, Shimosegawa T. Long-term effect of *Helicobacter pylori* eradication on the reversibility of acid secretion in profound hypochlorhydria. Aliment Pharmacol Ther. 2004;19:1181–8.
162. Moss S, Calam J. *Helicobacter pylori* and peptic ulcers: the present position. Gut. 1992;33:289–92.
163. Lichtenberger LM, Dial EJ, Romero JJ, Lechago J, Jarboe LA, Wolfe MM. Role of luminal ammonia in the development of gastropathy and hypergastrinemia in the rat. Gastroenterology. 1995;108:320–9.
164. Kaneko H, Uchida K, Mitsuma T, Kotera H, Nagai H, Furusawa A, et al. Effect of a long-term oral ammonia administration on immunoreactive-somatostatin concentrations of rat stomach. Nihon Shokakibyo Gakkai zasshi. 1992;89:1477–83.
165. Graham DY, Opekun A, Lew GM, Klein PD, Walsh JH. *Helicobacter pylori*-associated exaggerated gastrin release in duodenal ulcer patients. The effect of bombesin infusion and urea ingestion. Gastroenterology. 1991;100:1571–5.
166. El Nujumi A, Dorrian C, Chittajallu R, Neithercut W, McColl K. Effect of inhibition of *Helicobacter pylori* urease activity by acetohydroxamic acid on serum gastrin in duodenal ulcer subjects. Gut. 1991;32:866–70.
167. Paoluzi OA. Del Vecchio Giovanna Blanco RC, Monteleone I, Monteleone G, Pallone F. Impairment of ghrelin synthesis in *Helicobacter pylori*-colonized stomach: New clues for the pathogenesis of *H. pylori*-related gastric inflammation. World J Gastroenterol. 2014;20:639–46.
168. Piotrowski J, Majka J, Slomiany A, Slomiany B. *Helicobacter pylori* lipopolysaccharide inhibition of gastric mucosal somatostatin receptor. Biochem Mol Biol Int. 1995;36:491–8.
169. Wang L, Basa NR, Shaikh A, Luckey A, Heber D, St-Pierre DH, et al. LPS inhibits fasted plasma ghrelin levels in rats: role of IL-1 and PGs and functional implications. Am J Physiol Gastrointest Liver Physiol. 2006;291:G611–G20.
170. Stengel A, Goebel M, Wang L, Reeve JR, Taché Y, Lambrecht NW. Lipopolysaccharide differentially decreases plasma acyl and desacyl ghrelin levels in rats: potential role of the circulating ghrelin-acylating enzyme GOAT. Peptides. 2010;31:1689–96.
171. Iyo T, Kaneko H, Konagaya T, Mori S, Kotera H, Uruma M, et al. Effect of intragastric ammonia on gastrin-, somatostatin-and somatostatin receptor subtype 2 positive-cells in rat antral mucosa. Life Sci. 1999;64:2497–504.
172. Retallack JE CS, Ring M, Moloche RM, Hezeko U, Finlay BB, et al. Differential regulation of SSTR mRNA expression in human antral cells infected with *Helicobacter pylori* (abstract). Gastroenterology. 1999;116:291.
173. Kusugami K, Ando T, Imada A, Ina K, Ohsuga M, Shimizu T, et al. Mucosal macrophage inflammatory protein-1alpha activity in Helicobacter. J Gastroenterol Hepatol. 1999;20(6).
174. Blaser MJ. Hypotheses on the pathogenesis and natural history of *Helicobacter pylori*-induced inflammation. Gastroenterology. 1992;102:720–7.

175. Zavros Y, Rathinavelu S, Kao JY, Todisco A, Del Valle J, Weinstock JV, et al. Treatment of *Helicobacter* gastritis with IL-4 requires somatostatin. Proc Natl Acad Sci. 2003;100:12944–9.
176. Lehmann FS, Golodner EH, Wang J, Chen M, Avedian D, Calam J, et al. Mononuclear cells and cytokines stimulate gastrin release from canine antral cells in primary culture. Am J Physiol Gastrointest Liver Physiol. 1996;270:G783–G8.
177. Beales I, Blaser MJ, Srinivasan S, Calam J, Perez-Perez GI, Yamada T, et al. Effect of *Helicobacter pylori* products and recombinant cytokines on gastrin release from cultured canine G cells. Gastroenterology. 1997;113:465–71.
178. Weigert N, Schaffer K, Schusdziarra V, Classen M, Schepp W. Gastrin secretion from primary cultures of rabbit antral G cells: stimulation by inflammatory cytokines. Gastroenterology. 1996;110:147–54.
179. Beales I, Post L, Calam J, Yamada T, DelValle J. Tumour necrosis factor alpha stimulates gastrin release from canine and human antral G cells: possible mechanism of the *Helicobacter pylori* – gastrin link. Eur J Clin Invest. 1996;26:609–11.
180. Beales I, Calam J, Post L, Srinivasan S, Yamada T, DelValle J. Effect of transforming growth factor alpha and interleukin 8 on somatostatin release from canine fundic D cells. Gastroenterology. 1997;112:136–43.
181. Prinz C, Neumayer N, Mahr S, Classen M, Schepp W. Functional impairment of rat enterochromaffin-like cells by interleukin 1 beta. Gastroenterology. 1997;112:364–75.
182. Witteman E, Mravunac M, Becx M, Hopman W, Verschoor J, Tytgat G, et al. Improvement of gastric inflammation and resolution of epithelial damage one year after eradication of *Helicobacter pylori*. J Clin Pathol. 1995;48:250–6.
183. Yamamoto S, Kaneko H, Konagaya T, Mori S, Kotera H, Hayakawa T, et al. Interactions among gastric somatostatin, interleukin-8 and mucosal inflammation in *Helicobacter pylori*-positive peptic ulcer patients. Helicobacter. 2001;6:136–45.
184. Konagaya T, Kusugami K, Kaneko H, Nishio Y, Osuga M, Shimizu T, et al. Negative correlation between somatostatin levels and interleukin-8 activity in gastric antral mucosa (abstract. Gut. 1997;41Supp:1–24.
185. Fantuzzi G, Faggioni R. Leptin in the regulation of immunity, inflammation, and hematopoiesis. J Leukoc Biol. 2000;68:437–46.
186. Grunfeld C, Zhao C, Fuller J, Pollack A, Moser A, Friedman J, et al. Endotoxin and cytokines induce expression of leptin, the ob gene product, in hamsters. J Clin Invest. 1996;97:2152–7.
187. Barbier M, Cherbut C, Aube A, Blottiere H, Galmiche J. Elevated plasma leptin concentrations in early stages of experimental intestinal inflammation in rats. Gut. 1998;43:783–90.
188. Roberts H, Hardie L, Chappell L, Mercer J. Parasite-induced anorexia: leptin, insulin and corticosterone responses to infection with the nematode, Nippostrongylus brasiliensis. Parasitology. 1999;118:117–23.
189. Lord GM, Matarese G, Howard JK, Baker RJ, Bloom SR, Lechler RI. Leptin modulates the T-cell immune response and reverses starvation-induced immunosuppression. Nature. 1998;394:897–901.
190. Sitaraman S, Liu X, Charrier L, Gu LH, Ziegler TR, Gewirtz A, et al. Colonic leptin: source of a novel proinflammatory cytokine involved in IBD. FASEB J. 2004;18:696–8.
191. Checchi S, Montanaro A, Pasqui L, Ciuoli C, Cevenini G, Sestini F, et al. Serum ghrelin as a marker of atrophic body gastritis in patients with parietal cell antibodies. J Clin Endocrinol Metab. 2007;92:4346–51.
192. Covacci A, Rappuoli R. Tyrosine-phosphorylated bacterial proteins trojan horses for the host cell. J Exp Med. 2000;191:587–92.
193. McColl KEL eOE, Gillen D, Ardill JES, Murray L, Crabtree JE. H. pylori induced hypergastrinemia is related to bacterial CagA status (abstract). Gastroenterology. 1997;112:215.
194. Queiroz D, Mendes E, Rocha G, Moura S, Oliveira A, Oliveira C. Somatostatin concentration (SC) and *Helicobacter pylori* (HP) vac A genotypes and cag A. Gut. 1996;39 Suppl 2:60.

195. Kim JH, Park HJ, Cho JS, Lee KS, Lee SI, Park IS, et al. Relationship of CagA to serum gastrin concentrations and antral G, D cell densities in *Helicobacter pylori* infection. Yonsei Med J. 1999;40:301–6.
196. Yamaoka Y, Kodama T, Kita M, Imanishi J, Kashima K, Graham D. Relation between clinical presentation, *Helicobacter pylori* density, interleukin 1β and 8 production, and cagA status. Gut. 1999;45:804–11.
197. Zaki M, Coudron PE, McCuen RW, Harrington L, Chu S, Schubert ML. H. pylori acutely inhibits gastric secretion by activating CGRP sensory neurons coupled to stimulation of somatostatin and inhibition of histamine secretion. Am J Phys Gastrointest Liver Physiol. 2013;304:G715–G22.
198. Kuipers E, Pena A, Festen H, Meuwissen S, Uyterlinde A, Roosendaal R, et al. Long-term sequelae of *Helicobacter pylori* gastritis. Lancet. 1995;345:1525–8.
199. Lee H-M, Wang G, Englander EW, Kojima M, Greeley Jr GH. Ghrelin, a new gastrointestinal endocrine peptide that stimulates insulin secretion: enteric distribution, ontogeny, influence of endocrine, and dietary manipulations. Endocrinology. 2002;143:185–90.
200. Masuda Y, Tanaka T, Inomata N, Ohnuma N, Tanaka S, Itoh Z, et al. Ghrelin stimulates gastric acid secretion and motility in rats. Biochem Biophy Res Commun. 2000;276:905–8.
201. de la Cour CD, Lindström E, Norlén P, Håkanson R. Ghrelin stimulates gastric emptying but is without effect on acid secretion and gastric endocrine cells. Regul Pept. 2004;120:23–32.
202. Suzuki H, Masaoka T, Hosoda H, Ota T, Minegishi Y, Nomura S, et al. *Helicobacter pylori* infection modifies gastric and plasma ghrelin dynamics in Mongolian gerbils. Gut. 2004;53:187–94.

Chapter 7
Symptom Generation

Juntaro Matsuzaki and Hidekazu Suzuki

Abstract Patients whose dyspepsia symptoms had disappeared after 12 months from *Helicobacter pylori* (*H. pylori*) eradication therapy were decided to be called *H. pylori*-associated dyspepsia (HpD), and were clearly distinguished from functional dyspepsia. *H. pylori* eradication is more effective than placebo with a number needed to treat (NNT) of 14 for *H. pylori*-positive dyspepsia. *H. pylori* is likely to be associated with the presence of postprandial distress symptoms rather than epigastric pain symptoms, although the evidence whether therapeutic responses to *H. pylori* eradication differ between subgroups of dyspepsia is still limited. The altered ghrelin secretion from the stomach, the presence or severity of microscopic duodenitis, and altered expression of muscle-specific microRNAs in the gastric smooth muscle layer would be possible mechanisms of HpD.

Keywords Dyspepsia • The Rome criteria • Ghrelin

7.1 *Helicobacter pylori*-Associated Dyspepsia

Most individuals infected with *Helicobacter pylori* (*H. pylori*) have few or no symptoms. However some may experience chronic dyspepsia symptoms even though they do not have peptic ulcer or gastric cancer. Dyspepsia is a term which includes a group of symptoms, such as epigastric pain, epigastric burning, uncomfortable postprandial fullness, and early satiation, which are thought to originate in the gastroduodenal region. By the Rome III definition, functional dyspepsia (FD) is diagnosed when no structural or biochemical explanation for a patient's symptoms is identified after appropriate investigations, regardless of the existence of *H. pylori* infection [1]. FD is one of the most common gastrointestinal diseases, which greatly impacts the quality of life. Since recent studies including systematic reviews and

J. Matsuzaki
Division of Molecular and Cellular Medicine, National Cancer Center Research Institute, 5-1-1 Tsukiji, Chuo-ku 104-0045, Tokyo, Japan

H. Suzuki (✉)
Medical Education Center, Keio University School of Medicine, 35 Shinanomachi, Shinjuku-ku, Tokyo, 160-8582 Tokyo, Japan
e-mail: hsuzuki.a6@keio.jp

H. Suzuki et al. (eds.), *Helicobacter pylori*, DOI 10.1007/978-4-431-55705-0_7

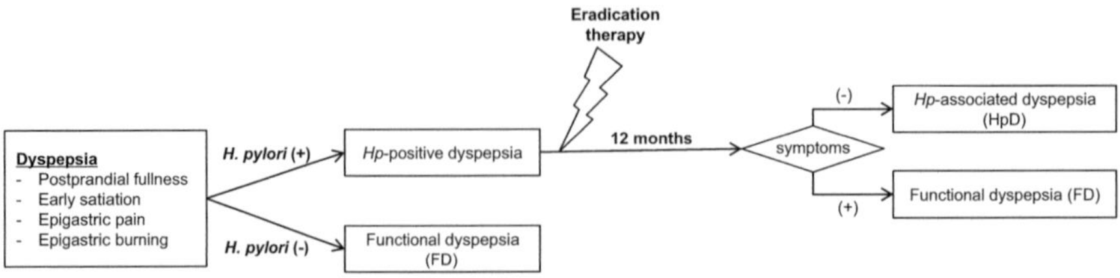

Fig. 7.1 *Helicobacter pylori*-associated dyspepsia (HpD) is defined as chronic dyspepsia which is improved for more than 12 months after *H. pylori* eradication. HpD should be separated from functional dyspepsia (FD)

meta-analysis showed the significant association between *H. pylori* infection and dyspepsia symptoms [2, 3], chronic dyspepsia symptoms which are thought to be caused by *H. pylori* infection are decided to be separated from FD and defined as *H. pylori*-associated dyspepsia (HpD) in the Kyoto Global Consensus Conference held on Jan. 30–Feb. 1, 2014 [4]. In this meeting, patients who remain symptom-free 12 months after eradication are considered to be cases of HpD, while patients who continue to experience dyspepsia even after *H. pylori* eradication will be considered as FD [5] (Fig. 7.1). In this section, we reviewed the epidemiology and the pathophysiology of HpD.

7.2 Epidemiology

An old meta-analysis showed that the prevalence of dyspepsia symptoms was greater in patients with *H. pylori* infection than in those without *H. pylori* infection, with an OR [odds ratio] of 2.3 (95 % CI [confidence interval] 1.9–2.7) [6]. More recent meta-analysis of 103 reports containing 312415 individuals showed that the prevalence of uninvestigated dyspepsia was higher in *H. pylori*-positive individuals (OR 1.18; 95 % CI 1.04–1.33) [7].

According to the Rome III criteria, FD patients were categorized into epigastric pain syndrome (EPS) and postprandial distress syndrome (PDS). Fang et al. reported the prevalence of *H. pylori* infection in FD patients diagnosed by the Rome III criteria in Taiwan [8]. In this study, 491 fulfilled the diagnostic criteria of FD among 2378 individuals. 298 (60.7 %) and 353 (71.9 %) individuals were diagnosed with EPS and PDS respectively, whereas 169 (34.4 %) had the overlap syndrome. *H. pylori* infection was positively associated with FD (OR 1.60; 99.5 % CI 1.03–2.48). *H. pylori* were associated with PDS alone (OR 1.86; 99.5 % CI 1.01–3.45), but not with EPS alone (OR 1.43; 99.5 % CI 0.72–2.84) or overlap syndrome (OR 1.12; 99.5 % CI 0.55–2.28). The density of *H. pylori*, severity of intestinal metaplasia, and infiltration of neutrophils and monocytes were not significantly different among the three subgroups. However, there was a trend of more moderate and marked gastric atrophy at the antrum in the subgroup of PDS alone. Among *H. pylori*-infected patients, a trend ($p = 0.044$) of more CagA-positive strains was observed in PDS alone (98.4 %), as compared with EPS alone (89.1 %) and the overlap syndrome (85.7 %). Piriyapong et al. also investigated the prevalence and impact of *H. pylori* infection on 300 patients with FD in Thailand and showed that *H. pylori* infection was significantly higher in PDS than EPS patients (27.1 % vs 16.7 %; OR 1.86; 95 % CI, 1.01–3.53) [9].

There is evidence of a small but statistically significant benefit in eradicating *H. pylori* in *H. pylori*-positive dyspepsia. In a Cochrane review of 21 placebo-controlled trials, the NNT (number needed to treat) for improvement in symptoms after eradicating *H. pylori* was 14, with no heterogeneity between studies and no evidence of funnel plot asymmetry [10]. Therefore, the eradication therapy is recommended as first-line therapy for *H. pylori*-positive dyspepsia. Zhao et al. reviewed 14 randomized controlled trials which contained information on the long-term (12 months or more) effects of *H. pylori* eradication on dyspeptic symptoms, and a subgroup analysis on geographical regions was conducted [11]. The improvements of dyspepsia symptoms in patients of eradication group were similarly better than in patients of control group in the European (OR 1.49; 95 % CI, 1.10–2.02), Asian (OR 1.54; 95 % CI, 1.07–2.21), and American populations (OR 1.43; 95 % CI, 1.12–1.83). Kim et al. reported that among the successfully eradicated dyspepsia patients ($n = 67$), male ($p = 0.013$) and higher initial BMI ($p = 0.016$) were associated with the improvement of dyspepsia at 1 year in Korean population [12]. Lan et al. reported that *H. pylori* eradication tended to be effective only in the EPS subgroup in China [13], although the other reports did not show the beneficial differences between subgroups of dyspepsia [14, 15].

7.3 Pathophysiology

Several hypotheses for the cause of HpD are considered [16], but pathophysiological conditions which were reported to be associated with dyspepsia symptoms have been limited (Fig. 7.2).

Ghrelin, which is produced and secreted by the A-like cells of the oxyntic glands of the stomach, has a well-established role in increasing appetite and food intake and in stimulating gastric emptying and acid secretion [17]. *H. pylori*-infected patients were shown to have lower gastric ghrelin mRNA expression than uninfected subjects [18]. Furthermore, the suppression of ghrelin expression is correlated with severity of glandular atrophy and chronic inflammation in the gastric corpus. Plasma ghrelin levels also decrease in *H. pylori*-infected patients [19, 20]. *H. pylori* infection may induce gastric motor dysfunction and reduce appetite with suppressed ghrelin secretion. Moreover, Lee et al. reported that preprandial ghrelin levels are significantly lower in PDS patients [21]. Shindo et al. also revealed that the maximum gastric emptying time for PDS is significantly higher with significant lower acyl-ghrelin levels in these patients [22]. Moreover, recent study showed that repeated ghrelin administrations have stimulatory effects

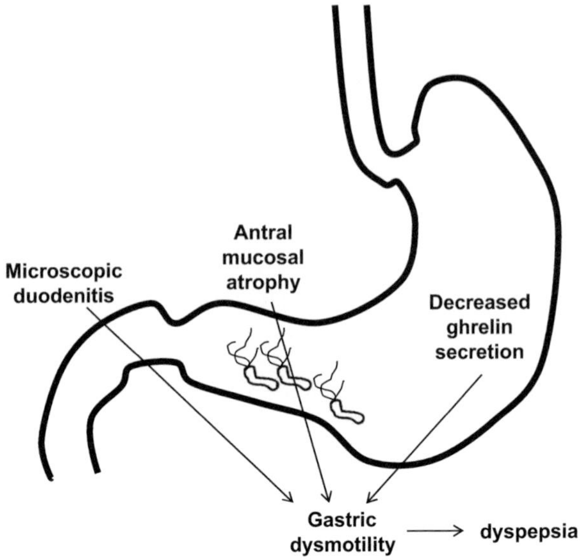

Fig. 7.2 Pathophysiological conditions associated with dyspepsia symptoms

on food intake in FD patients [23]. Taken together, *H. pylori* infection may reduce appetite with suppressed ghrelin secretion.

Inflammatory cell infiltration in the duodenal mucosa would be another possible cause of HpD. Mirbagheri et al. reported that *H. pylori* infection was significantly associated with presence and severity of microscopic duodenitis [24]. Although severity of dyspepsia symptoms was not higher in *H. pylori*-infected patients than *H. pylori*-noninfected patients, microscopic duodenitis significantly worsened the dyspepsia symptoms in the presence of *H. pylori* infection. Moreover, they also compared the symptom response to *H. pylori* eradication in dyspepsia patients in presence or absence of microscopic duodenitis [25]. Among 37 dyspepsia patients with *H. pylori*, microscopic duodenitis was observed in 16 (43.2 %). The improvement in severity of dyspepsia symptoms in the presence of microscopic duodenitis was significantly greater than when it was absent.

We previously reported the importance of muscle-specific microRNAs (miRNAs), such as *miR-1* and *miR-133*, in gastric motility disorders associated with *H. pylori* infection in mice [26]. These miRNAs were downregulated in the stomach, while the expression levels of histone deacetylase 4 (HDAC4) and serum response factor (SRF), which are target genes of *miR-1* and *miR-133*, were increased. Aberrant expression of HDAC4 and SRF induced gastric muscular hyperproliferation, thereby the gastric emptying was impaired.

7.4 Future Prospects

As described above, the evidence of the association between *H. pylori* infection and dyspepsia has been increasing. However, it is still unknown why most of individuals with *H. pylori* infection have no symptoms, while some of them have chronic dyspepsia symptoms. This point would be explained by distinct host-microbe interactions, but the evidence is insufficient for them. Tahara et al. reported the p22PHOX C242T polymorphism in host was inversely related to the risk of dyspepsia just in *H. pylori*-infected patients, but not in *H. pylori*-noninfected patients [27], although the reason for this difference is unknown. We therefore need to conduct further investigations about the complex interactions between *H. pylori* and the host to reveal mechanisms of HpD.

References

1. Tack J, Talley NJ, Camilleri M, Holtmann G, Hu P, Malagelada JR, et al. Functional gastroduodenal disorders. Gastroenterology. 2006;130(5):1466–79. doi:10.1053/j.gastro.2005.11.059. S0016-5085(06)00508-7 [pii].
2. Suzuki H, Matsuzaki J, Hibi T. What is the difference between *Helicobacter pylori*-associated dyspepsia and functional dyspepsia? J Neurogastroenterol Motil. 2011;17(2):124–30. doi:10.5056/jnm.2011.17.2.124.
3. Suzuki H, Moayyedi P. *Helicobacter pylori* infection in functional dyspepsia. Nat Rev Gastroenterol Hepatol. 2013;10(3):168–74. doi:10.1038/nrgastro.2013.9.
4. Sugano K, Tack J, Kuipers EJ, Graham DY, El-Omar EM, Miura S, et al. Kyoto global consensus report on *Helicobacter pylori* gastritis. Gut. 2015;64(9):1353–67. doi:10.1136/gutjnl-2015-309252.
5. Miwa H, Kusano M, Arisawa T, Oshima T, Kato M, Joh T, et al. Evidence-based clinical practice guidelines for functional dyspepsia. J Gastroenterol. 2015;50(2):125–39. doi:10.1007/s00535-014-1022-3.
6. Armstrong D. *Helicobacter pylori* infection and dyspepsia. Scand J Gastroenterol Suppl. 1996;215:38–47.
7. Ford AC, Marwaha A, Sood R, Moayyedi P. Global prevalence of, and risk factors for, uninvestigated dyspepsia: a meta-analysis. Gut. 2014. doi:10.1136/gutjnl-2014-307843.
8. Fang YJ, Liou JM, Chen CC, Lee JY, Hsu YC, Chen MJ, et al. Distinct aetiopathogenesis in subgroups of functional dyspepsia according to the Rome III criteria. Gut. 2014. doi:10.1136/gutjnl-2014-308114.
9. Piriyapong K, Tangaroonsanti A, Mahachai V, Vilaichone RK. *Helicobacter pylori* infection impacts on functional dyspepsia in Thailand. Asian Pac J Cancer Prev. 2014;15(24):10887–91.
10. Moayyedi P, Soo S, Deeks J, Delaney B, Harris A, Innes M, et al. Eradication of *Helicobacter pylori* for non-ulcer dyspepsia. Cochrane Database of Syst Rev. 2006;2, CD002096. doi:10.1002/14651858.CD002096.pub4.
11. Zhao B, Zhao J, Cheng WF, Shi WJ, Liu W, Pan XL, et al. Efficacy of *Helicobacter pylori* eradication therapy on functional dyspepsia: a meta-analysis of randomized controlled studies with 12-month follow-up. J Clin Gastroenterol. 2014;48(3):241–7. doi:10.1097/MCG.0b013e31829f2e25.
12. Kim SE, Park YS, Kim N, Kim MS, Jo HJ, Shin CM, et al. Effect of *Helicobacter pylori* eradication on functional dyspepsia. J Neurogastroenterol Motil. 2013;19(2):233–43. doi:10.5056/jnm.2013.19.2.233.

13. Lan L, Yu J, Chen YL, Zhong YL, Zhang H, Jia CH, et al. Symptom-based tendencies of *Helicobacter pylori* eradication in patients with functional dyspepsia. World J Gastroenterol. 2011;17(27):3242–7. doi:10.3748/wjg.v17.i27.3242.
14. Gwee KA, Teng L, Wong RK, Ho KY, Sutedja DS, Yeoh KG. The response of Asian patients with functional dyspepsia to eradication of *Helicobacter pylori* infection. Eur J Gastroenterol Hepatol. 2009;21(4):417–24. doi:10.1097/MEG.0b013e328317b89e.
15. Mazzoleni LE, Sander GB, Francesconi CF, Mazzoleni F, Uchoa DM, De Bona LR, et al. *Helicobacter pylori* eradication in functional dyspepsia: HEROES trial. Arch Intern Med. 2011;171(21):1929–36. doi:10.1001/archinternmed.2011.533.
16. Suzuki H, Mori H. *Helicobacter pylori*: *Helicobacter pylori* gastritis-a novel distinct disease entity. Nat Rev Gastroenterol Hepatol. 2015;12(10):556–7. doi:10.1038/nrgastro.2015.158.
17. Kojima M, Hosoda H, Date Y, Nakazato M, Matsuo H, Kangawa K. Ghrelin is a growth-hormone-releasing acylated peptide from stomach. Nature. 1999;402(6762):656–60. doi:10.1038/45230.
18. Tatsuguchi A, Miyake K, Gudis K, Futagami S, Tsukui T, Wada K, et al. Effect of *Helicobacter pylori* infection on ghrelin expression in human gastric mucosa. Am J Gastroenterol. 2004;99(11):2121–7. doi:10.1111/j.1572-0241.2004.30291.x.
19. Suzuki H, Masaoka T, Hosoda H, Nomura S, Ohara T, Kangawa K, et al. Plasma ghrelin concentration correlates with the levels of serum pepsinogen I and pepsinogen I/II ratio—a possible novel and non-invasive marker for gastric atrophy. Hepatogastroenterology. 2004;51(59):1249–54.
20. Isomoto H, Nakazato M, Ueno H, Date Y, Nishi Y, Mukae H, et al. Low plasma ghrelin levels in patients with *Helicobacter pylori*-associated gastritis. Am J Med. 2004;117(6):429–32. doi:10.1016/j.amjmed.2004.01.030.
21. Lee KJ, Cha DY, Cheon SJ, Yeo M, Cho SW. Plasma ghrelin levels and their relationship with gastric emptying in patients with dysmotility-like functional dyspepsia. Digestion. 2009;80(1):58–63. doi:10.1159/000215389.
22. Shindo T, Futagami S, Hiratsuka T, Horie A, Hamamoto T, Ueki N, et al. Comparison of gastric emptying and plasma ghrelin levels in patients with functional dyspepsia and non-erosive reflux disease. Digestion. 2009;79(2):65–72. doi:10.1159/000205740.
23. Akamizu T, Iwakura H, Ariyasu H, Hosoda H, Murayama T, Yokode M. Repeated administration of ghrelin to patients with functional dyspepsia: its effects on food intake and appetite. Eur J Endocrinol. 2008;158(4):491–8. doi:10.1530/EJE-07-0768. 158/4/491.
24. Mirbagheri SA, Khajavirad N, Rakhshani N, Ostovaneh MR, Hoseini SM, Hoseini V. Impact of *Helicobacter pylori* infection and microscopic duodenal histopathological changes on clinical symptoms of patients with functional dyspepsia. Dig Dis Sci. 2012;57(4):967–72. doi:10.1007/s10620-011-1960-z.
25. Mirbagheri SS, Mirbagheri SA, Nabavizadeh B, Entezari P, Ostovaneh MR, Hosseini SM, et al. Impact of microscopic duodenitis on symptomatic response to *Helicobacter pylori* eradication in functional dyspepsia. Dig Dis Sci. 2015;60(1):163–7. doi:10.1007/s10620-014-3285-1.
26. Saito Y, Suzuki H, Tsugawa H, Suzuki S, Matsuzaki J, Hirata K, et al. Dysfunctional gastric emptying with down-regulation of muscle-specific microRNAs in *Helicobacter pylori*-Infected mice. Gastroenterology. 2011;140:189–98. doi:10.1053/j.gastro.2010.08.044.
27. Tahara T, Shibata T, Wang F, Nakamura M, Sakata M, Nakano H, et al. A genetic variant of the p22PHOX component of NADPH oxidase C242T is associated with reduced risk of functional dyspepsia in *Helicobacter pylori*-infected Japanese individuals. Eur J Gastroenterol Hepatol. 2009;21(12):1363–8. doi:10.1097/MEG.0b013e32830e2871.

Chapter 8
Gastric Non-*Helicobacter pylori* Helicobacter: Its Significance in Human Gastric Diseases

Masahiko Nakamura, Anders Øverby, Somay Y. Murayama, Hidekazu Suzuki, Tetsufumi Takahashi, Shinichi Takahashi, and Hidenori Matsui

Abstract In the post-*H. pylori* eradication era, the clinical significance of gastric non-*H. pylori* helicobacters (NHPH; also referred to as *H. heilmannii*-like organisms and *H. heilmannii* sensu lato) is gradually increasing. This group of bacteria may inhabit the stomach of domestic and wild animals including cats, dogs, pigs, primates, rodents, cheetahs, and rabbits. NHPH are zoonotic microorganisms, meaning that they may transmit between animals and humans. They may be distinguished from *H. pylori* regarding their microbiology involving larger cells with more distinct spiral shape and bipolarity, localization in the stomach layer and regional distribution, urease activity and virulence factors, and relation to gastric diseases where gastric NHPH infection is often associated with milder gastritis than *H. pylori* but higher risk of gastric MALT lymphoma. At present, pure culture of NHPH species remains a challenge, but the full genome sequences of some of the species have been reported. Recent and ongoing prevalence studies indicate a higher clinical relevance of these bacteria than earlier impressions suggested. Current efforts in improving cultivation and detection methodology are contributing to an increased understanding of their microbiology, prevalence, and relevance to human diseases.

M. Nakamura (✉) • A. Øverby • T. Takahashi
Division of Pathophysiology, Research and Education Center for Clinical Pharmacy, School of Pharmaceutical Sciences, Kitasato University, 5-9-1 Shirokane, Minato-ku 108-8641, Tokyo, Japan
e-mail: nakamuram@pharm.kitasato-u.ac.jp

S.Y. Murayama
School of Pharmacy, Nihon University, Funabashi-shi, Chiba, Japan

H. Suzuki
Medical Education Center, Keio University School of Medicine, 35 Shinanomachi, Shinjuku-ku, Tokyo, 160-8582 Tokyo, Japan

S. Takahashi
3rd Department of Internal Medicine, Kyorin University, Mitaka, Tokyo, Japan

H. Matsui
Kitasato Institute for Life Sciences, Kitasato University, Minato-ku, Tokyo, Japan

H. Suzuki et al. (eds.), *Helicobacter pylori*, DOI 10.1007/978-4-431-55705-0_8

Keywords Non-*Helicobacter pylori* helicobacter • *Helicobacter suis* • *Helicobacter heilmannii* • Zoonosis

8.1 Introduction

Recently, many spiral bacteria belonging to the *Helicobacter* genus other than *H. pylori* have been described in many tissues, organs, and animals. They are generally longer than *H. pylori* with a size ranging from 4 to 10 μm, spiral shaped, highly motile with three to eight coils, up to 14 uni- or bipolar flagellae, and no periplasmic filaments [1]. These bacteria also differ from *H. pylori* when considering traits such as zoonosis, distribution/sublocations in the stomach, relation to gastric diseases, urease activity, and response to antibiotics. To date, 35 species belonging to the *Helicobacter* genus have been identified (Fig. 8.1) that may be divided into gastric helicobacters and enterohepatic helicobacters. In this chapter, we would like to focus on the gastric helicobacters and further aim at the gastric helicobacters other than *H. pylori*, the so-called non-*H. pylori* helicobacters (NHPH), and discuss these bacteria regarding their microbiology, pathogenesis, and significance for human health.

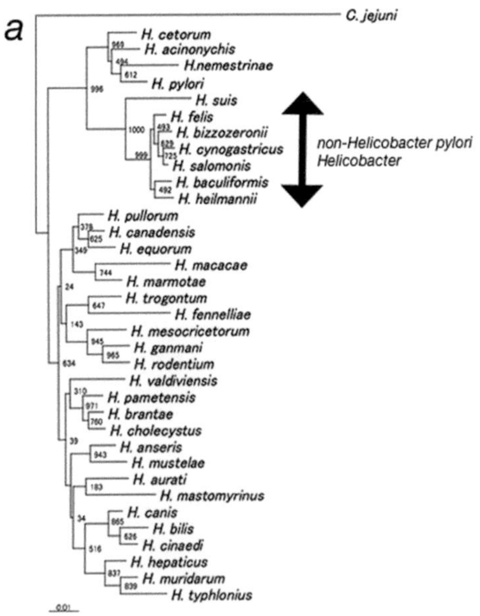

b *Type strain of 35 Helicobacter* species and its accession number of 16S rDNA

Species	Type strain	16S rRNA sequence accession no	bp
Helicobacter acinonychis	CCUG 29263	M88148	1485
Helicobacter anseris	MIT 04-9362	DQ415545	1473
Helicobacter aurati	MIT 97-5075c	AF297868	1455
Helicobacter baculiformis	M50	EF070342	1455
Helicobacter bilis	Hb1	U18766	1652
Helicobacter bizzozeronii	CCUG 35545	Y09404	1421
Helicobacter brantae	MIT 04-9366	DQ415546	1474
Helicobacter canadensis	NLEP-16143	AF262037	1457
Helicobacter canis	ATCC 51401	AY631945	1473
Helicobacter cetorum	MIT 99-5656	AF292378	1495
Helicobacter cholecystus	ATCC 700242	AY686606	1472
Helicobacter cinaedi	CCUG 18818	M88150	1473
Helicobacter cynogastricus	JKM4	DQ004689	1462
Helicobacter equorum	EqF1	DQ307735	1454
Helicobacter felis	ATCC 49179	M57398	1474
Helicobacter fennelliae	CCUG 18820	M88154	1833
Helicobacter ganmani	CCUG 43526	AF000221	1418
Helicobacter heilmannii	ASB1	HM625820	1403
Helicobacter hepaticus	Hh-2	U07574	1470
Helicobacter macacae	MIT 99-5501	AF333338	1654
Helicobacter marmotae	MIT 98-6070	AF333341	1473
Helicobacter mastomyrinus	MIT 97-5574	AY742307	1760
Helicobacter mesocricetorum	MU 97-1514	AF072471	1420
Helicobacter muridarum	ST1	M80205	1478
Helicobacter mustelae	ATCC 43772	M35048	1481
Helicobacter nemestrinae	ATCC 49396	X67854	1463
Helicobacter pametensis	B9A Seymour	M88147	1481
Helicobacter pullorum	ATCC 51801	AY631956	1473
Helicobacter pylori	ATCC 43504	U01330	1503
Helicobacter rodentium	MIT 95-1707	U96296	1423
Helicobacter salomonis	Inkinen	U89351	1441
Helicobacter suis	HS1	EF204589	1451
Helicobacter trogontum	LRB 8581	U65103	1422
Helicobacter typhlonius	MIT 97-6810	AF127912	1614
Helicobacter valdiviensis	WBE14	KF549903.	1383
Campilobacter jejuni	ATCC 33560	NR_118520	1435

Fig. 8.1 Phylogenetic tree of 35 *Helicobacter* species (**a**) and type strains (**b**)

8.2 History

At first, we would like to introduce the reader to the somewhat complicated story of NHPH and nomenclature as this has been subject to some confusion in discussions and in published papers. In fact, the first discovery of spiral-shaped microorganism in the stomach from an animal dates back to 1881, a study conducted by Rappin using light microscopy [2]. Bizzozero and Salomon also showed similar studies [3, 4], and in 1906 the German scientist Krienitz reported spiral-shaped microorganisms in the human stomach [5]. In 1919, Kasai and Kobayashi were the first to achieve an interventional study by using salvarsan, an arsene compound, to treat infection [6]. In 1962, Weber et al. found spiral-shaped bacteria in the gastric mucosa of cats and dogs by using electron microscopy [7]. As 1984 will be remembered for the paramount discovery of gastric *H. pylori* by Warren and Marshall [8], 3 years later was marked by a German report by Heilmann who detected gastric spiral bacteria in 39 patients (0.25 %) with upper gastrointestinal symptoms. The report was later published in an English journal [9]. Among the patients examined 34 patients had chronic active gastritis and 4 had chronic gastritis, and they were all treated with the bismuth treatment. Following this report, Dent and McNulty found spiral-shaped bacteria in 6 human cases and further reported that they were incapable of cultivating these bacteria in vitro [10, 11]. They designated these spiral-shaped bacteria *Gastrospirillum hominis*. In 1988, they proposed that these spiral-shaped bacteria could transmit from cat to human [12]. Three years later, Ito and Takahashi found urease-positive gastric bacteria in a cynomolgus monkey [13], a strain that would turn out to be used for later experimentation [14]. In 1993, Solnick et al. analyzed the 16S rRNA of this bacterium and discovered that this bacterium was phylogenetically close to *H. felis* [15]. *Gastrospirillum hominis* was from that point renamed to *H. heilmannii*. Further phylogenetic analyses revealed that *H. heilmannii* did not only encompass one species, but rather a group of bacteria. Further renaming resulted in *H. heilmannii* type 1 and 2, of which type 1 was identical to the type found in pigs also called *H. suis* [16]. Type 2 represents a group of bacteria which includes, for instance, *H. felis*, *H. bizzozeronii*, and *Candidatus H. heilmannii*. Challenges in in vitro cultivation have hampered identification and phylogenetic analyses of these bacteria, but recent efforts have resolved many of the species that belong to the *Helicobacter* genus. In order to avoid further confusion, Haesebrouck proposed to use the terms *H. heilmannii* sensu stricto and sensu lato [17, 18]. Another commonly applied term is *H. heilmannii*-like organisms (HHLO), which may be further subdivided into types 1, 2, and 4 [19]. More recently, the term non-*H. pylori* helicobacters (NHPH) has been applied as it covers all "the other" helicobacters than *H. pylori* without manifesting to one such species of historical reason [20]. The term is per definition the same as *H. heilmannii* s.l. and will be used throughout this chapter.

8.3 Characteristics of NHPH

8.3.1 Zoonosis

Unlike *H. pylori*, NHPH show a zoonotic infection pattern. Reported animal hosts for gastric NHPH include dogs, cats, primates, pigs, cheetahs, rodents, and rabbits [21]. Stolte et al. conducted a study in which the relationship between infection in humans and contact with domestic animals was made. In NHPH-positive cases, 70 % of the patients had a history of contact with pet animals, while in the general population, 37 % has a history of contact with pet animals [22]. In addition, the rate of coinfection with NHPH and *H. pylori* was found to be very low, suggesting that NHPH infection might conflict with and somehow prevent an *H. pylori* infection (although opposing reports do exist as mentioned later in this chapter). Another report demonstrated the transmission of NHPH from a cat to a veterinarian who was treating the cat [23]. The relationship between pet keeping and risk of infection for NHPH has also been demonstrated in a prevalence study conducted in Korea by Chung and colleagues [24]. Although transmission from animals to humans is a commonly accepted concept, studies that raise important questions also exist. For example, Priestnall et al. have reported that the NHPH most transmitted to human is HHLO type 1, or *H. suis*, which is a pig-specific species of NHPH. The type of NHPH predominant in cats and dogs are presumably HHLO type 2 and 4, suggesting that pet keeping is not necessarily the main transmission route of NHPH from animal to human [18]. A prevalence study conducted in China of more than 1500 patients all positive for *H. pylori* showed that of those who were coinfected with NHPH, about half was infected with *H. suis* [25]. An ongoing prevalence study in Japan by Øverby and Nakamura of gastric disease patients negative for *H. pylori* also suggests the pig-specific species *H. suis* to be the predominant type found in humans. Considering the number of people who are in contact with pet animals compared to the number of people who are in contact with live pigs, it is likely that there is a missing piece to this controversial puzzle of how humans are infected with NHPH. Flahou and Haesebrouck have shown that NHPH may be detected in minced meat from pork from the supermarket, suggesting that transmission through diet may also be a possibility [26].

8.3.2 Distribution in the Stomach

As for the localization in the gastric mucosa, it has been reported that the majority of the bacteria inhabit the mucus layer like *H. pylori* but that they also may reside in the fundic glandular tissue as shown, which is clearly detected in our histochemical observation in Fig. 8.2 [27]. A more precise pathological observation revealed that most of the cells residing in the fundic gland were actually in the intracellular canaliculi in the parietal cells [28]. However, by using electron microscopy, some

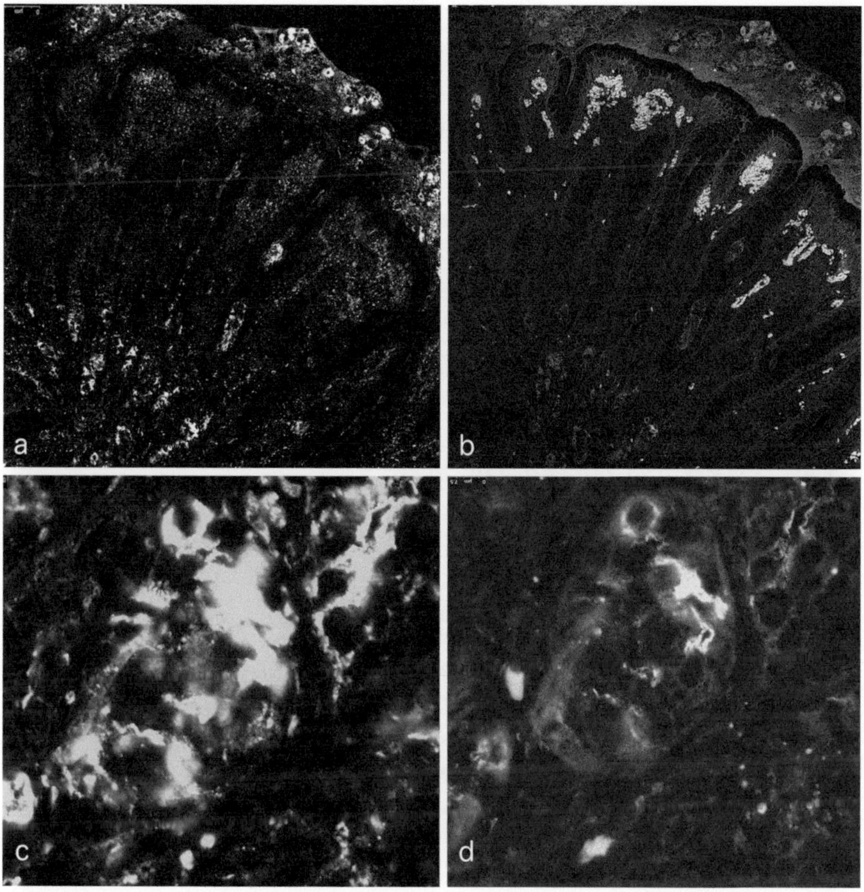

Fig. 8.2 Immunohistochemical observation of *Helicobacter suis* in the human fundic mucosa. (**a**, **b**) *Helicobacter suis* are localized in the base of the fundic mucosa as well as in the mucus layer above the surface epithelial cells. (**c**, **d**) In the base of the fundic glands, *Helicobacter suis* are localized mainly in the actin-rich parietal cells. (**a**, **c**) Stained with *Helicobacter* antibody to the *Helicobacter suis*-positive human gastric mucosa. (**b**, **d**) stained with FITC-phalloidin showing the actin-rich erythrocytes in the tip portion and parietal cells in the base of the fundic mucosa. (**a**, **b**) x200, (**c**, **d**) x800

of the bacteria were found to be localized in the lamina propria mucosae as well as in the intracellular canaliculus. This resulted in the adjacent parietal cells to become apoptotic suggesting a relation to gastric MALT lymphoma formation [23].

8.3.3 Prevalence of NHPH

Up until 2001, O'Rourke performed meta-analysis of 500 NHPH-positive cases [29–31], and she found that the infection rate is very diverse. In Western countries, the prevalence of NHPH in the general population was less than 0.5 %, while in East Europe and Asia, it varied from 1.2 % to 6.2 %. In a prevalence study conducted in China mentioned above with *H. pylori*-positive patients with symptoms of a gastric disease, close to 12 % was shown to coinhabit an NHPH infection [32]. In an ongoing study in Japan based on patients with a gastric disease but negative for *H. pylori*, more than half appear to be infected with NHPH (Øverby and Nakamura, unpublished data).

8.3.4 NHPH and Human Diseases

Stolte et al. compared the disease formation by *H. pylori* and NHPH and found that NHPH induced a milder gastritis than *H. pylori* but that NHPH was a stronger inducer of gastric MALT lymphoma than *H. pylori* [33, 34]. They used histochemistry, specific immunoabsorbent method, and PCR analysis using 16S rDNA, and five MALT lymphoma cases were *H. pylori* negative and NHPH positive. From the survey of the patients from 1988 to 1998, 1745 out of 263, 680 cases were *H. pylori* positive (0.66 %), while 8 out of 543 cases were NHPH positive (1.47 %) and confirmed the stronger relation of NHPH to the gastric MALT lymphoma formation. Okiyama et al. found 15 NHPH-positive cases out of 4074 serial cases, among these 11 patients constituted chronic gastritis cases, and 4 patients were diagnosed with gastric MALT lymphoma [35]. The relationship between NHPH infection and gastric cancer is still a controversional topic. Foschini et al. have reported that all NHPH-positive cases had gastritis, and 1 gastric cancer case was coinfected with NHPH and *H. pylori* [36]. The relation to the gastric dysplasia to NHPH infection was suggested in cases in Thailand [31]. Sasaki has recently reported a patient case positive for NHPH infection and with nodular gastritis, which is thought to be one of the precancerous lesions of the gastric cancer [37].

8.3.5 Diagnosis, Culture, and Genome Sequence Studies

Detection of NHPH may be performed with different approaches. However, PCR analysis of the 16S rDNA is the most reliable method but is also time-consuming. The pathological identification of gastric spiral bacteria may sometimes be inaccurate, as under certain conditions *H. pylori*, for instance, may display a different morphology than typically seen including a longer shape [38]. At present, serum serological test and fecal bacterial detection tests are not available for NHPH, somewhat limiting the detection methodology in patients. Difficulties in cultivation NHPH have hampered the identification of these microorganisms. Lee et al. reported culture of a spiral bacterium isolated from the antrum area of the stomach from a cat [39]. In 1996, Andersen et al. reported a presumably successful cultivation of *H. heilmannii* (s. s.) but later turned out to be identified as *H. bizzozeronii* [40, 41]. As a new diagnostic method, Trebesius et al. used fluorescent in situ hybridization of the 16S rDNA and reported five species of NHPH from human cases, most of which coincided with *H. suis* [42]. In 2003, Chisholm et al. invented a novel PCR method for NHPH detection from biopsy specimen and found 2.3 % positive cases from dyspeptic cases in New England, which was quite higher compared with their former report highlighting the importance of appropriate detection methodology [43]. In 2004, O'Rourke et al. analyzed 26 human and animal samples positive for NHPH infection and showed that 15 (28 %) of these were very infected with *H. suis*. The rest were identified as *H. felis*, *H. bizzozeronii*, *H. salomonis*, and *Candidatus H. heilmannii* [44]. In 2008, Baele et al. reported the first pure culture of *H. suis* isolated from pig stomach [45]. In 2011, the whole genome sequences of *H. suis* and *H. felis* were reported [46, 47]. When compared with *H. pylori*, the genome sequence of *H. suis* was shown to lack *cagA* but did contain *hpaA* and *horH*; *comB*, related to type IV secretory system and similar genes to *H. pylori* neutrophil-activating protein; γ-glutamyl transpeptidase (GGT); and flavodoxin vacuolating cytotoxin A gene. In *H. felis*, the genome sequence was shown to lack *cagPAI* and *vacA* but contain *comB*, GGT-encoding gene, immunomodulator (*napA*), collagenase and secretory serine protease htrA, and several chemotaxis sensors and restriction/modification system [48]. The sequence of *H. bizzozeronii* isolated from a gastritis patient [49] and the sequence of *H. heilmannii* isolated from cat [50] were subsequently reported.

8.4 Conclusion

Infection with the zoonotic NHPH is becoming an increasing issue in the clinic and eventually in the general population. Recent studies have shown NHPH to be linked to human gastric diseases especially gastric MALT lymphoma, require up-to-date sensitive methodology in order to be detected, and be more prevalent than what was previously thought. In the post-*H. pylori* eradication era, gastric infection with NHPH is likely to become a significant burden and should be subject to further investigation in order to resolve its issues linked with gastric MALT lymphoma and gastric cancer.

References

1. Stoffel MH, Friess AE, Burnens A, Schmassmann A, Neiger R. Distinction of gastric Helicobacter spp. in humans and domestic pets by scanning electron microscopy. Helicobacter. 2000;5:232–9.
2. Rappin J. Contre a l'etude de bacteri de la bouche a l'etat normal.1881;68. In: Breed RS, Murray EGD, Hitchens AP, editors. Bergey's manual of determinative bacteriology. 6th ed. Baltimore: Williams and Wilkins Co; 1948. p. 217.
3. Bizzozero G. Ueber die schlauchformigen drusen desmagendarmkanals und die beziehungen ihres epithels zu demoberflachenepithel der schleimhaut. Archivfur MikrofkopischeAnatomie Entwickiungs mechanik. 1893;42:82.
4. Salomon H. Ueber das spirillum des saugetiermagens und seinverhalten zu den belegzellen. Centralblatt fur Bakteriologie, Parasitenkunde V Infektionskrankheiten. 1896;XIX:433–43.
5. Krienitz W. Ueber das Auftreten von Spirochaten verscheidener Form im Mageninhalt bei Carcinoma ventriculi. Dtsch Med Wocknenschr. 1906;32:872.
6. Kasai K, Kobayashi R. The stomach spirochete occurring in mammals. J Parasitol. 1919;6:1–11.
7. Weber AF, Schmittdiel EF. Electron microscopic and bacteriologic studies of spirilla isolated from the fundic stomachs of cats and dogs. Am J Vet Res. 1962;23:422–7.
8. Marshall BJ, Warren JR. Unidentified curved bacilli in the stomach of patients with gastritis and peptic ulceration. Lancet. 1984;1(8390):1311–5.
9. Heilmann KL, Bochard F. Gastritis due to spiral shaped bacteria other than *Helicobacter pylori*: clinical, histological, and ultrastructural findings. Gut. 1991;32:137–40.
10. Dent JC, McNulty CA, Uff JC, Wilkinson SP, Gear MW. Spiral organisms in the gastric antrum. Lancet. 1987;11:96.
11. McNulty CA, Dent JC, Curry A, Uff JS, Ford GA, Gear MW, Wilkinson SP. New spiral bacterium in gastric mucosa. J Clin Pathol. 1989;42:585–91.
12. Dye KR, Marshall BJ, Fnerson HF, Onerrant RT, McCall RW. Gastritis in a human due to infection with an organism resembling the cat gastric spirillum. Gastroenterology. 1988;94: A108.
13. Itoh T, Yanagawa M, Singaki N, Masubuchi N, Takahashi S, Saito S. Isolation of *Helicobacter heilmannii* like organism from the stomachs of cynomolgus monkey and colonization of them in mice. Gastroenterology. 1994;106:A99.
14. Nakamura M, Murayama SY, Serizawa H, Sekiya Y, Eguchi M, Takahashi S, Nishikawa K, Takahashi T, Matsumoto T, Yamada H, Hibi T, Tsuchimoto K, Matsui H. "*Candidatus* Helicobacter heilmannii" from a cynomolgus monkey induces gastric mucosa-associated lymphoid tissue lymphomas in C57BL/6 mice. Infect Immun. 2007;75:1214–22.
15. Solnick JV, O'Rourke J, Lee A, Paster BJ, Dewhirst FE, Tompkins LS. An uncultured gastric spiral organism is a newly identified Helicobacter in humans. J Infect Dis. 1993;168:379–85.
16. Mendes EN, Queiroz DM, Rocha GA, Moura SB, Leite VH, Fonseca ME. Ultrastructure of a spiral micro-organism from pig gastric mucosa ("*Gastrospirillum suis*"). J Med Microbiol. 1990;33:61–6.
17. De Groote D, Ducatelle R, van Doorn LJ, Tilmant K, Quint WGV, Verschuurn A, Haesebrouck F. Detection of "*Candidatus* Helicobacter suis" in gastric samples of pig by PCR: comparison with other invasive diagnostic techniques. J Clin Microbiol. 2000;38:1131–5.
18. Haesebrouck F, Pasmans F, Flahou B, Smet A, Vandamme P, Ducatelle R. Non-Helicobacter pylori Helicobacter species in the human gastric mucosa: a proposal to introduce the terms *H. heilmannii sensu lato* and *sensu stricto*. Helicobacter. 2011;16:339–40.
19. Priestnall SL, Wiinberg B, Spohr A, Neuhaus B, Kuffer M, Wiedmann M, Simpson W. Evaluation of "*Helicobacter heilmannii*" subtypes in the gastric mucosas of cats and dogs. J Clin Microbiol. 2004;42:2144–51.

20. Fox JG. The non-*H pylori* helicobacters: their expanding role in gastrointestinal and systemic diseases. Gut. 2002;50:273–83.
21. Van den Bulck K, Baele M, Hermans K, Ducatelle R, Haesebrouck F, Decostere A. First report on the occurrence of "*Helicobacter heilmannii*" in the stomach of rabbits. Vet Res Commun. 2005;29:271–27.
22. Stolte M, Wellens E, Bethke B, Ritter M, Eidt H. *Helicobacter heilmannii* (formerly *Gastrospirillum hominis*) gastritis: an infection transmitted by animals? Scand J Gastroenterol. 1994;29:1061–4.
23. Lavelle JP, Landas S, Mitros FA, Conklin JL. Acute gastritis associated with spiral organisms from cats. Dig Dis Sci. 1994;39:744–50.
24. Chung T-H, Kim H-D, Lee Y-S, Hwang C-Y. Determination of the prevalence of *Helicobacter heilmannii*-like organisms type 2 (HHLO-2) infection in humans and dogs using non-invasive genus/species-specific PCR in Korea. J Vet Med Sci. 2014;76:73–9.
25. Liu J, He L, Haesebrouck F, Gong Y, Flahou B, Cao Q, Zhang J. Prevalence of coinfection with gastric Non-Helicobacter pylori Helicobacter (NHPH) species in Helicobacter pylori-infected patients suffering from gastric disease in Beijing, China. Helicobacter. 2015;20:284–90. doi:10.1111/hel.12201.
26. De Cooman L, Flahou B, Houf K, Smet A, Ducatelle R, Pasmans F, Haesebrougk F. Survival of Helicobacter suis bacteria in retail pig meat. Int J Food Microbiol. 2013;166(1):164–7. doi:10.1016/j.ijfoodmicro.2013.05.020.
27. Carnot P, Lelievre A. Morphologie du product d'excretion des cellules bordants. Comptes Rendus Soc Biol. 1909;66:311–3.
28. Regard C. Sur une curieuse lacalisation de spirilles parasites dans les canalisations glandulaires de la gastrique normale, chez le chien et le chat. Soc Biol. 1909;66:229–31.
29. Kubonova K, Trupl J, Jancula L, Polák E, Vráblik V. Presence of spiral bacteria ('*Gastrospirillum hominis*') in the gastric mucosa. Eur J Clin Microbial Infect Dis. 1991;10:459–60.
30. Yang HT, Goliger JA, Song M, Zhou D. High prevalence of *Helicobacter heilmannii* infection in China. Dig Dis Sci. 1998;43:1493.
31. Yali Z, Yamada N, Wen M, Matsuhisa T, Miki M. *Gastrospirillum hominis* and Helicobacter pylori infection in Thai individuals – comparison of histopathological changes of gastric mucosa. Pathol Int. 1998;48:507–11.
32. Kato S, Ozawa K, Sekine H, Ohyauchi M, Shimosegawa T, Minoura T, Iinuma K. *Helicobacter heilmannii* infection in a child after successful eradication of *Helicobacter pylori*: case report and review of literature. J Gastroenterol. 2005;40:94–7.
33. Stolte M, Kroher G, Meining A, Morgner A, Bayerdörffer E, Bethke B. A comparison of Helicobacter pylori and H. heilmannii gastritis. A matched control study involving 404 patients. Scand J Gastroenterol. 1997;32:28–33.
34. Morgner A, Lehn N, Andersen LPP, Thiede C, Bennedsen M, Trebesius K, Neubauer B, et al. *Helicobacter heilmannii*-associated primary gastric low-grade MALT lymphoma: complete remission after curing the infection. Gastroenterology. 2000;118:821–8.
35. Okiyama Y, Matsuzawa K, Hidaka E, Sano K, Akamatsu T, Ota H. *Helicobacter heilmannii* infection: clinical, endoscopic and histopathological features in Japanese patients. Pathol Int. 2005;55:398–404.
36. Foschini MP, Pieri F, Cerasoli S, Accardo P, Formica G, Biasiucci A, Donzelli C, et al. *Helicobacter heilmannii*: anatomo-clinical study of 14 new cases. Pathologica. 1999;91:18–24.
37. Sasaki M, Goji S, Tamura Y, Nakamura M, Matsui H, Murayama SY, Ebi M, Ogasawara N, Funai Y, Kasugai K. H*elicobacter suis*-infected nodular gastritis and a review of diagnostic sensitivity for *Helicobacter heilmannii*-like organisms. Case Rep Gastroenterol. 2015;9:179–87.

38. Vinette KM, Gibney KM, Proujansky R, Fawcett PT. Growth of *Helicobacter pylori* in a long spiral form does not alter expression of immunodominant proteins. BMC Microbiol. 2002;2:24.
39. Lee A, Dent J, Hazell S, McNulty C. Origin of spiral organisms in human gastric antrum. Lancet. 1988;1(8580):300–1.
40. Andersen LP, Norgaard A, Holck S, Blom J, Elsborg L. Isolation of a *Helicobacter heilmannii*-like organism from the human stomach. Eur J Clin Microbiol Infect Dis. 1996;15:95–6.
41. Jalava K, On SLW, Harrington CS, Andersen LP, Hanninen ML, Vandamme P. A cultured strain of "Helicobacter heilmannii", a human gastric pathogen, identified as H. bizzozeronii: evidence for zoonotic potential of Helicobacter. Emerg Infect Dis. 2001;7:1036–8.
42. Trebesius K, Adler K, Vieth M, Stolte M, Haas R. Specific detection and prevalence of *Helicobacter heilmannii*-like organisms in the human gastric mucosa by fluorescent in situ hybridization and partial 16S ribosomal DNA sequencing. J Clin Microbiol. 2001;39:1510–6. doi:10.1128/JCM.39.4.1510-1516.2001.
43. Chisholm SA, Owen RJ. Development and application of a novel screening PCR assay for direct detection of '*Helicobacter heilmannii*'-like organisms in human gastric biopsies in Southeast England. Diagn Microbiol Infect Dis. 2003;46(1):1–7.
44. O'Rourke JL, Solnick JV, Neilan BA, Seidel K, Hayter R, Hansen LM, Lee A. Description of "*Candidatus* Helicobacter heilmannii" based on DNA sequence analysis of 16S rRNA and urease genes. Int J Syst Evol Microbiol. 2004;54:2203–11.
45. Baele M, Decostere A, Vandamme P, Ceelen L, Hellemans A, Chiers K, Ducatelle R, Haesebrouck F. Isolation and characterization of *Helicobacter suis* sp. nov. from pig stomachs. Int J Syst Evol Microbiol. 2008;58:1350–8.
46. Vermoote M, Vandekerckhove TTM, Flahou B, Pasmans F, Smet A, De Groote D, Van Criekinge W, et al. Genome sequence of *Helicobacter suis* supports its role in gastric pathology. Vet Res. 2011;42:51.
47. Arnold IC, Zigova Z, Holden M, Lawley TD, Rad R, Dougan G, Falkow S, et al. Comparative whole genome sequence analysis of the carcinogenic bacterial model pathogen *Helicobacter felis*. Genome Biol Evol. 2011;3:302–8.
48. Schott T, Rossi M, Hänninen M-L. Genome sequence of *Helicobacter bizzozeronii* strain CIII-1, an isolate from human gastric mucosa. J Bacteriol. 2011;193:4565–6.
49. Smet A, Van Nieuwerburgh F, Ledesma J, Flahou B, Deforce D, Ducatelle R, Haesebrouck F. Genome sequence of *Helicobacter heilmannii* sensu stricto ASB1 isolated from the gastric mucosa of a kitten with severe gastritis. Genome Announc. 2013;1:e00033–12.
50. Montecucco C, Rappuoli R. Living dangerously: how *Helicobacter pylori* survives in the human stomach. Nat Rev Mol Cell Biol. 2001;2:457–66.

Part III
Diagnosis

Chapter 9
Urea Breath Test and Rapid Urease Test

Akiko Shiotani, Maria Pina Dore, and David Y. Graham

Abstract The urea breath test (UBT) and the rapid urease test (RUT) are the most commonly used diagnostic methods for *H. pylori* infection and rely on detecting the presence of urease produced by *H. pylori*. The sensitivity of the UBT is excellent (often exceeding 95 %). The majority of false-positive results after eradication occur when the results are near the cutoff value and are most frequent in areas where atrophic gastritis is common and citric acid is not used as an adjuvant. The sensitivity of RUT is slightly less (approximately 80–95 %). Increasing the number and size of biopsy fragments, collecting them from the antrum and corpus, or sampling gastric mucus more widely instead of biopsy samples achieves better results with the RUT. Negative results of the either test should not be taken as evidence of the absence of the infection especially in PPI users as well as in patients with bleeding or a history of partial gastrectomy.

Keywords Urea breath test • Rapid urease test • Diagnosis • *Helicobacter pylori* • Proton pump inhibitor • Gastrectomy • Citric acid • Bleeding • False-negative results

9.1 Introduction

The vast numbers of infected patients worldwide and the severe clinical outcomes have made *Helicobacter pylori* (*H. pylori*) infection a public health issue. Because at least 20 % of those with clinically latent *H. pylori* infections eventually develop serious clinical diseases, the infection should be eradicated whenever it is discovered unless there are compelling reasons [1, 2]. If a simple and effective therapy or

A. Shiotani (✉)
Department of Internal Medicine, Kawasaki Medical School, 577 Matsushima, Kurashiki, Okayama 701-0192, Japan
e-mail: shiotani@med.kawasaki-m.ac.jp

M.P. Dore
Department of Clinical and Experimental Medicine, University of Sassari, Sassari, Italy

D.Y. Graham
Department of Medicine, Michael E. DeBakey VAMC and Baylor College of Medicine, Houston, TX, USA

H. Suzuki et al. (eds.), *Helicobacter pylori*, DOI 10.1007/978-4-431-55705-0_9

Table 9.1 Accuracy of diagnostic tests for *H. pylori* infection

	Rapid urease test		Urea breath test		Stool antigen test [5]	
	Sensitivity	Specificity	Sensitivity	Specificity	Sensitivity	Specificity
Before eradication	○	◎	◎	◎	○	◎
After eradication	X	○	◎	○	◎	◎
PPI use	X	◎	△	◎	△	◎
Upper GI bleeding	X	△	○	○	△	○
Gastrectomy	△	◎	△	○	○	◎

◎ more than 95 %, ○ 85 ~ 95 %, △ more than 75 %, X 75 % and less

vaccine were available, *H. pylori* would likely be targeted for a worldwide elimination campaign. The recommended approach is a three-part strategy of test, treat, and confirm cure. The decision to treat must be based on an accurate diagnosis, and the test of choice among a variety of available tests depends on the clinical condition and the indication for testing. Noninvasive testing includes the urea breath test (UBT), stool antigen test, and serologic tests. Biopsy-based tests include the rapid urease test (RUT), histological evaluation, culture, and molecular tests using the polymerase chain reaction (PCR) [3, 4].

Tests differ in specificity and sensitivity, and choice of which test to use is often influenced by the pretest probability of infection (local prevalence of infection). Additional important factors are availability, cost, clinical setting, and factors that might influence the accuracy of testing such as post-eradication, the use of proton pump inhibitors (PPIs), bismuth or antibiotics, upper gastrointestinal (GI) bleeding, past history of partial gastrectomy, etc. (Table 9.1). The choice between noninvasive and invasive testing is dependent on whether there are other indications to perform GI endoscopy. In this paper, we review the RUT and UBT which are the most commonly used diagnostic methods for *H. pylori* infection. Both rely on detecting the presence of urease produced by *H. pylori*.

9.2 Urea Breath Test (UBT)

The UBT is a preferred test for the diagnosis and confirmation of cure because it is a noninvasive, simple, and accurate method for diagnosis of active infections [3]. Testing requires the patient to ingest a small quantity of urea in which the carbon is labeled with either the stable isotope ^{13}C or the radioactive isotope ^{14}C [3, 6]. If *H. pylori* is present, *H. pylori* breaks down orally ingested ^{13}C- or ^{14}C-labeled urea into CO_2 and ammonia. $^{13}CO_2$ or $^{14}CO_2$ diffuses into the blood, is exhaled via the lungs, and can be measured in the exhaled air. Both ^{13}C- and ^{14}C-UBTs are sensitive and specific for *H. pylori* detection; however, ^{13}C is generally preferred because it is not radioactive and thus avoids the potential problems associated with the use of radioactive substances, especially in children and in pregnant women.

Most previous studies report similar high sensitivity and specificity of the UBT exceeding 95 % [7]. Positive and negative UBT results tend to cluster outside of the range between 2‰ and 5‰ such that change in cutoff value within this range would be expected to have little effect on the clinical accuracy of the test (see below) [7, 8].

The simplicity, good tolerance, and economy of the citric acid test meal probably make its systematic use advisable. Some protocols use a test meal to delay gastric emptying and attempt to spread the urea within the stomach. Over time the choice of test meal switched from an actual test meal such as a pudding to citric acid. A study comparing a pudding test meal, ascorbic acid, and two doses of citric acid in 11 volunteers showed that the increased intragastric urease activity associated with citric acid could not be attributed only to gastric emptying. Citric acid and malic acid both enhance urease activity possibly via an effect on UreI, a proton-gated urea channel, thus making urea more accessible to the intra-bacterial urease [9]. The dose of urea varies from essentially none with the ^{14}C test to mg quantities of ^{13}C-urea. Originally, the ^{13}C-UBT was studied with a dose of ^{13}C-urea of 5 mg/kg of body weight [10]. Subsequently, doses of 125 and 100 mg were validated, and more recently 75 mg, 50 mg, or less and possibly even 10 mg of ^{13}C-urea have proved to be sufficient [11–15]. The UBT protocol containing citric acid test meal can be performed with relatively low doses (<100 mg) of urea: 75 mg or even 50 mg seem to be sufficient. With the most widely used protocol (with citric acid and 75 mg of urea), excellent accuracy is obtained when breath samples are collected as early as 10–15 min after urea ingestion [16].

It is frequently asked whether fasting before UBT is required. Most have found no significant differences between tests performed under fasting and non-fasting conditions. Thus fasting prior to testing seems to be not necessary and non-fasting may be more applicable in the routine setting. In the USA, fasting from solid food for one hour is required. Because the issue remains controversial, it would seem prudent to perform UBT in fasting condition from solid food for one hour until new data will definitively clarify this issue [7, 17–20].

9.2.1 Assessment for Eradication

Overall, the UBT is generally the best noninvasive method for detection of active *H. pylori* infections and for confirmation of cure of the infection. The excellent sensitivity of the UBT, especially after eradication therapy, may be explained by the fact that the UBT is more likely to produce positive results than biopsy-based tests including RUT in cases of moderate colonization or patchy distribution of *H. pylori*. The sensitivity and specificity were previously reported to be greater than 95 %, and it is widely recommended as the primary method for confirmatory testing after *H. pylori* eradication therapy [7, 12].

The accuracy of UBT for eradication assessment is markedly affected by two factors: the timing of the assessment and the cutoff value used for the assessment. After eradication treatment, the values resulting from the ^{13}C-UBT have converged around the cutoff value. A cutoff value of 5‰ for ^{13}C-UBT is widely used for eradication assessment worldwide. In Korea and Japan, the optimum cutoff value of the ^{13}C-UBT performed using the UBiTkit (Otsuka Pharmaceutical; measurement at 20 min after the administration of 100 mg of urea) for the detection of *H. pylori* infection before and after eradication is 2.5‰ [15]. The cutoff value of 2.5‰ was based on a multicenter trial conducted in patients prior to treatment for eradication of *H. pylori* in Japan [15], while the cutoff value is 2.4‰ in the USA.

Early experience with a low proportion of false-positive results has been recently challenged by studies from Spain and Korea [21, 22]. In Korea they reported a low specificity of 47.1 % instead of high sensitivity of 99.3 % using the film-coated urea tablet without citric acid after *H. pylori* eradication [22]. False-positive results after eradication tend to occur in studies using cutoff in patients with atrophic gastritis. The majority of false-positive results occur when the results are near the cutoff value and are most frequent in areas where atrophic gastritis is common and citric acid is not used. In those locations we recommend that tests with results between the cutoffs (e.g., 2.5‰ and 10‰) should be considered indeterminate and confirmed with a different test. In Korea this accounted for up to 10 % of cases of post-*H. pylori* eradication therapy. It is unknown if the addition of citric acid which increases the urease activity of *H. pylori* and inhibits non-*H. pylori* ureases would prevent this problem. After failure of *H. pylori* therapy, the infection may recover slowly such that testing should be delayed for at least 4 weeks after the end of therapy and 2 weeks after treatment with PPI to allow any remaining *H. pylori* to achieve a sufficient density to be detected [23].

The stool antigen test is also useful for confirming *H. pylori* eradication; however, stool testing is inconvenient and has low diagnostic accuracy when polyclonal antibodies are used. Both the UBT and stool antigen test have similar accuracy if one ensures one is using a validated stool antigen test based on the use of monoclonal antibodies [24].

9.2.2 Diagnosis in Children

Citric acid has demonstrated good performance as an adjuvant to the UBT in adults and is well accepted in children when a sweetener is included as in the USA. Apple, orange, or grape juices are often used as alternatives in UBT of children although their suitability has not been evaluated critically. The standard ^{13}C-UBT is less accurate for the diagnosis of *H. pylori* infection in young children, especially under 6 years old, because the test is assessed by measuring the relative amounts of natural and $^{13}CO_2$. Attempts to adjust based on cutoff value, pretest meal, and urea dose are gross attempts to this adjustment and are not recommended as CO_2 is highly dependent on size which varies greatly in different populations. The problem can be overcome by adjusting for the CO_2 production rate [25].

A systematic review and meta-analysis showed that stool antigen tests using a monoclonal antibody are also highly accurate for the diagnosis of *H. pylori* infection in children [26]. The specificity of UBT was less than 90 % in young children aged 6–30 months in the developed countries, while both monoclonal stool antigen test and UBT have proven to be reliable in South American developing countries [27].

9.3 Rapid Urease Test (RUT)

The diagnosis of *H. pylori* can be established by endoscopy to obtain specimens for RUTs, histology, or culture. Choosing among these tests depends upon the clinical circumstance, the accuracy of the tests, and their relative costs and especially their local availability. Endoscopy is generally not indicated only to establish *H. pylori* status, and rather, the primary use of endoscopy is to evaluate for the presence of upper GI diseases, many of which are consequences of *H. pylori* infections, or to obtain biopsies for culture to perform antimicrobial susceptibility testing. As noted above, endoscopy is expensive and not without risk, and it is prudent to combine visual inspection with diagnostic testing for *H. pylori*. Japan is one of the countries with the highest incidence of gastric cancer, and the Japanese Minister of Health, Labour and Welfare has recently approved the application of medical insurance of *H. pylori* eradication in patients with chronic gastritis diagnosed by endoscopy based on the strategy for the elimination of gastric cancer deaths. Therefore in Japan, endoscopy for screening of gastric cancer and evaluation of gastric cancer risk is required prior to *H. pylori* eradication for chronic gastritis.

The RUT is inexpensive, rapid, widely available, and highly specific. In 1998, American College of Gastroenterology suggested that the RUT is the test of first choice when endoscopy is indicated [28]. Additional biopsy specimens should also be taken from normal appearing mucosa and placed into formalin.

RUTs are based on the fact that *H. pylori* urease splits urea into ammonia and CO_2. When a gastric biopsy containing *H. pylori* is placed into a urea-containing medium, the ammonia produced by bacterial urease will increase the pH, and this is detected by the color change of a pH indicator. In regions where cost is an important factor in terms of whether testing can be done, any laboratory can produce "home-made" tests for pennies. These tests are made from a solution containing 2 g urea, 10 mL of 0.5 % (w/v) phenol red, and 20 mg sodium azide in 100 mL of 0.01 M sodium phosphate buffer, pH6.5. Approximately ½ to 1 mL of this solution is placed in a dram vial and the biopsy is immersed in this solution. The pH indicator is initially yellow, and the addition of *H. pylori*-positive biopsy specimens will change the solution from yellow to pink. Many commercial RUTs are available, including tests that use urea-impregnated agar (e.g., *hp*fast, GI Supply, Camp Hill,

PA; CLO test, TriMEd Specialties, Inc., Lenexa, Kans), liquid-based tests (Helicocheck, Otsuka Pharmaceutical, Tokyo, Japan), or dry filter-paper tests using a urea-impregnated semipermeable membrane (PyloriTek, Serim Research Co., Elkhart, IN). They typically provide a result within 1–24 h, depending on the format of the test and the bacterial density in the biopsy specimens. The speed of the reaction in the agar tests can be increased by the use of a warmer (e.g., Helicoview, GI Supply, PA), the use of large biopsy specimens, or adding several specimens to the agar [29–31]. A study comparing the two types of RUTs, a liquid-based test and a dry filter-paper test, showed they had significantly faster reaction times than an agar test such as the CLO test [32]. Some RUTs marketed in Europe are reported to give accurate results within minutes. However, clinical experience has not shown rapid or ultrarapid results to be advantageous clinically.

The specificity for these RUTs varies from 90 % to 100 %, but their sensitivity is slightly less (approximately 80–95 %) [21, 30, 31, 33, 34]. Positive results should generally prompt *H. pylori* treatment. However, false-positive results may occur if the test is not interpreted within 24 h because of the growth of urease-containing mouth bacteria. We recommend RUTs be discarded after 24 h. Importantly, a negative RUT should not be taken as evidence of the absence of the infection especially in PPI users and those with atrophic gastritis (see below).

9.4 False-Negative Results

Because the RUT samples such a tiny fraction of the gastric mucosa, there is a high possibility of false-negative results with RUT due to the presence of a low-level, suppressed, or patchy infection. This problem with false-negative tests is responsible for the abovementioned dictum not to accept a negative RUT as the sole criterion for either the absence of the infection. Extensive atrophy or intestinal metaplasia leads to patchy distribution of the infection and may be associated with low *H. pylori* density and false-negative RUT results. Corpus-predominant gastritis or achlorhydria can also lead to false-negative UBTs, as like RUT. The number of bacteria present in the biopsy specimen is the main cause in reducing the test's sensitivity. It is estimated that densities lower than 10^4–10^5 organisms may result in false-negative tests [35]. The false-negative result could also be caused by the use of *H. pylori*- and urease-suppressive therapies, such as PPIs, antibiotics, or bismuth compounds [36, 37].

Preimmersion of the biopsy forceps in formalin does not adversely affect viability of the organisms, although it has been suggested that formalin contamination of forceps used to collect the biopsy may possibly contribute to reduced sensitivity [38–41].

9.4.1 PPI Users

False-negative rates greater than 30 % have been reported when PPI is used just prior or at the time of testing, and this problem is present with histology, culture, RUT, UBT, and stool antigen testing. Gatta et al. [13] reported that the sensitivity of both UBT and stool antigen test was significantly decreased (UBT range, 77.1–85.4 %; stool test, 83 %) after 14 days of PPI treatment, while it was unchanged in those that took antacids. PPIs should be stopped 2 weeks (at least 1 week) before the UBT [42]. H_2 receptor antagonists have no effect on *H. pylori* and can be continued up to the day of testing for histology, RUT, and stool antigen testing. However, high pH caused by H_2 receptor antagonists may reduce the accuracy of the UBT, especially the ^{14}C-UBT. Citric acid can overcome the problem with the ^{13}C-UBT.

9.4.2 Diagnosis in Patients with Upper GI Bleeding

The false-negative result in RUT or culture may often occur in bleeding patients [43, 44]. The biological mechanisms for the false-negative results in the setting of acute upper GI bleeding are poorly understood and may indeed vary depending on the test. Blood adversely affects the performance of the RUT. One suggestion is the pH-buffering effect of the blood or serum albumin rather than a direct inhibition on the urease activity. The color change of the pH indicator was progressively suppressed by higher concentrations of serum albumin regardless of the presence of anti-*H. pylori* antibody [45]. Velayos et al. [46] investigated the accuracy of UBT performed immediately after emergency endoscopy in patients with peptic ulcer bleeding by comparing the results with those of UBT performed after hospital discharge. The sensitivity and specificity of the early UBT were 86 and 66 %, respectively, with a negative predictive value of 50 %. In contrast, Tu et al. [47] compared invasive and noninvasive methods for detecting *H. pylori* infection in bleeding peptic ulcer and reported a higher sensitivity of UBT compared to RUT (95 % vs. 46 %). In reality, there is no compelling reason to diagnose *H. pylori* in a bleeding patient. Since *H. pylori* infections are acquired in childhood, most patients have been infected for decades. There is thus no rush to diagnose and diagnostic testing can be delayed until after the acute problems have settled.

9.4.3 Diagnosis in Patients with History of Partial Gastrectomy

The studies using UBT performed in patients with a history of partial gastrectomy indicated that the sensitivity is decreased in patients with partial gastrectomy. One problem is that bile inhibits *H. pylori* and thus the density of the organism is often low. The accuracy of UBT in patients' post-distal gastrectomy was reported to be lower than that of RUT [48]. The UBT may become unreliable because of rapid gastric emptying and/or enterogastric alkaline reflux. A cutoff of 4.0‰ using the commercially available test has been recommended [49]. A meta-analysis indicated that the sensitivity and specificity of UBT were 0.77 (95 % CI, 0.72–0.82) and 0.89 (95 % CI, 0.85–0.93) compared to the RUT (0.79 (95 % CI, 0.72–0.84) and 0.94 (95 % CI, 0.90–0.97)) and histology (0.93 (95 % CI, 0.88–0.97) and 0.85 (95 % CI, 0.73–0.93)) [50]. In contrast, a Japanese trial used the UBT protocol that included ingestion of 100 mg ^{13}C-urea, the use of mouthwash, and the body in a horizontal position on the left side. Using that approach the sensitivity of ^{13}C-UBT in patients with a remnant stomach was improved to 95.7 % [51]. Overall, histology is often the best choice for the diagnosis of *H. pylori* infection after partial gastrectomy [50]. However, stool antigen test may also be reliable, and the specificity of stool antigen test in Japanese patients who underwent distal gastrectomy was reported to be 90.5 % in contrast of a standard UBT of only 59 % [52].

9.5 False-Positive Results

In contrast, urease-producing bacteria (*Streptococcus*, *Staphylococcus*, *Gardnerella*, *Lactococcus*, and *Enterococcus*) could cause false-positive results of urease-dependent tests [53–55]. Although some members of the microbiota in the oropharynx produce urease which is swallowed in the saliva, most non-*H. pylori* urease enzymes have a pKa greater than seven and are thus inactive in the acidic conditions of the stomach. However, in the achlorhydric patient, their presence may prove to be a problem. A previous Japanese study demonstrated that five bacterial species with urease activity (*Proteus mirabilis*, *Citrobacter freundii*, *Klebsiella pneumoniae*, *Enterobacter cloacae*, and *Staphylococcus aureus*) were subsequently isolated from the oral cavity and/or stomach and all of the patients with a false-positive UBT result were suffering from atrophic gastritis [55]. The use of the citric acid adjuvant will also lower the intragastric pH below the pH optimum of non-*H. pylori* ureases and thus reduce the chance of false-positive results seen in patients with hypochlorhydria and overgrowth of non-*H. pylori* urease-containing organisms. False-positive RUT is infrequent but may occur if the sample is kept beyond 24 h [56].

9.6 Improvement of the UBT

Mouthwashing prior to a standard UBT has been recommended to reduce the interaction of the urea with mouth urease-containing organisms to reduce frequency of false-positive UBT results; however, it has not been critically examined

[57]. Alternative methods have been to use a straw to drink the urea solution or the use of film-coated ^{13}C-urea tablets to decrease the interaction of the urea and urease-positive bacteria in the oral cavity. The simplest method is to use citric acid and to delay the first breath sample to at least 10 min to allow the low pH of the citric acid solution to inhibit non-*H. pylori* ureases and to enhance *H. pylori* urease activity. One potential problem with urea tablets or capsules is that they can potentially empty from the stomach without exposing the *H. pylori* to the labeled urea. Clearly, more head-to-head studies are needed to compare different delivery methods.

9.7 Improvement of the RUT

Increasing the number and size of biopsy fragments and/or collecting them from various regions of the stomach (e.g., combining biopsies for the antrum and body in one well) achieves better results with the RUT [58]. Combining tissues not only increased the detection of *H. pylori* compared with testing separate specimens but also produces faster results [59]. Obtaining two samples one from the antrum and one from the corpus avoiding areas of ulceration and obvious intestinal metaplasia is sufficient to obtain optimal results and provide the highest yield [56].

The tissue sample contained in the agar of an RUT can be used for other purposes. For example, the sample can be removed from the agar gel of positive tests and used for molecular testing for *H. pylori* and/or for the presence of clarithromycin resistance [60]. Since the sample contains host tissue, it could be used for other testing such as the CYP2C19 genotype of the host as well as 23S rRNA of *H. pylori* [61].

To increase the sensitivity of RUT, authors have recommended sampling gastric mucus more widely instead of biopsy samples. We assessed whether adherent gastric mucus to biopsy forceps instead of biopsy samples was suitable for diagnosis. Gastric mucus was obtained by gently scraping gastric mucosa from the antrum greater curvature to corpus greater curvature using biopsy forceps and put into RUT tube. The accuracy of RUT using gastric mucus was superior to that using combined two biopsy specimens in 494 subjects including 300 *H. pylori*-positive patients. The sensitivity and specificity of using gastric mucus were 0.93 (95 % CI, 0.90–0.96) and 0.93 (95 % CI, 0.89–0.96) compared to using biopsy specimens (0.87 (95 % CI, 0.80–0.94) and 0.92 (95 % CI, 0.86–0.99)). (The data is not published.) An alternative is to use an endoscopy brush to collect mucus. As noted previously, the use of the RUT seems to have declined in Western countries possibly because of the desire to obtain the additional information available with histology.

References

1. Graham DY. Can therapy even be denied for *Helicobacter pylori* infection? Gastroenterology. 1997;113(6 Suppl):S113–7.
2. Shiotani A, Nurgalieva ZZ, Yamaoka Y, Graham DY. *Helicobacter pylori*. Med Clin North Am. 2000;84(5):1125–36. viii.
3. Graham DY, Klein PD. Accurate diagnosis of *Helicobacter pylori*. 13C-urea breath test. Gastroenterol Clin North Am. 2000;29(4):885–93. x.
4. Qureshi WA, Graham DY. Diagnosis and management of *Helicobacter pylori* infection. Clin Cornerstone. 1999;1(5):18–28.
5. Gisbert JP, Pajares JM. Stool antigen test for the diagnosis of *Helicobacter pylori* infection: a systematic review. Helicobacter. 2004;9(4):347–68. doi:10.1111/j.1083-4389.2004.00235.x.
6. Peura DA, Pambianco DJ, Dye KR, Lind C, Frierson HF, Hoffman SR, et al. Microdose 14C-urea breath test offers diagnosis of *Helicobacter pylori* in 10 minutes. Am J Gastroenterol. 1996;91(2):233–8.
7. Gisbert JP, Pajares JM. Review article: 13C-urea breath test in the diagnosis of *Helicobacter pylori* infection – a critical review. Aliment Pharmacol Ther. 2004;20(10):1001–17. doi:10.1111/j.1365-2036.2004.02203.x.
8. Graham DY, Opekun AR, Jogi M, Yamaoka Y, Lu H, Reddy R, et al. False negative urea breath tests with H2-receptor antagonists: interactions between *Helicobacter pylori* density and pH. Helicobacter. 2004;9(1):17–27.
9. Shiotani A, Saeed A, Yamaoka Y, Osato MS, Klein PD, Graham DY. Citric acid-enhanced *Helicobacter pylori* urease activity in vivo is unrelated to gastric emptying. Aliment Pharmacol Ther. 2001;15(11):1763–7.
10. Graham DY, Klein PD, Evans Jr DJ, Evans DG, Alpert LC, Opekun AR, et al. Campylobacter pylori detected noninvasively by the 13C-urea breath test. Lancet. 1987;1(8543):1174–7.
11. Bielanski W, Konturek SJ. New approach to 13C-urea breath test: capsule-based modification with low-dose of 13C-urea in the diagnosis of *Helicobacter pylori* infection. J Physiol Pharmacol. 1996;47(3):545–53.
12. Gatta L, Ricci C, Tampieri A, Osborn J, Perna F, Bernabucci V, et al. Accuracy of breath tests using low doses of 13C-urea to diagnose *Helicobacter pylori* infection: a randomised controlled trial. Gut. 2006;55(4):457–62. doi:10.1136/gut.2005.078626.
13. Gatta L, Vakil N, Ricci C, Osborn JF, Tampieri A, Perna F, et al. Effect of proton pump inhibitors and antacid therapy on 13C urea breath tests and stool test for *Helicobacter pylori* infection. Am J Gastroenterol. 2004;99(5):823–9. doi:10.1111/j.1572-0241.2004.30162.x.
14. Graham DY, Malaty HM, Cole RA, Martin RF, Klein PD. Simplified 13C-urea breath test for detection of *Helicobacter pylori* infection. Am J Gastroenterol. 2001;96(6):1741–5. doi:10.1111/j.1572-0241.2001.03867.x.
15. Ohara S, Kato M, Asaka M, Toyota T. Studies of 13C-urea breath test for diagnosis of *Helicobacter pylori* infection in Japan. J Gastroenterol. 1998;33(1):6–13.
16. Lopes AI, Vale FF, Oleastro M. *Helicobacter pylori* infection - recent developments in diagnosis. World J Gastroenterol. 2014;20(28):9299–313. doi:10.3748/wjg.v20.i28.9299.
17. Epple HJ, Kirstein FW, Bojarski C, Frege J, Fromm M, Riecken EO, et al. 13C-urea breath test in *Helicobacter pylori* diagnosis and eradication. Correlation to histology, origin of 'false' results, and influence of food intake. Scand J Gastroenterol. 1997;32(4):308–14.
18. Moayyedi P, Braunholtz D, Heminbrough E, Clough M, Tompkins DS, Mapstone NP, et al. Do patients need to fast for a 13C-urea breath test? Eur J Gastroenterol Hepatol. 1997;9(3):275–7.
19. Ng FH, Lai KC, Wong BC, Wong WM, Wong SY, Chow KC, et al. [13C]-urea breath test without prior fasting and without test meal is accurate for the detection of *Helicobacter pylori* infection in Chinese. J Gastroenterol Hepatol. 2002;17(8):834–8.
20. Wang WM, Lee SC, Wu DC, Chen LT, Liu CS, Peng CF, et al. Simplified 13C-urea breath test for the diagnosis of *Helicobacter pylori* infection – the availability of without fasting and without test meal. Kaohsiung J Med Sci. 2000;16(12):607–13.

21. Calvet X, Sanchez-Delgado J, Montserrat A, Lario S, Ramirez-Lazaro MJ, Quesada M, et al. Accuracy of diagnostic tests for *Helicobacter pylori*: a reappraisal. Clin Infect Dis. 2009;48(10):1385–91. doi:10.1086/598198.
22. Kwon YH, Kim N, Lee JY, Choi YJ, Yoon K, Hwang JJ, et al. The diagnostic validity of citric acid-free, high dose c-urea breath test after *Helicobacter pylori* eradication in Korea. Helicobacter. 2015. doi:10.1111/hel.12189.
23. Laine L, Estrada R, Trujillo M, Knigge K, Fennerty MB. Effect of proton-pump inhibitor therapy on diagnostic testing for *Helicobacter pylori*. Ann Intern Med. 1998;129(7):547–50.
24. Shimoyama T. Stool antigen tests for the management of *Helicobacter pylori* infection. World J Gastroenterol. 2013;19(45):8188–91. doi:10.3748/wjg.v19.i45.8188.
25. Klein PD, Malaty HM, Czinn SJ, Emmons SC, Martin RF, Graham DY. Normalizing results of 13C-urea breath testing for CO_2 production rates in children. J Pediatr Gastroenterol Nutr. 1999;29(3):297–301.
26. Leal YA, Cedillo-Rivera R, Simon JA, Velazquez JR, Flores LL, Torres J. Utility of stool sample-based tests for the diagnosis of *Helicobacter pylori* infection in children. J Pediatr Gastroenterol Nutr. 2011;52(6):718–28. doi:10.1097/MPG.0b013e3182077d33.
27. Queiroz DM, Saito M, Rocha GA, Rocha AM, Melo FF, Checkley W, et al. *Helicobacter pylori* infection in infants and toddlers in South America: concordance between [13C]urea breath test and monoclonal *H. pylori* stool antigen test. J Clin Microbiol. 2013;51 (11):3735–40. doi:10.1128/JCM.01752-13.
28. Howden CW, Hunt RH. Guidelines for the management of *Helicobacter pylori* infection. Ad Hoc committee on practice parameters of the American college of gastroenterology. Am J Gastroenterol. 1998;93(12):2330–8. doi:10.1111/j.1572-0241.1998.00684.x.
29. Weston AP, Campbell DR, Hassanein RS, Cherian R, Dixon A, McGregor DH. Prospective, multivariate evaluation of CLOtest performance. Am J Gastroenterol. 1997;92(8):1310–5.
30. Yousfi MM, El-Zimaity HM, Cole RA, Genta RM, Graham DY. Detection of *Helicobacter pylori* by rapid urease tests: is biopsy size a critical variable? Gastrointest Endosc. 1996;43 (3):222–4.
31. Yousfi MM, El-Zimaity HM, Cole RA, Genta RM, Graham DY. Does using a warmer influence the results of rapid urease testing for *Helicobacter pylori*? Gastrointest Endosc. 1996;43(3):260–1.
32. Tseng CA, Wang WM, Wu DC. Comparison of the clinical feasibility of three rapid urease tests in the diagnosis of *Helicobacter pylori* infection. Dig Dis Sci. 2005;50(3):449–52.
33. Laine L, Lewin D, Naritoku W, Estrada R, Cohen H. Prospective comparison of commercially available rapid urease tests for the diagnosis of *Helicobacter pylori*. Gastrointest Endosc. 1996;44(5):523–6.
34. Yousfi MM, El-Zimaity HM, Cole RA, Genta RM, Graham DY. Comparison of agar gel (CLOtest) or reagent strip (PyloriTek) rapid urease tests for detection of *Helicobacter pylori* infection. Am J Gastroenterol. 1997;92(6):997–9.
35. Ricci C, Holton J, Vaira D. Diagnosis of *Helicobacter pylori*: invasive and non-invasive tests. Best Pract Res Clin Gastroenterol. 2007;21(2):299–313. doi:10.1016/j.bpg.2006.11.002.
36. Lewis JD, Kroser J, Bevan J, Furth EE, Metz DC. Urease-based tests for *Helicobacter pylori* gastritis. Accurate for diagnosis but poor correlation with disease severity. J Clin Gastroenterol. 1997;25(2):415–20.
37. Capurso G, Carnuccio A, Lahner E, Panzuto F, Baccini F, Delle Fave G, et al. Corpus-predominant gastritis as a risk factor for false-negative 13C-urea breath test results. Aliment Pharmacol Ther. 2006;24(10):1453–60. doi:10.1111/j.1365-2036.2006.03143.x.
38. Castellote J, Guardiola J, Porta F, Falco A. Rapid urease test: effect of preimmersion of biopsy forceps in formalin. Gastrointest Endosc. 2001;53(7):744–6. doi:10.1067/mge.2001.114786.
39. Ozaslan E, Koseoglu T, Purnak T, Yildiz A. A forgotten cause of false negative rapid urease test: formalin contamination of the sample. Hepatogastroenterology. 2010;57:99–100. 2 p preceding table of contents.

40. Wettstein A, Loy C, Frommer DJ. Effect of immersion of biopsy forceps in formalin on tissue urease activity. J Gastroenterol Hepatol. 1999;14(10):984–6.
41. Yousfi MM, Reddy R, Osato MS, Graham DY. Culture of *Helicobacter pylori*: effect of preimmersion of biopsy forceps in formalin. Helicobacter. 1996;1(1):62–4.
42. Graham DY, Opekun AR, Hammoud F, Yamaoka Y, Reddy R, Osato MS, et al. Studies regarding the mechanism of false negative urea breath tests with proton pump inhibitors. Am J Gastroenterol. 2003;98(5):1005–9. doi:10.1111/j.1572-0241.2003.07426.x.
43. Choi YJ, Kim N, Lim J, Jo SY, Shin CM, Lee HS, et al. Accuracy of diagnostic tests for *Helicobacter pylori* in patients with peptic ulcer bleeding. Helicobacter. 2012;17(2):77–85. doi:10.1111/j.1523-5378.2011.00915.x.
44. Vaira D, Menegatti M, Miglioli M. What is the role of *Helicobacter pylori* in complicated ulcer disease? Gastroenterology. 1997;113(6 Suppl):S78–84.
45. Leung WK, Sung JJ, Siu KL, Chan FK, Ling TK, Cheng AF. False-negative biopsy urease test in bleeding ulcers caused by the buffering effects of blood. Am J Gastroenterol. 1998;93 (10):1914–8. doi:10.1111/j.1572-0241.1998.00457.x.
46. Velayos B, Fernandez-Salazar L, Pons-Renedo F, Munoz MF, Almaraz A, Aller R, et al. Accuracy of urea breath test performed immediately after emergency endoscopy in peptic ulcer bleeding. Dig Dis Sci. 2012;57(7):1880–6. doi:10.1007/s10620-012-2096-5.
47. Tu TC, Lee CL, Wu CH, Chen TK, Chan CC, Huang SH, et al. Comparison of invasive and noninvasive tests for detecting *Helicobacter pylori* infection in bleeding peptic ulcers. Gastrointest Endosc. 1999;49(3 Pt 1):302–6.
48. Adamopoulos AB, Stergiou GS, Sakizlis GN, Tiniakos DG, Nasothimiou EG, Sioutis DK, et al. Diagnostic value of rapid urease test and urea breath test for *Helicobacter pylori* detection in patients with Billroth II gastrectomy: a prospective controlled trial. Dig Liver Dis. 2009;41(1):4–8. doi:10.1016/j.dld.2008.05.010.
49. Konturek PC, Konturek SJ, Hahn EG. Duodenal alkaline secretion: its mechanisms and role in mucosal protection against gastric acid. Dig Liver Dis. 2004;36(8):505–12. doi:10.1016/j.dld.2004.03.008.
50. Tian XY, Zhu H, Zhao J, She Q, Zhang GX. Diagnostic performance of urea breath test, rapid urea test, and histology for *Helicobacter pylori* infection in patients with partial gastrectomy: a meta-analysis. J Clin Gastroenterol. 2012;46(4):285–92. doi:10.1097/MCG.0b013e318249c4cd.
51. Kubota K, Hiki N, Shimizu N, Shimoyama S, Noguchi C, Tange T, et al. Utility of [13C] urea breath test for *Helicobacter pylori* detection in partial gastrectomy patients. Dig Dis Sci. 2003;48(11):2135–8.
52. Yan J, Yamaguchi T, Odaka T, Suzuki T, Ohyama N, Hara T, et al. Stool antigen test is a reliable method to detect *Helicobacter pylori* in the gastric remnant after distal gastrectomy for gastric cancer. J Clin Gastroenterol. 2010;44(1):73–4. doi:10.1097/MCG.0b013e3181aae65e.
53. Brandi G, Biavati B, Calabrese C, Granata M, Nannetti A, Mattarelli P, et al. Urease-positive bacteria other than *Helicobacter pylori* in human gastric juice and mucosa. Am J Gastroenterol. 2006;101(8):1756–61. doi:10.1111/j.1572-0241.2006.00698.x.
54. Gurbuz AK, Ozel AM, Narin Y, Yazgan Y, Baloglu H, Demirturk L. Is the remarkable contradiction between histology and 14C urea breath test in the detection of *Helicobacter pylori* due to false-negative histology or false-positive 14C urea breath test? J Int Med Res. 2005;33(6):632–40.
55. Osaki T, Mabe K, Hanawa T, Kamiya S. Urease-positive bacteria in the stomach induce a false-positive reaction in a urea breath test for diagnosis of *Helicobacter pylori* infection. J Med Microbiol. 2008;57(Pt 7):814–9. doi:10.1099/jmm.0.47768-0.
56. Uotani T, Graham DY. Diagnosis of *Helicobacter pylori* using the rapid urease test. Ann Transl Med. 2015;3(1):9. doi:10.3978/j.issn.2305-5839.2014.12.04.
57. Lee TH, Yang JC, Lee SC, Farn SS, Wang TH. Effect of mouth washing on the. J Gastroenterol Hepatol. 2001;16(3):261–3.

58. Hsu WH, Wang SS, Kuo CH, Chen CY, Chang CW, Hu HM, et al. Dual specimens increase the diagnostic accuracy and reduce the reaction duration of rapid urease test. World J Gastroenterol. 2010;16(23):2926–30.
59. Moon SW, Kim TH, Kim HS, Ju JH, Ahn YJ, Jang HJ, et al. United rapid urease test is superior than separate test in detecting *Helicobacter pylori* at the gastric antrum and body specimens. Clin Endosc. 2012;45(4):392–6. doi:10.5946/ce.2012.45.4.392.
60. Li Y, Rimbara E, Thirumurthi S, Trespalacios A, Reddy R, Sabounchi S, et al. Detection of clarithromycin resistance in *Helicobacter pylori* following noncryogenic storage of rapid urease tests for 30 days. J Dig Dis. 2012;13(1):54–9. doi:10.1111/j.1751-2980.2011.00549.x.
61. Furuta T, Sagehashi Y, Shirai N, Sugimoto M, Nakamura A, Kodaira M, et al. Influence of CYP2C19 polymorphism and *Helicobacter pylori* genotype determined from gastric tissue samples on response to triple therapy for *H. pylori* infection. Clin Gastroenterol Hepatol. 2005;3(6):564–73.

Chapter 10
Endoscopic Findings of *H. pylori* Infection

Mototsugu Kato

Abstract Chronic gastritis has usually been diagnosed by histological examination because of discrepancy between histological findings and endoscopic findings. However, recent advances in endoscopy have gradually clarified endoscopic findings that correspond to histological changes. According to the infectious condition of *Helicobacter pylori*, the gastric mucosa is divided into three states, normal mucosa without a history of *H. pylori* infection (non-gastritis), current *H. pylori* infection (active gastritis), and past history of *H. pylori* infection (inactive gastritis). The Kyoto classification of endoscopic gastritis is a novel classification system; 19 endoscopic findings related to gastritis are characterized according to infectious condition of *H. pylori*. Regular arrangement of collecting venules, diffuse redness, and maplike redness are highly specified characteristic features in each of the non-gastritis, active gastritis, and inactive gastritis states. Since the risk of gastric cancer is different among the three states of *H. pylori*, the diagnosis of gastric mucosa by endoscopic examination is important for screening of gastric cancer. Adequate evaluation of gastric mucosa by endoscopic examination requires appropriate training.

Keywords Sydney System • Kyoto classification • Chronic gastritis • Atrophy • Intestinal metaplasia

10.1 Role of Endoscopy in Diagnosis of Gastritis

H. pylori infection is the most common factor causing chronic inflammation of the gastric mucosa [1, 2]. Infiltration of mononuclear cells and polynuclear cells in the gastric mucosa induces atrophic change and intestinal metaplasia during long-term persistent infection with *H. pylori*. A wide variety of upper gastrointestinal tract diseases, such as gastric ulcer, duodenal ulcer, gastric adenocarcinoma, gastric mucosa-associated lymphoid tissue lymphoma, and gastric hyperplastic polyps,

---

M. Kato (✉)
Division of Endoscopy, Hokkaido University Hospital, North 14, West 5, Kita-ku, Sapporo, Hokkaido 060-8648, Japan
e-mail: m-kato@med.hokudai.ac.jp

H. Suzuki et al. (eds.), *Helicobacter pylori*, DOI 10.1007/978-4-431-55705-0_10

Table 10.1 Relationship between histological and endoscopic findings

Histological findings	Endoscopic findings
Hyperemia	Erythema
Edema	Mucosal swelling
Epithelial defect	Erosion, Ulcer
Hemorrhage	Bleeding spot
Infiltration of polymorphonuclear cells and mononuclear cells	Diffuse redness
	Disappearance of RAC
Atrophy	Visibility of vascular pattern
	Rugal atrophy
Intestinal metaplasia	Whitish elevated lesion (specific type)
	Methylene blue stained
	Light blue crest (by IEE)
	White opaque substance (by IEE)

occur from the background of chronic gastritis [3–7]. Chronic gastritis has usually been diagnosed by histological examination. It has long been believed that endoscopic findings correlate poorly with histopathological findings of chronic gastritis [8, 9]. The Sydney System based on this concept was divided into the histological division and endoscopic division [10, 11]. However, recent advances in endoscopy have resulted in improvement in the diagnosis of chronic gastritis without the need for histological assessment of biopsied specimens. Various endoscopic features associated with gastric inflammation that often accompanies structural mucosal changes are assessed. Endoscopic findings that correspond to histological changes have gradually become clear by using not only white light endoscopy but also image-enhanced endoscopy (IEE) (Table 10.1).

According to the infectious condition of *H. pylori*, the gastric mucosa is divided into three states, normal mucosa without a history of *H. pylori* infection (non-gastritis), current *H. pylori* infection (active gastritis), and past history of *H. pylori* infection (inactive gastritis). The frequency of gastric cancer from normal mucosa without a history of *H. pylori* infection is the lowest [12, 13]. Although successful eradication of *H. pylori* reduces the incidence of gastric cancer, the risk of gastric cancer continues for a long time after eradication of *H. pylori* [14, 15]. The risks of gastric cancer are different among the three mucosal conditions. It is important to diagnose *H. pylori* status during endoscopic examination for screening of gastric cancer.

The Kyoto classification of endoscopic gastritis is a novel classification system established in Japan [16]. In this classification system, nineteen endoscopic findings related to gastritis are characterized according to topography and infectious condition of *H. pylori* (Table 10.2). A subset of these endoscopic findings is described in the endoscopic division of the Sydney System [11]. Regular arrangement of collecting venules (RAC), diffuse redness, and maplike redness are highly specified

Table 10.2 Kyoto classification of gastritis

			○	Often detected
			×	Not detected
			△	Sometimes detected
Topography	Terminology	Infected	Noninfected	After eradication
Angstrom	Atrophy	○	×	○~×
	Diffuse redness	○	×	×
	Foveolar-hyperplastic polyp	○	×	○~×
	Maplike redness	×	×	○
	Xanthoma	○	×	○
	Hematin	△	○	○
	Red streak	△	○	○
	Intestinal metaplasia	○	×	○~△
	Mucosal swelling	○	×	×
	Patchy redness	○	○	○
	Depressed erosion	○	○	○
Corpus	Enlarged fold, tortuous fold	○	×	×
	Sticky mucus	○	×	×
Corpus ~Fornix	Fundic gland polyp	×	○	○
	Spotty redness	○	×	△~×
	Multiple white and flat elevated lesions	△	○	○
Lower body ~angle	Regular arrangement of collecting venules (RAC)	×	○	×~△
Antrum	Nodularity	○	×	△~×
	Raised erosion	△	○	○

characteristic features in each of the non-gastritis, active gastritis, and inactive gastritis states.

10.2 Normal Mucosa with No History of *H. pylori* Infection

By conventional white light endoscopic observation, spotty redness is visible everywhere on fundic gland mucosa with no history of *H. pylori* infection. This endoscopic finding is called RAC (Fig. 10.1a). RAC was reported to have high sensitivity and high specificity for the *H. pylori*-negative normal stomach [17]. In the normal fundic gland mucosa, the microsurface structure consists of round or oval crypt openings that are identical to gastric glands. Pin-like dark spots in crypt openings are identical to the center of the gastric gland. The subepithelial capillary loops surrounding the crypts form a regular honeycomb-like network. Collecting venules that are joined with subepithelial capillaries drain directly away from the

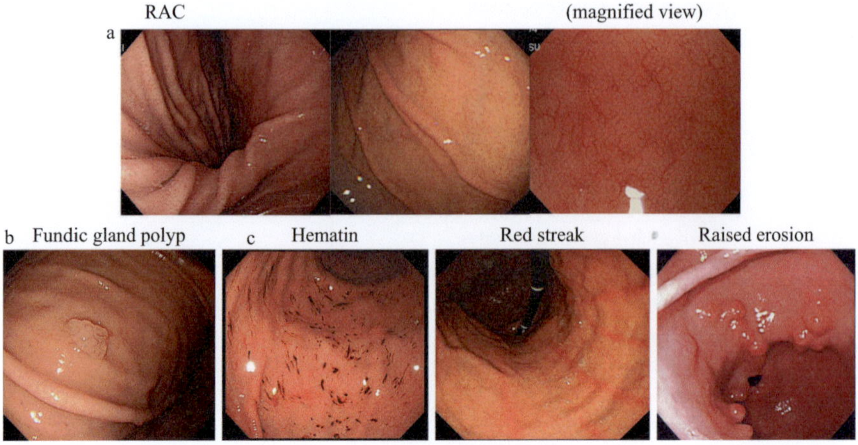

Fig. 10.1 Endoscopic findings in normal mucosa with no history of *H. pylori* infection

mucosal surface toward the submucosa. Collecting venules are arranged every 350 nm which is equivalent to the width of 10–12 gastric crypts. Collecting venules are recognized by magnifying endoscopy as starfish-like vascular structures in a regular arrangement. Since the visibility of RAC is affected by inflammation and atrophic change in the gastric mucosa, RAC is not observed through the surface of mucosa with *H. pylori* infection [18]. However, the visibility of RAC in the area of upper and middle gastric body is not often disturbed in the case with antrum-predominant gastritis. Therefore, positivity of RAC in angle is important to suggest strongly normal mucosa with no history of *H. pylori* infection. RAC is not visible in the pyloric gland mucosa without *H. pylori* infection because of a different crypt structure.

Fundic gland polyps are characterized as multiple, small, sessile polyps usually scattered in the fundic gland region (Fig. 10.1b). They have the same color as that of the gastric mucosa. Sporadic fundic gland polyps in cases without familial adenomatous polyposis have been reported to have an association with negativity of *H. pylori* infection [19]. It was shown that prolonged use of proton pump inhibitors increased the risk of fundic gland polyp development in *H. pylori*-negative subjects [20, 21]. The presence of RAC strongly suggests *H. pylori*-negative mucosa including mucosa with successfully eradicated *H. pylori*.

Three endoscopic findings, hematin (bleeding spot), red streak (linear erythema), and raised erosion, are observed predominantly in normal mucosa, but sometimes in *H. pylori*-infected mucosa (Fig. 10.1c). Hematin is characterized as punctuated or ecchymotic reddish or brown-blackish flecks present in the gastric wall. Red streaks are defined as reddish longitudinal streaks in the antrum and corpus. Raised erosion is characterized as elevated mucosa with white excavation at the center. These endoscopic findings suggest normal mucosa, but definite diagnosis of normal mucosa is not determined by only these findings.

10.3 Gastric Mucosa with Current *H. pylori* Infection

Many findings using high-resolution white light endoscopy and IEE have been reported to be associated with histopathologic findings related to *H. pylori* infection [22–26]. Congestion and dilation of the subepithelial capillary network in gastric mucosa with *H. pylori* infection change the color of the mucosal surface to red. The hemoglobin index (IHB) of gastric mucosa determined by endoscopic measurement was reported to be increased in *H. pylori* gastritis [27]. Diffuse redness refers to uniform redness involving the entire fundic gland mucosa (Fig. 10.2a). RAC is invisible in mucosa with diffuse redness. The vascular structure on the surface mucosa with *H. pylori* infection observed by magnifying endoscopy is limited to subepithelial capillary network without visibility of collecting venules. Therefore, the presence of diffuse redness strongly suggests mucosa with current *H. pylori* infection [22, 23].

Endoscopic mucosal atrophy that develops in *H. pylori* gastritis or autoimmune gastritis is diagnosed by visibility of the vascular pattern and rugal atrophy (Fig. 10.2b). Extension of mucosal atrophy is evaluated according to the location of the endoscopic atrophic border. Extension pattern of atrophy is classified into closed type (C-1, C-2, C-3) and open type (O-1, O-2, O-3) by the Kimura-Takemoto classification [28]. To evaluate the relationship between endoscopic atrophy and histological atrophy, all of the subjects with high-stage OLGA (Operative Link on Gastritis Assessment) gastritis clustered in moderate-to-severe endoscopic gastric atrophy [29]. A 10-year prospective follow-up study with annual endoscopic examinations showed that atrophic change was extended in half of the *H. pylori*-positive subjects [30].

Intestinal metaplasia develops histologically during progression of mucosal atrophic change. A subset of intestinal metaplasia (IM) termed special type of IM is visible by conventional endoscopic observation as grayish-whitish, slightly opalescent patches [31] (Fig. 10.2c-1). Chromoendoscopy with methylene blue staining has been established to detect nonspecific IM as well as special type of IM [32] (Fig. 10.2c-2). Magnifying endoscopy using narrow band imaging or blue light imaging can detect IM as a light blue crest (LBC) without methylene blue staining (Fig. 10.2c-3). LBC is defined as a fine, blue, white line on the crest and gyri. LBC is caused by the difference in reflectance of light at the brush border [33]. LBC is a highly accurate sign of the presence of histological IM. Special type of IM is observed as a white opaque substance (WOS) by narrow band imaging [34]. The cause of WOS is thought to be lipid droplets in the epithelial mucosa with intestinal function (Fig. 10.2c-4).

Some endoscopic findings such as mucosal swelling (edema), enlarged and tortuous folds (rugal hyperplasia), sticky mucus, spotty redness, nodularity, xanthoma, and hyperplastic polyps are associated with *H. pylori* infection (Fig. 10.2d). Mucosal edema is characterized by soft, thick, and swollen gastric mucosa. Indigo carmine staining is useful for diagnosing mucosal edema as swollen areae gastricae, in which the inter-area groove is narrow [35]. An enlarged fold and

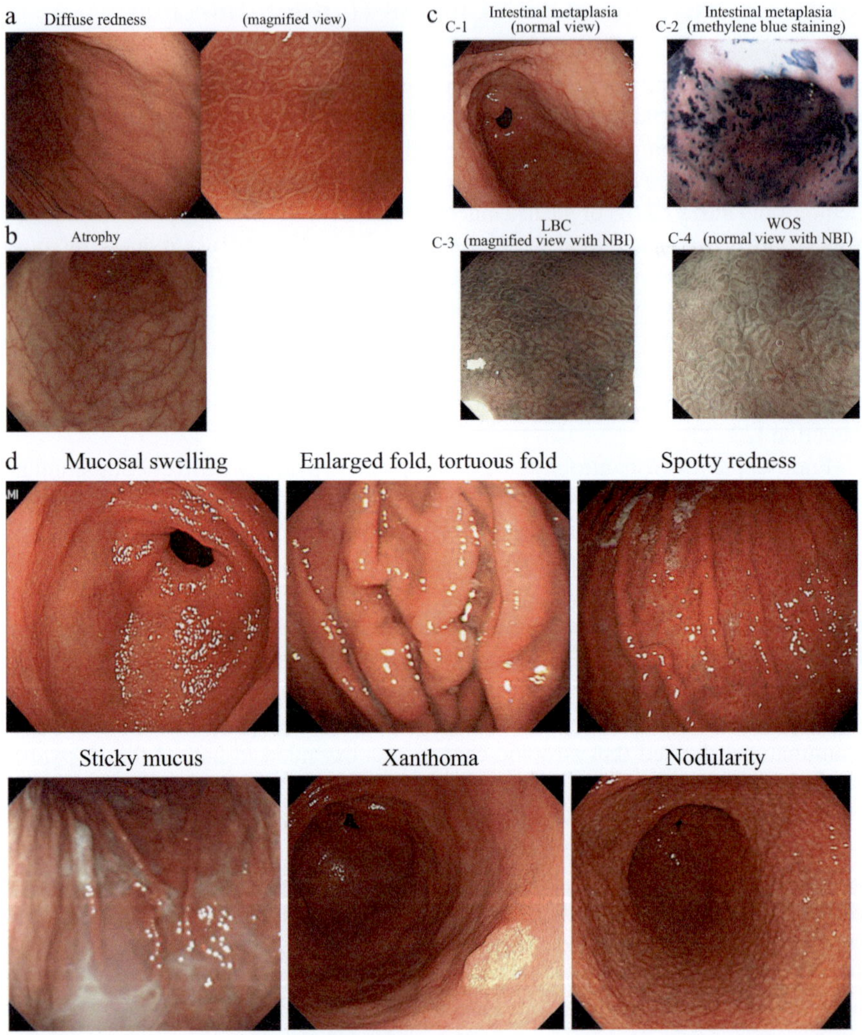

Fig. 10.2 Endoscopic findings in mucosa with current *H. pylori* infection

tortuous fold are fold changes induced by mucosal inflammation. A normal fold is straight, smooth, and approximately 5 mm in diameter. Sticky mucus means grayish or yellowish mucus adhered to the mucosal surface prior to washing with water. Spotty redness is observed as multiple tiny reddish spots in the fundic gland region. Spotty redness should be strictly differentiated from patchy erythema. Nodularity is characterized by the appearance of multiple whitish elevated lesions mainly in pyloric gland mucosa [36]. Lymphoid follicle is histologically proved to be included in a nodular lesion. Nodular gastritis induced by *H. pylori* infection has been shown to be a high risk for diffuse-type gastric cancer in young women

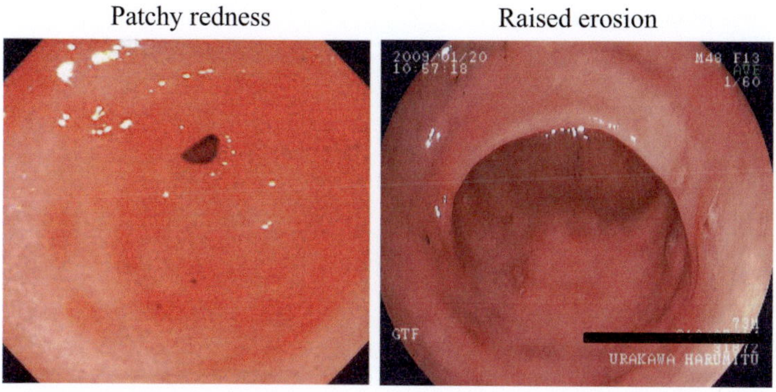

Fig. 10.3 Endoscopic findings related to non-*H. pylori*

[37]. Xanthoma is characterized as yellow-white, well-demarcated, single or multiple nodules or plaques of various sizes.

Patchy erythema and flat erosion are observed in all types of mucosa irrespective of *H. pylori* infection (Fig. 10.3). These two findings occur mostly in individuals taking aspirin or nonsteroidal anti-inflammatory drugs. Patchy erythema is defined as localized reddish maculae of various sizes. Flat erosion is characterized by mucosal defects and whitish patches of various sizes.

10.4 Gastric Mucosa with a Past History of *H. pylori* Infection

Successful eradication of *H. pylori* dramatically improves histopathologic findings of gastritis and may prevent various diseases associated with *H. pylori* infection. Differences in endoscopic findings before and after *H. pylori* eradication treatment were investigated by conventional endoscopy and magnifying endoscopy with narrow band imaging [38–42].

A multicenter prospective trial was conducted to elucidate short-term changes in conventional white light endoscopic features after cure of *H. pylori* infection [38]. Spotty redness of fundic gland mucosa improved significantly when eradication was successful. The frequency of gastric flat erosion was increased after cure of *H. pylori* infection. There were significant differences in diffuse redness and enlarged and tortuous folds between the group with successful eradication and the group with failed eradication. Appearance of flat erosion of the stomach is related to rapid recovery of acid output in corpus-predominant gastritis after successful *H. pylori* eradication [43]. A 5-year follow-up study using conventional endoscopy showed no consistent alteration in the atrophic border [39]. The disappearance of nodularity, recovery of RAC, and appearance of fundic gland polyps have been observed during long-term follow-up after successful eradication [40]. In

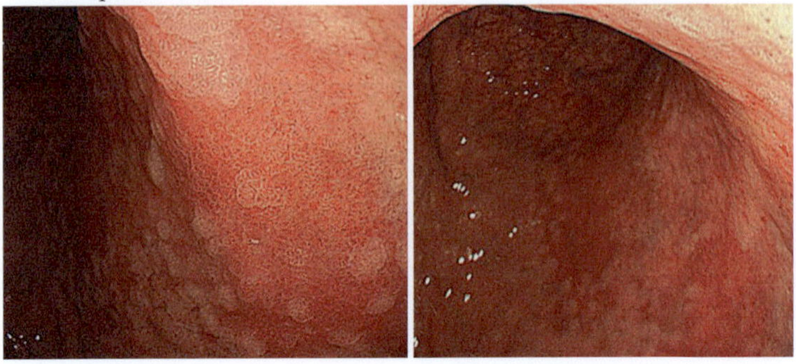

Fig. 10.4 Endoscopic findings in mucosa with past history of *H. pylori* infection

magnifying endoscopic observation with narrow band imaging, disappearance of erythema and swelling of the areas between gastric pits, circular change of enlarged or elongated pits, pinhole-like changes of white pits, and recovery of RAC after successful eradication have been reported [41, 42].

Maplike redness (mottled patchy erythema) was reported to be the only useful endoscopic finding for predicting gastric mucosa with successful eradication of *H. pylori* [44] (Fig. 10.4). Maplike redness is characterized as slightly depressed redness that is distinguishable from whitish mucosa. The area of maplike redness is atrophic mucosa including IM [45]. The mechanism of the appearance of maplike redness is thought to be strengthening of the contrast between non-atrophic mucosa and atrophic mucosa after diffuse redness has disappeared by successful eradication. In addition, coexistence of atrophic change that induced by *H. pylori* and fundic gland polyps or RAC that are associated with *H. pylori*-negative suggest mucosa with a past history of *H. pylori* infection.

10.5 Conclusion

Diagnosis of gastric mucosa by endoscopic examination is important for assessing the risk of gastric cancer development. Gastric cancer risk is different among the three types of gastric mucosal, normal mucosa without a history of *H. pylori* infection (non-gastritis), with current *H. pylori* infection (active gastritis), and with a past history of *H. pylori* infection (inactive gastritis). Atrophy, intestinal metaplasia, nodularity, and enlarged and tortuous folds have been reported to be associated with gastric cancer risk [46–52]. Atrophic change and intestinal metaplasia can be accurately detected by conventional endoscopy and image-enhanced endoscopy. However, adequate evaluation of gastric mucosa by endoscopic examination requires appropriate training [53].

References

1. Marshall BJ, Armstrong JA, McGechie DB, et al. Attempt to fulfill Koch's postulate for pyloric campylobacter. Med J Aust. 1985;142:436–9.
2. Dixon MF, Genta RM, Yardley JH, et al. and the participants in the International Workshop on the Histopathology of Gastritis, Houston 1994. Classification and grading of gastritis. The updated Sydney System. Am J Surg Pathol. 1996;6:1161–81.
3. Marshall BJ, Goodwin CS, Warren JR, et al. Prospective double-blind trial of duodenal ulcer relapse after eradication of Campylobacter pylori. Lancet. 1988;ii:1437–42.
4. Malfertheiner P, Leodolter A, Peitz U. Cure of *Helicobacter pylori*-associated ulcer disease through eradication. Baillieres Best Pract Res Clin Gastroenterol. 2000;14:119–32.
5. International agency for research on cancer, World Health Organization. Schistosomes, liver flukes and *Helicobacter pylori*. IARC Monogr Eval Carcinog Risk Hum. 1994;61:177–241.
6. Ohkusa T, Takashimizu I, Fujiki K, et al. Disappearance of hyperplastic polyps in the stomach after eradication of *Helicobacter pylori*. A randomized, clinical trial. Ann Intern Med. 1998;129:712–5.
7. Wotherspoon AC, Doglioni C, Diss TC, et al. Regression of primary low-grade B-cell gastric lymphoma of mucosa-associated lymphoid tissue type after eradication of *Helicobacter pylori*. Lancet. 1993;342:575–7.
8. Laine L, Cohen H, Sloane R, Marin-Sorensen M, Weinstein WM. Interobserver agreement and predictive value of endoscopic findings for *H. pylori* and gastritis in normal volunteers. Gastrointest Endosc. 1995;42:420–3.
9. Bah A, Saraga E, Armstrong D, et al. Endoscopic features of *Helicobacter pylori*-related gastritis. Endoscopy. 1995;27:593–6.
10. Price AB. The Sydney system: histological division. J Gastroenterol Hepatol. 1991;6:209–22.
11. Tytgat GNJ. The Sydney System: endoscopic division. Endoscopic appearances in gastritis/duodenitis. J Gastroenterol Hepatol. 1991;6:223–34.
12. Haruma K, editor. The Kyoto classification of gastritis. Tokyo: Nippon Medical Center; 2014. (in Japanese)
13. Matsuo T, Ito M, Takata S, et al. Low prevalence of Helicobacter pylori-negative gastric cancer among Japanese. Helicobacter. 2011;16:415–9.
14. Ono S, Kato M, Suzuki M, et al. Frequency of Helicobacter pylori-negative gastric cancer and gastric mucosal atrophy in a Japanese endoscopic submucosal dissection series including histological, endoscopic and serological atrophy. Digestion. 2012;86:59–65.
15. Ford AC, Forman D, Hunt RH, et al. Helicobacter pylori eradication therapy to prevent gastric cancer in healthy asymptomatic infected individuals: systematic review and meta-analysis of randomised controlled trials. BMJ. 2014;348:g3174.
16. Yoon SB, Park JM, Lim CH, et al. Effect of Helicobacter pylori eradication on metachronous gastric cancer after endoscopic resection of gastric tumors: a meta-analysis. Helicobacter. 2014;19:243–8.
17. Yagi K, Nakamura A, Sekine A. Characteristic endoscopic and magnified endoscopic findings in the normal stomach without *Helicobacter pylori* infection. J Gastroenterol Hepatol. 2002;17:39–45.
18. Kato M, Nakagawa S, Shimizu Y, et al. The efficacy of magnifying endoscopy with adaptive IHb enhancement for diagnosis of *H. pylori* induced gastritis. Dig Endosc. 2002;14:S72–75.
19. Sakai N, Tatsuta M, Hirasawa R, et al. Low prevalence of Helicobacter pylori infection in patients with hamartomatous fundic polyps. Dig Dis Sci. 1998;43:766–72.
20. El-Zimaity HM, Jackson FW, Graham DY. Fundic gland polyps developing during omeprazole therapy. Am J Gastroenterol. 1997;92:1858–60.
21. Hongo M, Fujimoto K, Gastric Polyps Study Group. Incidence and risk factor of fundic gland polyp and hyperplastic polyp in long-term proton pump inhibitor therapy: a prospective study in Japan. J Gastroenterol. 2010;45:618–24.

22. Kato T, Yagi N, Kamada T, et al. Diagnosis of Helicobacter pylori infection in gastric mucosa by endoscopic features: a multicenter prospective study. Dig Endosc. 2013;25:508–18.
23. Cho JH, Chang YW, Jang JY, et al. Close observation of gastric mucosal pattern by standard endoscopy can predict Helicobacter pylori infection status. J Gastroenterol Hepatol. 2013;28:279–84.
24. Nomura S, Terao S, Adachi K, et al. Endoscopic diagnosis of gastric mucosal activity and inflammation. Dig Endosc. 2013;25:136–46.
25. Nakagawa S, Kato M, Shimizu Y, et al. Relationship between histopathologic gastritis and mucosal microvascularity: observations with magnifying endoscopy. Gastrointest Endosc. 2003;58:71–5.
26. Tahara T, Shibata T, Nakamura M, et al. Gastric mucosal pattern by using magnifying narrow-band imaging endoscopy clearly distinguishes histological and serological severity of chronic gastritis. Gastrointest Endosc. 2009;70:246–53.
27. Uchiyama K, Ida K, Okuda J, et al. Correlations of hemoglobin index (IHb) of gastric mucosa with Helicobacter pylori (*H. pylori*) infection and inflammation of gastric mucosa. Scand J Gastroenterol. 2004;39:1054–60.
28. Kimura K, Takemoto T. An endoscopic recognition of the atrophic border and its significance in chronic gastritis. Endoscopy. 1969;3:87–97.
29. Quach DT, Le HM, Hiyama T, et al. Relationship between endoscopic and histologic gastric atrophy and intestinal metaplasia. Helicobacter. 2013;18:151–7.
30. Sakaki N, Kozawa H, Egawa N, et al. Ten-year prospective follow-up study on the relationship between Helicobacter pylori infection and progression of atrophic gastritis, particularly assessed by endoscopic findings. Aliment Pharmacol Ther. 2002;16 Suppl 2:198–203.
31. Takemoto T. Endoscopic diagnosis of chronic gastritis. Diagn Treat. 1966;54:1274–85 (in Japanese).
32. Ida K, Hashimoto Y, Kawai K. In vivo staining of gastric mucosa: its application to endoscopic diagnosis of intestinal metaplasia. Endoscopy. 1975;7:18–24.
33. Uedo N, Ishihara R, Iishi H, et al. A new method of diagnosing gastric intestinal metaplasia: narrow-band imaging with magnifying endoscopy. Endoscopy. 2006;38:819–24.
34. Yao K, Iwashita A, Nambu M, et al. Nature of white opaque substance in gastric epithelial neoplasia as visualized by magnifying endoscopy with narrow-band imaging. Dig Endosc. 2012;24:419–25.
35. Ida K, Hashimoto Y, Takeda S, et al. Endoscopic diagnosis of gastric cancer with dye scattering. Am J Gastroenterol. 1975;63:316–20.
36. Nakashima R, Nagata N, Watanabe K, et al. Histological features of nodular gastritis and its endoscopic classification. J Dig Dis. 2011;12:436–42.
37. Kamada T, Hata J, Tanaka A, et al. Nodular gastritis and gastric cancer. Dig Endosc. 2006;18:79–83.
38. Kato M, Terao S, Adachi K, et al. Changes in endoscopic findings of gastritis after cure of H. pylori infection: multicenter prospective trial. Dig Endosc. 2013;25:264–73.
39. Oda Y, Miwa J, Kaise M, et al. Five-year follow-up study on histological and endoscopic alterations in the gastric mucosa after Helicobacter pylori eradication. Dig Endosc. 2004;16:213–8.
40. Chen MJ, Shih SC, Wang TE, et al. Endoscopic patterns and histopathological features after eradication therapy in Helicobacter pylori-associated nodular gastritis. Dig Dis Sci. 2008;53:1893–7.
41. Yagi K, Nakamura A, Sekine A. Magnifying endoscopy of the gastric body: a comparison of the findings before and after eradication of Helicobacter pylori. Dig Endosc. 2002;14:S76–82.
42. Okubo M, Tahara T, Shibata T, et al. Changes in gastric mucosal patterns seen by magnifying NBI during *H. pylori* eradication. J Gastroenterol. 2011;46:175–82.
43. Feldman M, Cryer B, Sammer D, et al. Influence of *H. pylori* infection on meal-stimulated gastric acid secretion and gastroesophageal acid reflux. Am J Physiol. 1999;277:1159–64.

44. Watanabe K, Nagata N, Nakashima R, et al. Predictive findings for Helicobacter pylori-uninfected, -infected and -eradicated gastric mucosa: validation study. World J Gastroenterol. 2013;19:4374–9.
45. Ngata N, Shimbo T, Akiyama J, et al. Predictability of gastric intestinal metaplasia by mottled patchy erythema seen on endoscopy. Gastroenterol Res. 2011;4:203–9.
46. Cheli R, Santi L, Ciancamerla G, et al. A clinical and statistical follow-up study of atrophic gastritis. Am J Dig Dis. 1973;18:1061–5.
47. Kato I, Tominaga S, Ito T, et al. Atrophic gastritis and stomach cancer risk: cross-section analyses. Jpn J Cancer Res. 1992;83:1041–6.
48. Uemura N, Okamoto S, Yamamoto S, et al. Helicobacter pylori infection and the development of gastric cancer. N Engl J Med. 2001;345:784–9.
49. Tsai YC, Hsiao WH, Yang HB, et al. The corpus-predominant gastritis index may serve as an early marker of Helicobacter pylori-infected patients at risk of gastric cancer. Aliment Pharmacol Ther. 2013;37:969–78.
50. Masuyama H, Yoshitake N, Sasaki T, et al. Relationship between the degree of endoscopic atrophy of the gastric mucosa and carcinogenic risk. Digestion. 2015;91:30–6.
51. Kamada T, Tanaka A, Yamanaka Y, et al. Nodular gastritis with Helicobacter pylori infection is strongly associated with diffuse-type gastric cancer in young patients. Dig Endosc. 2007;19:180–4.
52. Nishibayashi H, Kanayama S, Kiyohara T, et al. Helicobacter pylori-induced enlarged-fold gastritis is associated with increased mutagenicity of gastric juice, increased oxidative DNA damage, and an increased risk of gastric carcinoma. J Gastroenterol Hepatol. 2003;18:1384–91.
53. Sugano K, Tack J, Kuipers EJ, et al. Kyoto global consensus report on Helicobacter pylori gastritis. Gut. 2015;64:1353–67.

Chapter 11
Stratification of Gastric Cancer Risk by *H. pylori* Infection

Kazuhiko Inoue

Abstract Infection by *Helicobacter pylori* (*H. pylori*) has been clearly shown to be strongly associated with the development of gastric cancer and is even considered a necessary condition for its development. In addition, it has been shown that among individuals with *H. pylori* infection, those who have advanced atrophy of the gastric mucosa are at high risk for gastric cancer. It is possible to evaluate the state of *H. pylori* infection by a serum antibody test for *H. pylori* and to evaluate the degree of gastric mucosal atrophy by the pepsinogen (PG) method. Furthermore, it is possible to stratify the risk for gastric cancer by using the ABC classification, which combines serum *H. pylori* antibody testing and the PG method. In this classification, a group of individuals who are *H. pylori* antibody (−) and PG method (−) (group A) is considered to be at low risk; a group with *H. pylori* antibody (+) and PG method (−) (group B) is considered to be at average risk; and groups with *H. pylori* antibody (+) and PG method (+) (group C) or *H. pylori* antibody (−) and PG method (+) (group D) are considered to be at high risk. The ABC classification is a simple, objective blood test, which is being widely used because it has the major advantage of providing consistent results regardless of the evaluating clinicians. The use of appropriate imaging studies based on the ABC stratification of gastric cancer risk would be an effective strategy toward elimination of gastric cancer.

Keywords Gastric cancer • Screening • Endoscopy • Radiography • *Helicobacter pylori* • Pepsinogen • ABC classification

K. Inoue (✉)
Department of General Medicine, Kawasaki Medical School, 577 Matsushima, Kurashiki, Okayama 701-0192, Japan
e-mail: inouekimkk@nifty.com

H. Suzuki et al. (eds.), *Helicobacter pylori*, DOI 10.1007/978-4-431-55705-0_11

11.1 *H. pylori* Infection and Gastric Cancer

11.1.1 H. pylori *Infection and the Development of Gastric Cancer*

H. pylori, discovered by Warren and Marshall [1] in 1983, has been shown to be strongly associated not only with peptic ulcer but also with the development of gastric cancer in subsequent numerous studies.

Based on three epidemiologic studies reported in 1991 [2–4], the International Agency for Research on Cancer, a subordinate agency of the World Health Organization, made the comment that *H. pylori* is a "definitive carcinogen" in gastric cancer [5]. Therefore, *H. pylori* is being a high-risk carcinogen in gastric cancer at the same level as smoking is considered a risk for lung cancer.

Uemura et al. [6] conducted a cohort study of screening for gastric cancer by yearly upper gastrointestinal endoscopy (endoscopy) in individuals with or without *H. pylori* infection, which was defined strictly by various methods. The authors detected gastric cancer in 0.4–0.5 %/year in individuals with *H. pylori* infection but none in those without *H. pylori* infection. The authors also reported that *H. pylori* infection was strongly associated with the development of not only the intestinal but also the diffuse type of gastric cancer. In addition, Matsuo et al. [7] analyzed 3161 gastric cancer cases for the prevalence of *H. pylori*-negative gastric cancer, defined by all of the following criteria: negative serum *H. pylori* antibody; negative *H. pylori* by microscopic observations; no histologic gastritis by microscopic examinations; no gastric mucosal atrophy on endoscopy; and negative ^{13}C-urea breath test or rapid urease test. The authors detected only 21 cases of *H. pylori*-negative gastric cancer (0.66 %, 95 % confidence interval (CI) 0.41–1.01). Ono et al. [8] also investigated the status of *H. pylori* infection in 240 early gastric cancers treated by endoscopic submucosal dissection using histological examination, *H. pylori* culture, ^{13}C-urea breath test, and serum *H. pylori* antibodies and found only one case (0.42 %) of *H. pylori*-negative gastric cancer. These findings suggest that the incidence of gastric cancer in individuals without *H. pylori* infection is very low, although it is not nil.

In animal experiments, Watanabe et al. [9] demonstrated that *H. pylori* infection induces gastric cancer in Mongolian gerbils. Furthermore, although excessive salt intake has previously been suggested to increase the risk for gastric cancer, Kato et al. [10] reported that high salt diets promote gastric chemical carcinogenesis in Mongolian gerbils with *H. pylori* infection but not in those without *H. pylori* infection.

Taken together, these findings support the evidence that *H. pylori* infection is required for the development of gastric cancer.

11.1.2 The Risk for Gastric Cancer Among Individuals with H. pylori *Infection*

The finding by Uemura et al. [6] that the incidence of gastric cancer among individuals with *H. pylori* infection was 0.4–0.5 %/year indicates that not all individuals with *H. pylori* infection may develop gastric cancer. The question is then, what kind of individuals with *H. pylori* infection are more likely to develop gastric cancer? Correa [11] has proposed a hypothesis for a carcinogenic pathway starting from *H. pylori* infection causing superficial gastritis, followed by atrophic gastritis, intestinal metaplasia, and finally gastric cancer. In addition, Uemura et al. [6] reported an increased risk for gastric cancer in severe atrophy of the gastric mucosa, associated with histological findings of intestinal metaplasia and corpus-predominant gastritis. Masuyama et al. [12] also reported that the gastric cancer detection rate increased significantly with the progression of endoscopic atrophy of gastric mucosa. Furthermore, our previous study on the background gastric mucosa in patients with gastric cancer, detected through an endoscopic screening during a complete medical checkup [13], showed that more than 70 % of cases with the diffuse type of gastric cancer as well as most cases with the intestinal type of gastric cancer were associated with endoscopic findings of open-type gastric mucosal atrophy, according to the Kimura and Takemoto classification [14]. In endoscopic surveillance during complete medical checkups over a period of more than 10 years, the incidence of gastric cancer was higher in a group of individuals with severe gastric mucosal atrophy detected in the first year of the 10-year period. These findings clearly demonstrate that advanced gastric mucosal atrophy increases the risk for gastric cancer among individuals with *H. pylori* infection.

Other than gastric mucosal atrophy, nodularity [15] and enlarged fold [16] have attracted attention as factors associated with increased risk for gastric cancer, especially the diffuse type. Therefore, in assessing the risk for gastric cancer we should focus not only on gastric mucosal atrophy but also on severe inflammation in the gastric mucosa.

11.1.3 H. pylori *Eradication and Expectation of Gastric Cancer Prevention*

Uemura et al. [17] investigated the effect of *H. pylori* eradication on subsequent development of cancer after endoscopic resection of early gastric cancer. They detected a second cancer in 6 (9 %) of the 67 patients who did not receive eradication therapy but none in the 65 patients who received eradication therapy during the follow-up period of 5 years. This was the first report on the prevention of gastric cancer development by *H. pylori* eradication in humans. Subsequently, a multicenter randomized control trial was done in Japan, which demonstrated that eradication of *H. pylori* after endoscopic resection of early gastric cancer significantly reduced the incidence of metachronous gastric cancer, with the hazard ratio of 0.339 (95 % CI 0.157–0.729) [18]. Furthermore, Yohn et al. [19] performed a meta-analysis and demonstrated a hazard ratio for the development of metachronous gastric cancer of 0.42 (95 % CI 0.32–0.58) after eradication of *H. pylori*. A meta-analysis in patients with *H. pylori*-associated gastritis alone without gastric cancer, peptic ulcer, or upper gastrointestinal symptoms also showed a significant reduction in the incidence of subsequent gastric cancer by *H. pylori* eradication [20]. However, the odds ratio was 0.66 (95 % CI 0.46–0.95), suggesting that the effect of *H. pylori* eradication in this patient population might be limited as compared to that in patients with early gastric cancer treated by endoscopic resection.

11.2 Methods to Evaluate the Risk for Gastric Cancer

11.2.1 Tests for H. pylori *Infection*

Diagnostic tests for *H. pylori* infection include microscopic examination, culture, and rapid urease test (which require endoscopy and biopsy), stool antigen, urine/serum antibody tests and ^{13}C-urea breath test (which do not require endoscopy). For use in screening, tests must be noninvasive (requiring no endoscopy), simple and convenient, cheap, and able to analyze multiple samples simultaneously. Now, serum antibody tests are considered most ideal for this purpose.

11.2.2 Methods to Evaluate Gastric Mucosal Atrophy

Atrophy of the gastric mucosa is characterized by thinning of the crypt epithelium and is best evaluated by histological examination, a gold standard method. In addition, endoscopic evaluation of the degree of gastric mucosal atrophy (Kimura and Takemoto classification) [14] is also very useful in clinical practice of gastroenterology. However, these morphological tests are invasive and therefore not suitable for screening. Furthermore, gastric analysis and 24-h gastric pH monitoring can be used for the evaluation of gastric mucosal atrophy, based on the fact that acid secretion decreases with progression of gastric mucosal atrophy; however, these methods are also invasive and not suitable for screening.

Serum pepsinogen (PG) reflects the atrophy and inflammation of the gastric mucosa, and a radioimmunoassay to measure the PG levels has been established by Samloff et al. [21, 22]. The serum PG method, proposed by Miki et al. [23, 24] is a simple blood test to detect gastric mucosal atrophy of more than a certain level, and it is suitable for screening. PG is classified into PG I, secreted mainly from chief cells in the gastric fundal glands, and PG II, secreted not only from the gastric fundal glands but also from the pyloric, cardiac, and Brunner's glands. The PG method is based on the observation that PG I secretion and the I/II ratio decrease in atrophic gastritis because the area of gastric fundal glands decreases by progression of gastric mucosal atrophy; a combination of serum PG I level of $\leqq$70 ng/ml and PG I/II ratio of $\leqq$3.0 is usually regarded as positive for atrophy.

11.2.3 About ABC Classification

ABC classification [25–27] is a combination of measurements of serum *H. pylori* antibody and serum PG method to evaluate the degree of health of the stomach and the risk for gastric cancer. In this classification, a group of individuals who are *H. pylori* antibody (−) and PG method (−) is defined as group A; a group with *H. pylori* antibody (+) and PG method (−) is defined as group B; a group with *H. pylori* antibody (+) and PG method (+) is defined as group C; and a group with *H. pylori* antibody (−) and PG method (+) is defined as group D. If highly sensitive *H. pylori* antibodies are used, the number of individuals classified into group D would be small; therefore, group C and group D are often combined as group C in the setting of mass screening (Table 11.1).

Theoretically, group A is generally considered to comprise a group of individuals who do not have *H. pylori* infection; group B is considered to have *H. pylori* infection but no severe gastric mucosal atrophy; group C is considered to have advanced gastric mucosal atrophy due to *H. pylori* infection; and group D is considered to comprise those in whom *H. pylori* cannot be detected due to very advanced gastric mucosal atrophy. Because chronic infection of *H. pylori* is established in childhood, an adult classified in group A by the ABC classification is unlikely to transit to group B, C, or D. Of course, it is possible for individuals in group B to change into group C after several years (Fig. 11.1).

Table 11.1 ABC classification. ABC classification is a combination of measurements of serum *H. pylori* antibody and serum pepsinogen method to evaluate the degree of health of the stomach and the risk for gastric cancer

<table>
<tr><td colspan="2" rowspan="2"></td><td colspan="2">H.pylori antibody</td></tr>
<tr><td>(−)</td><td>(+)</td></tr>
<tr><td rowspan="2">PG method</td><td>(−)</td><td>Group A</td><td>Group B</td></tr>
<tr><td>(+)</td><td colspan="2">Group C
(Group D) | (Group C)</td></tr>
</table>

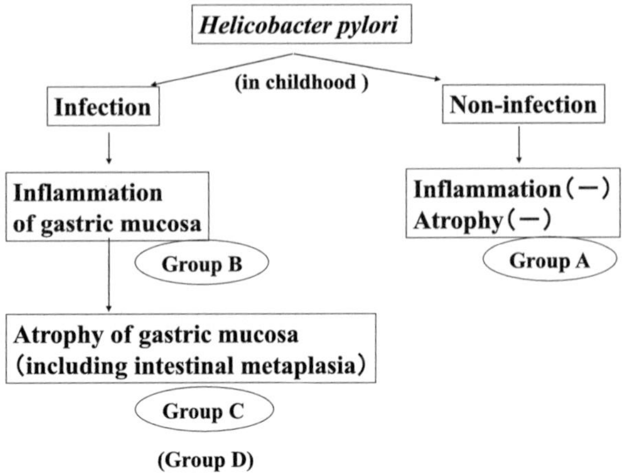

Fig. 11.1 *H. pylori* infection, gastric mucosal condition, and ABC classification. The gastric mucosal condition can approximately be predicted by ABC classification. Theoretically, group A exhibits no *H. pylori* infection, group B has *H. pylori* infection with associated inflammation and mild mucosal atrophy, and group C shows advanced mucosal atrophy due to *H. pylori* infection

11.3 Screening Systems for Gastric Cancer Using ABC Classification as Gateway

11.3.1 Usefulness of ABC Classification

In individuals who underwent complete medical checkups from 1995 to 2008, the incidence of gastric cancer detected by endoscopy on the same day was the highest (1.87 %, 39/2089) in groups C and D, followed by group B (0.21 %, 7/3395), whereas none of the individuals in group A developed gastric cancer (Table 11.2) [27]. Because it might be possible that the age distributions varied between groups, we reanalyzed the data by stratifying the individuals by sex and age. We found that the incidence of gastric cancer was significantly higher in men over 40 years of age in groups C and D. Furthermore, the incidence of gastric cancer was higher in women in their 50s in groups C and D. In addition, the incidence of gastric cancer detected by follow-up endoscopy performed after ABC classification during complete medical checkups was highest in groups C and D, followed by groups B and A. In follow-up studies by Ohata et al.[28] using radiology in workplace medical checkups and by Watabe et al.[29] using endoscopy in individuals who received complete medical checkups, the incidence of gastric cancer was reported to be highest in group D, followed by groups C, B, and A. These findings suggest that group A is associated with low risk, group B with average risk, and groups C and D with high risk for gastric cancer.

In addition, ABC classification can be used to evaluate the incidence of not only gastric cancer but also other upper gastrointestinal diseases [26, 27]. For example, the risks for gastric adenoma or hyperplastic polyp are also highest in groups C and D, followed by group B. The risk for peptic ulcer is highest in group B, followed by groups C and D. Similarly, the risk for peptic ulcer is also very low in group A. On the other hand, the risk for reflux esophagitis is highest in group A, followed by group B (Table 11.3).

Table 11.2 Incidence of gastric cancer detected by endoscopy in a medical check-up. The detection rate of gastric cancer was the highest in Group C (+D), followed by Group B. No gastric cancer was detected in Group A

		H. pylori antibody	
		(−)	(+)
PG method	(−)	Group A: 0 %	Group B: 0.21 %[b]
		(0/2802)	(7/3395)
	(+)	Group C (D): 1.87 %[a]	
		(39/2089)	

[a] $p < 0.01$ (v.s. Group A, Group B)
[b] $p < 0.05$ (v.s. Group A)

Table 11.3 ABC classification and risk of the upper gastrointestinal diseases. By ABC classification, we can evaluate a risk of various upper gastrointestinal diseases as well as gastric cancer

	Group A	Group B	Group C(D)
Gastric cancer	Low	Middle	High
Gastric adenoma	Low	Low-middle	High
Hyperplastic polyp	Low	Middle	High
Peptic ulcer	Low	High	Middle
Reflux esophagitis	High	Middle	Low

H. pylori infection and gastric mucosal atrophy can be evaluated by endoscopy and upper gastrointestinal radiography (radiography). According to the Management of Precancerous Conditions and Lesions in the Stomach (MAPA), published in Europe in 2012 [30] endoscopic observation with magnification endoscopy and narrow band imaging is recommended in all cases. In addition, when gastric mucosal atrophy or intestinal metaplasia is found, staging gastritis in multiple biopsied specimens obtained from each of the gastric antrum and the gastric corpus using the operative link for gastritis assessment (OLGA) system [31] is recommended. However, not all clinicians who are involved in endoscopy can necessarily evaluate these findings at the same level. In addition, performing endoscopy in all persons is not realistic in the setting of mass screening.

ABC classification is based on a simple and cheap blood test that is easy to carry out by anyone; it also has the advantage of providing consistent, reproducible results and is therefore suitable for mass screening.

11.3.2 Screening for Gastric Cancer Using ABC Classification

Although the risk for gastric cancer can be evaluated by ABC classification, the diagnosis is never done by this test. Imaging studies by radiology or endoscopy are required for the diagnosis. Modalities and intervals of imaging studies should be determined based on the risk assessment by ABC classification. In Japan, mass screening by radiology has been performed and shown to be effective in reducing the mortality of gastric cancer [32–34], thereby contributing greatly to the preventive strategy. However, even when screening by radiology has been done, endoscopy has been used in further examinations. In addition, the use of endoscopy is rapidly spreading and is now becoming easier to access than radiology. For individuals who are judged to be at high risk for gastric cancer according to ABC classification, routine examination by endoscopy of good quality is desirable. On the other hand, for individuals who are judged to be at low risk for gastric cancer, or more precisely for those who are found to have no *H. pylori* infection, imaging studies with endoscopy or radiology would be unnecessary.

When used as a population-based screening, the quality control of tests is of course important.

11.3.3 A Word of Caution About ABC Classification

The major problem in ABC classification is a cross-contamination of group A individuals from previously or persistently *H. pylori*-infected persons. If *H. pylori* eradication is successfully done, both serum PG I and PG II levels are decreased, and the PG I/II ratio is increased because the reduction rate of PG II is larger than that of PG I. Furthermore, in most instances the PG I/II ratio is ≧3 after *H. pylori* eradication even in individuals with advanced gastric mucosal atrophy; therefore, the PG method could be negative in the vast majority of cases, if this test is done after *H. pylori* eradication. In addition, the serum *H. pylori* antibody titer is decreased after successful eradication and is converted to negative over time. Therefore, if ABC classification is done after eradication, most cases may be classified falsely into 'group A'. Although the risk for gastric cancer is decreased to 1/3–1/2 by *H. pylori* eradication, it still remains. The risk for gastric cancer in patients who underwent *H. pylori* eradication therapy is apparently different from that in individuals with no *H. pylori* infection; therefore, these individuals should not be included in the same group. When ABC classification is used in mass screening, a possible previous history of *H. pylori* eradication should be examined. In patients who have received eradication therapy as group E (eradication group), ABC classification should not be done, but instead periodical image studies should be recommended.

Furthermore, the medical history alone may not be able to pick up some individuals who have received eradication therapy or those in whom *H. pylori* was eradicated by chance. Ideally, imaging studies, such as endoscopy, should be done at least once even in individuals in group A; however, it is practically difficult to perform the imaging studies in all of these individuals because of a lack of capacity, including man power. It is desirable to pick up individuals in whom previous or persistent infection of *H. pylori* cannot be excluded by actual values of serum PG and serum *H. pylori* antibody titer.

11.4 Toward Elimination of Gastric Cancer

In order to reduce the mortality of gastric cancer, early detection followed by early therapy is important and, along these lines, cancer screening as a secondary preventive measure is needed. The use of ABC classification as a gateway in evaluation of gastric cancer risk enables the promotion of efficient cancer screening. If *H. pylori* infection is detected according to the ABC classification, endoscopy-based imaging studies and, in addition, eradication of *H. pylori* should be ideally done to reduce the risk for gastric cancer. Then, even after eradication therapy, periodical surveillance with imaging studies is essential. The eradication therapy is sometimes termed as the primary prevention, but *H. pylori* infection prophylaxis must be considered the true primary prevention. It seems important to regard the eradication therapy as an intermediate measure between the primary prevention and secondary prevention in order to avoid neglecting the surveillance after eradication therapy.

Acknowledgements and Disclosures The author states that there is no involvement of funding sources.

The author states that there is no conflict of interest.

References

1. Warren JR, Marshall BJ. Unidentified curved bacilli on gastric epithelium in active gastritis. Lancet. 1983;321:1273–5.
2. Forman D, Newell DG, Fullerton F, Yarnell JW, Stacey AR, Wald N, et al. Association between infection with *Helicobacter pylori* and risk of gastric cancer. BMJ. 1991;302:1302–5.
3. Parsonnet J, Friedman GD, Vandersteen DP, Chang Y, Vogelman JH, Orentreich N, et al. *Helicobacter pylori* infection and risk of gastric carcinoma. N Engl J Med. 1991;325:1127–31.
4. Nomura A, Stemmerman GN, Chyou P, Kato I, Perez-Perez GI, Blaser MJ. *Helicobacter pylori* infection and gastric carcinoma among Japanese Americans in Hawaii. N Engl J Med. 1991;325:1132–6.
5. WHO. Schistosomes, liver flukes and *Helicobacter pylori*: IARC working group on the evaluation of carcinogenic risks to humans, IARC monographs on the evaluation of carcinogenetic risks to humans, IARC scientific publication no. 61, vol. 61. Lyon: IARC; 1994. p. 218–20.
6. Uemura N, Okamoto S, Yamamoto S, Matsumura N, Yamaguchi S, Yamakido M, et al. *Helicobacter pylori* infection and the development of gastric cancer. N Engl J Med. 2001;345:784–9.
7. Matsuo T, Ito M, Takata S, Tanaka S, Yoshihara M, Chayama K. Low prevalence of *Helicobacter pylori*-negative gastric cancer among Japanese. Helicobacter. 2001;16:415–9. doi:10.1111/j.1523-5378.2011.06893.x.
8. Ono S, Kato M, Suzuki M, Ishigaki S, Takahashi M, Haneda M, et al. Frequency of *Helicobacter pylori*-negative gastric cancer and gastric mucosal atrophy in a Japanese endoscopic submucosal dissection series including histological, endoscopic and serological atrophy. Digestion. 2012;86:59–65. doi:10.1159/000339176. Epub 2012 Jun 20.

9. Watanabe T, Tada M, Nagai H, Sasaki S, Nakao M. *Helicobacter pylori* infection induces gastric cancer in Mongolian gerbils. Gastroenterology. 1998;115:642–8.
10. Kato S, Tsukamoto T, Mizoshita T, Tanaka H, Kumagai T, Ota H, et al. High salt diets dose-dependently promote gastric chemical carcinogenesis in *Helicobacter pylori*-infected Mongolian gerbils associated with a shift in mucin production from glandular to surface mucous cells. Int J Cancer. 2006;119:1558–66.
11. Correa P. Human gastric carcinogenesis: a multistep and multifactorial process; First American cancer Society award lecture on cancer epidemiology and prevention. Cancer Res. 1992;52:6735–40.
12. Masuyama H, Yoshitake N, Sasai T, Nakamura T, Masuyama A, Zuiki T, et al. Relationship between the degree of the endoscopic atrophy of the gastric mucosa and carcinogenic risk. Digestion. 2015;91:30–6.
13. Inoue K, Fujisawa T, Chinuki D, Kushiyama Y. Characteristics of gastric mucosa in gastric cancer improvement – examination from a clinical survey by endoscopy. Stomach Intestine. 2009;44:1367–73 (in Japanese with English abstract).
14. Kimura K, Takemoto T. An endoscopic recognition of the atrophic border and its significance in chronic gastritis. Endoscopy. 1969;1:87–96.
15. Kamada T, Hata J, Tanaka A, Kusunoki H, Miyamoto M, Inoue K, et al. Nodular gastritis and gastric cancer. Dig Endosc. 2006;18:79–83. doi:10.1111/j.0915-5635.2006.00588.x.
16. Nishibayashi H, Kanayama S, Kiyohara T, Yamamoto K, Miyazaki Y, Yasunaga Y, et al. *Helicobacter pylori*-induced enlarged-fold gastritis is associated with increased mutagenicity of gastric juice, increased oxidative DNA damage, and an increased risk of gastric carcinoma. J Gastroenterol Hepatol. 2003;18:1384–91.
17. Uemura N, Mukai T, Okamoto S, Yamaguchi S, Mashiba H, Taniyama K, et al. Effect of *Helicobacter pylori* infection on subsequent development of cancer after endoscopic resection of gastric cancer. Cancer Epidemiol Biomarkers Prev. 1997;6:639–42.
18. Fukase K, Kato M, Kikuchi S, Inoue K, Uemura N, Okamoto S, et al. Effect of eradication of *Helicobacter pylori* on incidence of metachronous gastric carcinoma after endoscopic resection of early cancer: an open-label, randomized controlled trial. Lancet. 2008;372:392–7. doi:10.1016/S0140-6736(08)61159-9.
19. Yohn SB, Park JM, Lim CH, Cho YK, Choi MG. Effect of *Helicobacter pylori* eradication on metachronous gastric cancer after endoscopic resection of gastric tumors: a meta-analysis. Helicobacter. 2014;19:243–8. doi:10.1111/hel.12146.
20. Ford AC, Forman D, Hunt RH, Yuan Y, Moayyedi P, et al. *Helicobacter pylori* eradication therapy to prevent gastric cancer in healthy asymptomatic infected individuals: systematic review and meta-analysis of randomised controlled trials. BMJ. 2014;20:348:g3174. 10.1136/bmj.g3174.
21. Samloff IM. Pepsinogens I, and II; purification from gastric mucosa and radioimmunoassay in serum. Gastroenterology. 1982;82:26–33.
22. Samloff IM, Varis K, Ihamaki KT, Siurala M, Rotter JI. Relationship among serum pepsinogen I, serum pepsinogen II, and gastric mucosal histology. A study in relatives of patients with pernicious anemia. Gastroenterology. 1982;83:204–9.
23. Miki K, Ichinose M, Shimizu A, Huang SC, Oka H, Furihata C, et al. Serum pepsinogens as a screening test of extensive chronic gastritis. Gastroentereol Jpn. 1987;22:133–41.
24. Miki K, Ichinose M, Ishikawa K, Yahagi N, Matsushima M, Kakei N, et al. Clinical application of serum pepsinogen I and II levels for mass screening to detect gastric cancer. Jpn J Cancer Res. 1993;84:1086–90.
25. Inoue K. Evaluation of serum pepsinogens and anti-*H.pylori* antibodies for screening of gastric cancer. Clin Gastroenterol. 2002;17:1591–8 (in Japanese with English abstract).
26. Inoue K, Fujisawa T, Haruma H. Assessment of degree of health of the stomach by concomitant measurement of serum pepsinogen and serum *Helicobacter pylori* antibodies. Int J Biol Markers. 2010;25:207–12.

27. Inoue K, Fujisawa T, Nishi T, et al. Utilities and problems of the ABC classification – including examination of the reference value of pepsinogens. Helicobacter Res. 2011;15:422–7 (in Japanese).
28. Ohata H, Kitauchi S, Yoshimura N, Mugitani K, Iwane M, Nakamura H, et al. Progression of chronic atrophic gastritis with *Helicobacter pylori* infection increases risk of gastric cancer. Int J Cancer. 2004;109:138–43.
29. Watabe H, Mitsushima T, Yamaji Y, Okamoto M, Wada R, Kokubo T, et al. Predicting the development of gastric cancer from combining *Helicobacter pylori* antibodies and serum pepsinogen status: a prospective endoscopic cohort study. Gut. 2005;54:764–8.
30. Dinis-Ribeiro M, Areia M, de Vries AC, Marcos-Pinto R, Monteiro-Soares M, O'Connor A, et al. Management of precancerous conditions and lesions in the stomach (MAPS): guideline from the European Society of Gastrointestinal Endoscopy (ESGE), European Helicobacter Study Group (EHSG), European Society of Pathology (ESP), and the Sociedade Portuguesa de Endoscopia Digestiva (SPED). Endoscopy. 2012;44:74–94. doi:10.1055/s-0031-1291491. Epub 2011 Dec 23.
31. Rugge M, Meggio A, Pennelli G, Piscioli F, Giacomelli L, De Pretis G, et al. Gastritis staging in clinical practice: the OLGA staging system. Gut. 2007;56:631–6.
32. Ohshima A, Hirata N, Ubukata T, Fujimoto I. Evaluation of a mass screening program for stomach with a case-control study design. Int J Cancer. 1986;38:829–33.
33. Fukao A, Tsubono Y, Tsuji I, Hisamichi S, Sugahara N, Takao A. The evaluation of screening for gastric cancer in Miyagi Prefecture, Japan; a population-based case-control study. Int J Cancer. 1995;60:45–8.
34. Abe Y, Mitsushima T, Nagatani Y, Ikuma H, Minamihara Y. Epidemiological evaluation of the protective effect for dying of stomach cancer by screening programme for stomach cancer with applying a method of case-control study. A study of efficient screening programma for stomach cancer. J Gastroenterol. 1995;92:836–45 (in Japanese).

Chapter 12
Prevention of Gastric Cancer by *Helicobacter pylori* Eradication: Current Evidence and Future Prospects

Jyh-Ming Liou, Jaw-Town Lin, and Ming-Shiang Wu

Abstract Ecological studies showed higher cumulative incidence of gastric cancer in countries with higher prevalence of *H. pylori* infection. Meta-analysis of case–control studies nested within prospective cohort showed that *H. pylori* infection was associated with a 5.9-fold increased risk of non-cardia gastric cancer using blood samples collected more than 10 years. *H. pylori* was associated with increased risk of both diffuse type and intestinal type gastric cancer. Cag-A seropositivity was associated with higher risk of gastric cancer. Prospective cohort studies showed that gastric cancer developed in 1–3 % of *H. pylori* infected subjects. Gastric cancer was successfully induced in Mongolian gerbils and INS-GAS transgenic mice after inoculation of *H. pylori*. The incidence of gastric dysplasia and gastric cancer could be reduced in mice treated with eradication therapy. Recent meta-analysis of randomized control trials showed a significant reduction in the risk of gastric cancer in *H. pylori* infected subjects who received *H. pylori* eradication therapy. Based on these evidences, it is well agreed that *H. pylori* is a causal risk factor of gastric cancer. However, large well-designed randomized trials are highly anticipated to assess the effectiveness of the screen and treat strategy as well as the changes in the antibiotic resistance and risks in the development of gastroesophageal reflux disease, obesity, and allergic diseases after *H. pylori* eradication.

J.-M. Liou
Department of Internal Medicine, National Taiwan University Hospital, National Taiwan University College of Medicine, Taipei, Taiwan

J.-T. Lin
Department of Internal Medicine, National Taiwan University Hospital, National Taiwan University College of Medicine, Taipei, Taiwan

School of Medicine, Fu Jen Catholic University, New Taipei, Taiwan

M.-S. Wu (✉)
Department of Internal Medicine, National Taiwan University Hospital, National Taiwan University College of Medicine, Taipei, Taiwan

Departments of Internal Medicine and Primary Care Medicine, National Taiwan University Hospital, National Taiwan University, College of Medicine, No. 7, Chung-Shan S. Road, Taipei, Taiwan
e-mail: mingshiang@ntu.edu.tw

H. Suzuki et al. (eds.), *Helicobacter pylori*, DOI 10.1007/978-4-431-55705-0_12

Keywords *Helicobacter pylori* • Gastric cancer • Prevention • Eradication • Trial

12.1 Introduction

Gastric cancer remains the third most common cause of cancer related mortality worldwide [1]. After the discovery of *Helicobacter pylori (H. pylori)* in 1982, [2] many researchers have investigated the association of this bacterium with the gastric cancer, including adenocarcinoma and lymphoma [3–5]. In this chapter, we will summarize the landmark studies and trials in humans and in animals that provide important evidence to the causal association of *H. pylori* and gastric cancer. Whether gastric cancer can be prevented through screening and eradication of *H. pylori* will also be discussed. We will start with the ecological association of prevalence of *H. pylori* and incidence of gastric cancer in different countries. Then we will show the results from observational studies in human (case–control studies, nested case–control studies, and cohort studies). We will next present the results from interventional trials in human and in animal models. Finally, we will discuss the effectiveness, regimen to be used, the recurrence rate, and potential harms of the mass screening and eradication strategy to prevent gastric cancer in the community.

12.2 Ecological Association of Prevalence of *Helicobacter pylori* and Incidence of Gastric Cancer in Different Countries

The EUROGAST study group conducted a multicenter ecological study in 13 European countries to investigate the association between *H. pylori* infection and gastric cancer [6]. Random selection of 200 population-based subjects aged 25–34 and 55–64 years (50 males and 50 females from each age group) was done in each country [6]. They found significant associations between the cumulative incidence and mortality rates of gastric cancer with prevalence of *H. pylori* using the linear regression analysis [6]. They further estimated a sixfold increased risk of gastric cancer among populations with 100 % *H. pylori* infection compared to those without *H. pylori* infection [6].

The Asian population accounted for 72.9 % (527,074/723,027) of gastric cancer related mortality according to the GLOBOCAN 2012 database [1]. About 700,000 Asian people develops gastric cancer annually [1]. The age standardized incidence rate (ASR) of gastric cancer in Asia is shown in Fig. 12.1. The ASR of gastric cancer is higher than 20 per 100,000 in China, Japan, Korea, and Mongolia, and is lower than 10 per 100,000 in Hong Kong, India, Indonesia, Malaysia, Pakistan, Philippines, Singapore, and Thailand [1]. Generally, the ASR of gastric cancer correlates with the prevalence of *H. pylori* in Asia (Fig. 12.1). However, the ASR of gastric cancer is low in India, Pakistan, and Philippines despite the high prevalence

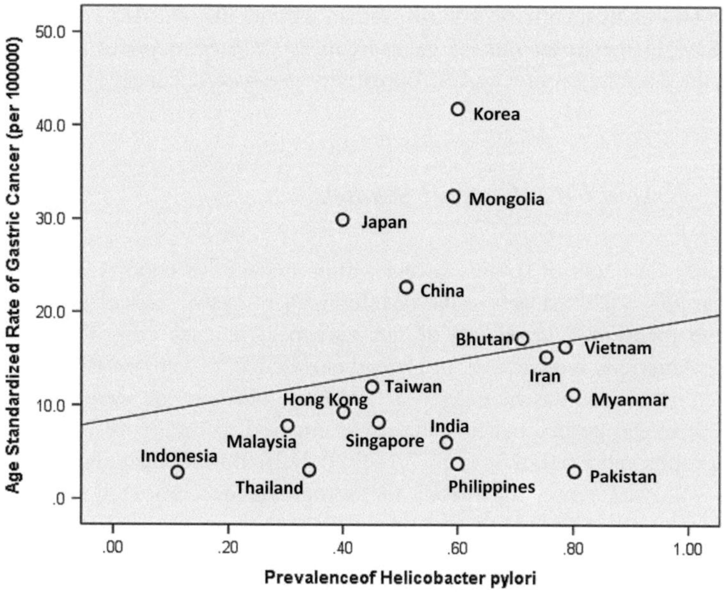

Fig. 12.1 Ecological association of prevalence of *Helicobacter pylori* and incidence of gastric cancer incidence in Asia-Pacific region [7, 8] (Note. Data were retrieved from the World Health Organization Internal Association of Cancer Registries (http://globocan.iarc.fr/Pages/age-specific_table_sel.aspx), the Taiwan Cancer Registry (http://cph.ntu.edu.tw/main.php?Page¼A5B2))

of *H. pylori* infection [7, 8]. Differences in dietary habits and genetic predisposing factors might account for this so called Asian Enigma, although underdiagnosis of gastric cancer and other competing causes of mortality might also contribute to the discrepancies [7, 8].

12.3 Association of *H. pylori* and Gastric Cancer in Observational Studies

12.3.1 Retrospective Case Control Studies

After the discovery of *H. pylori*, many retrospective case control studies have been conducted to assess its association with gastric cancer [9, 10]. The seroprevalence of *H. pylori* was significantly higher among the 112 incidence gastric cancer patients compared with the 103 matched controls (odds ratio 2.6, 95 % CI 1.4–5.0) [9]. However, conflicting results have been reported. Although some studies showed positive association of *H. pylori* and gastric cancer, others failed to confirm the association. One of the reasons for the contradictory results was the loss of *H. pylori* infection in the atrophic stomach at the time of gastric cancer diagnosis. Therefore, these studies tended to underestimate the risk of gastric cancer. Meta-analysis of case control studies showed that *H. pylori* was associated with 1.81-fold increased risk of gastric cancer [10]. Subsequent studies further showed the association of CagA seropositivity and gastric cancer [11].

12.3.2 Nested Case Control Studies

Conducting case control studies nested within prospective cohorts which used the blood samples collected before the development of gastric cancer may overcome the above mentioned limitation of the retrospective case control studies. In a Japanese American male cohort in Hawaii enrolled from 1967 to 1970, 109 of the 5908 men developed gastric cancer by 1989 [5]. The seroprevalence of *H. pylori* was 94 % in the gastric cancer patients, compared to the 76 % in the matched controls (odds ratio 6.0, 95 % CI 2.1–17.3) [12]. In another cohort of 128,992 persons followed since mid-1960, the seroprevalence was 84 % among the 186 patients with gastric cancer, compared to the 61 % in the 186 matched controls (odd ratio 3.6, 95 % CI 1.8–7.3) [13]. Based on the results from nested case control studies, *H. pylori* was classified as a class I human carcinogen by the International Agency for Research on Cancer in 1994.

Subsequent meta-analysis of 12 nested case control studies including 1228 gastric cancer cases showed that the association with *H. pylori* was restricted to non-cardia gastric cancers (odds ratio 3.0), but not the cardia gastric cancer. It was associated with increased risk of intestinal type and the diffuse type gastric cancer at non-carida portions [14]. They further showed a stronger association (OR 5.9) when the blood samples used for serology testing was collected more than 10 years before the diagnosis of gastric cancer [14]. Several nested case–control studies have been reported after that meta-analysis. Therefore, we performed an updated meta-analysis and the results were shown in Table 12.1 and Fig. 12.2 [13–28]. Taken together with the assumption that the prevalence of *H. pylori* infection were 35 % and 80 % in developed and in developing countries, respectively, it was estimated that about 65–80 % of gastric cancer could be attributed to *H. pylori* infection [14].

12.3.3 Cohort Studies

Hansson et al. estimated the risk of gastric cancer in 57,936 patients with gastric ulcer and duodenal ulcer in a large hospital-based retrospective cohort registered between 1965 and 1983 in the Swedish Inpatient Register [29]. Gastric cancer developed in 782 of the 29,287 patients with gastric ulcer (standardized incidence ratio 4.3; 95 % CI 4.0–4.6) after an average follow-up of 8.3 years [29]. In contrast, gastric cancer developed only in 136 of the 24,456 patients with duodenal ulcer (standardized incidence ratio 0.9, 95 % CI 0.7–1.1) after an average follow-up of 10.1 years [29]. The standardized incidence ratio (SIR) for gastric cancer among

Table 12.1 Association of *Helicobacter pylori* and gastric cancer in nested case control studies

Study	Country	Cohort enrolled year	Cohort size	Median Follow-up (years)	Mean age of GC at diagnosis	Male % among cases	No of HP (+) cases	Number of GC cases	No of HP (+) controls	Number of controls
Forman 1991 and Wald 1997 [15, 16]	UK	1975–1982	21,500	8.7	54 (39–59)	100 %	38	56	72	174
Parsonnet 1991 [13]	USA	1964–1969	128,992	15	68 (44–90)	69 %	94	111	67	111
Nomura 1991 [12]	USA	1967–1970	5908	13.8	72 (56–85)	100 %	103	109	93	109
Lin 1995 [17]	Taiwan	1984–1986	9775	2	63 (43–80)	100 %	20	29	129	146
Aromaa 1996 [18]	Finland	1968–1972	39,268	9.5	62 (32–85)	62 %	73	84	121	146
Webb 1996 [19, 20]	China	1986–1989	18,244	4.8	63 (49–76)	100 %	168	188	451	548
Siman 1997 [21]	Sweden	1972–1992	32,906	5.1	56 (38–70)	93 %	46	56	110	224
Watanabe 1997 [22]	Japan	1987	2858	3.6	69 (46–86)	58 %	41	45	170	225
Hansen 1999 [23]	Norway	1972–1986	101,601	12	56 (34–68)	75 %	166	208	619	983
Limburg 2001 [24]	China	1985	29,584	3.6	61 (36–75)	62 %	113	181	99	192
Shin 2005 [25]	Korea	1993–1994	10,699	2.6	63	66 %	72	86	278	344
Kamangar 2006 [26]	Finland	1985–1988	29,133	5.8	59 (50–70)	100 %	195	234	176	234
Sasazuki 2006 [27]	Japan	1990 and 1993	123,576	n/r	57 (40–69)	67 %	478	511	383	511
Palli 2007 [28]	Europe	1992–1998	360,000	6.1	n/r	55 %	195	233	625	910

GC gastric cancer, *HP Helicobacter pylori*, *n/r* not reported

Study name	Odds ratio	Lower limit	Upper limit	Z-Value	p-Value	Relative weight
Forman 1991 and Wald 1997	2.991	1.582	5.654	3.372	0.001	4.97
Parsonnet 1991	3.631	1.912	6.897	3.940	0.000	4.89
Nomura 1991	2.953	1.109	7.863	2.168	0.030	2.10
Lin 1995	1.568	0.683	3.599	1.060	0.289	2.92
Aromaa 1996	1.371	0.637	2.950	0.807	0.419	3.43
Webb 1996	1.807	1.082	3.017	2.260	0.024	7.66
Siman 1997	4.767	2.292	9.915	4.180	0.000	3.76
Watanabe 1997	3.316	1.137	9.675	2.194	0.028	1.76
Hansen 1999	2.324	1.618	3.339	4.561	0.000	15.33
Limburg 2001	1.561	1.033	2.359	2.113	0.035	11.81
Shin 2005	1.221	0.649	2.298	0.619	0.536	5.04
Kamangar 2006	1.648	1.046	2.595	2.155	0.031	9.76
Sasazuki 2006	4.841	3.227	7.262	7.622	0.000	12.24
Palli 2007	2.340	1.609	3.404	4.447	0.000	14.34
	2.328	2.020	2.683	11.669	0.000	

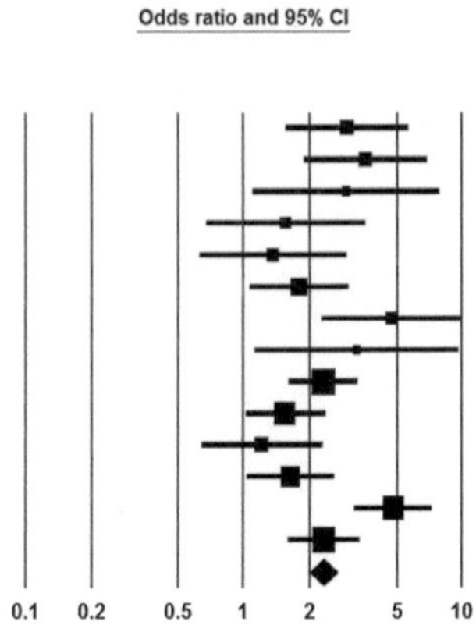

Fig. 12.2 Updated meta-analysis of nested case–control studies on the *H. pylori* infection and the gastric cancer risk [13–28]

patients with gastric ulcer and duodenal ulcer were 1.8 (95 % CI 1.6–2.0) and 0.6 (95 % CI 0.4–0.7), respectively [29]. This study indicated that compared to the general population, patients with gastric ulcer have increased risk of gastric cancer, whereas those with duodenal ulcer have reduced risk of gastric cancer.

Uemura et al. prospectively enrolled 1246 *H. pylori* infected and 280 non-infected Japanese patients who had peptic ulcer disease, gastric hyperplasia, or non-ulcer dyspepsia between 1990 and 1993 [30]. The *H. pylori* status was determined by histology, rapid urease test, and serology testing in all subjects. The mean ages at baseline were 52.3 and 52.7 years among *H. pylori* positive and negative patients, respectively. After a mean follow-up of 7.8 years, gastric cancer developed in 2.9 % (36/1246) among *H. pylori* positive patients. Interestingly, none (0/280) of the *H. pylori* negative subjects developed gastric cancer during the follow-up periods [30]. Gastric cancer developed in 4.7 % (21/445) of patients with non-ulcer dyspepsia, in 3.4 % (10/297) of patients with gastric ulcers, and in 2.2 % (5/229) of patients with gastric hyperplastic polyps [30]. However, none of the patients with duodenal ulcer (0/275) developed gastric cancer during follow-up periods [30]. In another prospective cohort study, Hsu et al. followed 618 *H. pylori* infected and 607 *H. pylori* negative patients with non-ulcer dyspepsia, gastric ulcer, and duodenal ulcer between 1990 and 1998 [31]. After a mean follow-up of 6.3 years, gastric cancer developed in 1.1 % (7/618) and 0 % (0/607) among *H. pylori* infected and negative patients, respectively [31]. They further showed that the presence of intestinal metaplasia at baseline was an independent risk factor of subsequent gastric cancer.

12.4 Animal Models

12.4.1 H. pylori *Induced Gastric Cancer in Animal Models*

Watanabe et al. inoculated *H. pylori* orally in 55 five-week-old male Mongolian gerbils. Another 30 gerbils were selected as controls. *H. pylori* were constantly detected in all of the inoculated gerbils throughout the study period [32]. Severe active chronic gastritis, ulcer, and intestinal metaplasia were observed in substantial number of gerbils examined since 26 weeks after inoculation of *H. pylori* [32]. Gastric adenocarcinoma was detected in 37 % (10/27) of gerbils 62 weeks after inoculation of *H. pylori* [32]. Wang et al. inoculated *H. felis* in insulin-gastrin (INS-GAS) transgenic mice. Gastric cancer developed in 75 % (6/8) of the INS-GAS mice that were greater than 20 months old [33]. Lee et al. further showed that inoculation of *H. pylori* may induce severe dysplasia and gastric cancer in INS-GAS transgenic mice 28 weeks after inoculation [34]. These studies provide important evidence that gastric cancer could be induced in animals after long-term *Helicobacter* infection.

12.4.2 Eradication Trials in Animal Models

It is also important to evaluate whether the elimination of *H. pylori* from infected animals could reduce the risk of gastric cancer. Lee et al. treated *H. pylori* infected transgenic INS-GAS mice with triple therapy containing omeprazole, clarithromycin, and metronidazole at 8, 12, or 22 weeks after inoculation of *H. pylori* infection [34]. They found that the severity of gastric dysplasia was significantly reduced in the mice treated with triple therapy compared to control mice [34]. More interestingly, they observed that gastric intraepithelia dysplasia was completely prevented in mice treated with 7-day triple therapy as early as 8 weeks after inoculation of *H. pylori* [34]. Romero-Gallo et al. conducted a similar study in Mongolian gerbils [35]. Thirty five gerbils were inoculated with *H. pylori* strain 7.13, a prototype strain which can induce gastric cancer [35]. These gerbils were treated with 14-day triple therapy containing lansoprazole, amoxicillin, and clarithromycin at 4 (n = 20) or 8 weeks (n = 15) after inoculation of *H. pylori.* Another 23 *H. pylori* inoculated gerbils not treated with antibiotics served as control groups. They found that gastric dysplasia or cancer developed in >60 % of the gerbils with persistent *H. pylori* infection, compared to none in the eradicated group [35]. These studies provided evidence that early eradication of *H. pylori* in animals may prevent the development of gastric dysplasia or cancer [34–36].

12.5 Eradication Trials in Human

12.5.1 Effect of H. pylori *Eradication on the Regression of Gastric Precancerous Lesions*

12.5.1.1 Cohort Studies

Several cohort studies confirmed that eradication of *H. pylori* may reduce the acute and chronic inflammation of gastric mucosa [37, 38]. However, whether the gastric precancerous lesions can be regressed after *H. pylori* eradication remains controversial [37, 38]. Ohkusa et al. showed that glandular atrophy in the corpus and intestinal metaplasia in the antrum improved 12–15 months after successful *H. pylori* eradication [37]. In a community mass eradication program, Lee et al. also showed that the incidence rate of gastric atrophy reduced from 8.2/100-person-years to 3.5/100-person-years after *H. pylori* eradication [38]. However, Lee et al. showed that the incidence rate of gastric intestinal metaplasia was not reduced after *H. pylori* eradication [38]. More well designed studies, preferably randomized trials with sufficient follow-up period and adequate biopsy number are needed to assess whether the gastric intestinal metaplasia could be regressed after *H. pylori* eradication.

12.5.1.2 Randomized Trials

In a factorial randomized trial in Columbia, 852 subjects with gastric precancerous lesions were randomized to receive *H. pylori* eradication therapy, ascorbic acid supplement, or β-carotene supplement ($2^3 = 8$ groups) [39]. The primary outcome was the risk of progression of gastric precancerous lesions. They found that patients treated with *H. pylori* eradication therapy were more likely (relative risk 8.7, 95 % CI 2.7–28.2) to have regression of the gastric precancerous lesions [39]. Sung et al. also showed that acute and chronic inflammations were significantly reduced after *H. pylori* eradication compared to the untreated group 1-year later [40]. However, neither gastric atrophy nor intestinal metaplasia showed significant improvement in the treated group [40]. You et al. showed that the risk of severe atrophic gastritis, intestinal metaplasia, dysplasia, and gastric cancer were significantly reduced (OR 0.60, 95 % 0.47–0.75) in the group treated with *H. pylori* eradication, compared to the untreated group [41]. Wong et al. showed that the precancerous lesions were more likely to be regressed in the group treated with *H. pylori* eradication (OR 1.8, 95 % CI 1.2–22. 8), compared to the placebo group [42]. These collectively indicated that eradication therapy provide beneficial effect on the regression of precancerous lesions compared to the untreated group.

12.5.2 *Effect of* H. pylori *Eradication in the Prevention of Gastric Cancer*

12.5.2.1 Cohort Studies

In a nationwide cohort study using the Taiwan National Health Insurance Database (NHID), Wu et al. included 80,255 patients who were hospitalized between 1997 and 2004 with a primary diagnosis of peptic ulcer disease and received *H. pylori* eradication therapy [43]. These patients were classified as "early eradication" (within 1 year) and "late eradication" (after 1 year). They found that "late eradication" was associated with increased risk of gastric cancer compared to the general population (standardized incidence ratio, SIRs 1.36, 95 % CI 1.24–1.49), whereas the SIR of the "early eradication" population was similar to that of the general population (SIR 1.05, 95 % CI 0.96–1.14) [43]. They further showed that early eradication was an independent protective factor for gastric cancer (hazards ratio 0.77).

12.5.2.2 Randomized Trials

Eight randomized trials that compare the *H. pylori* eradication and placebo on the primary prevention of gastric cancer and its precursor lesions have been reported [39–42, 44–51]. However, none of the included subjects developed gastric cancer during the follow-up periods in two of these trials [50, 51]. A randomized trial assessed the effect of *H. pylori* eradication compared to no treatment in the secondary prevention of metachronous gastric cancer in patients who received endoscopic resection for early gastric cancer [52]. The demographic characteristics of these trials were summarized in Table 12.2 [39–42, 44–49, 52]. It is noteworthy that the primary outcome was the incidence of gastric cancer in only one of these trials. Wong et al. showed insignificant reduction in the risk of gastric cancer at the end of study [42]. New cases of gastric cancer developed in 7 and 11 subjects who received *H. pylori* eradication and placebo, respectively (Hazard risk 0.63, 95 % CI 0.24–1.62) [42]. In the subgroup analysis in subjects who did not have gastric precancerous lesions at baseline, none of the subjects in the treated group developed gastric cancer compared to 6 subjects in the placebo group ($p = 0.02$) [42]. The result indicated that eradication therapy should be given as early as possible in the primary prevention of gastric cancer. Other trials used the gastric precancerous lesion as the primary outcome. In a 15-year follow-up report, Ma et al. showed that a total of 34 and 52 subjects in the treated and untreated groups developed gastric cancer, respectively [48]. In a recent meta-analysis, Ford et al. demonstrated a significant reduction in the gastric cancer incidence in healthy *H. pylori* infected subjects who received *H. pylori* eradication compared to the untreated group [53].

Table 12.2 Randomized trials comparing the efficacy of *H. pylori* eradication on the prevention of gastric cancer and its precursor lesions

Study	Country	Settings	Subjects	Mean age (years)	Gender Male %	% with precancerous lesions at baseline	Primary outcome	Follow-up periods (years)
Correa et al. [39]	Columbia	community	Healthy subjects	51.1 (26–69)	46.1 %	100 %	Gastric precancerous lesions	6.0
Leung and Zhou et al. [40, 44, 45]	China	community	Healthy subjects	52 (35–75)	47.8 %	33.7 %	Gastric precancerous lesions	10
Wong et al. [42]	China	community	Healthy subjects	42.2 (35–65)	54 %	37.7 %	Gastric cancer incidence	7.5
Saito et al. [46]	Japan	community	Healthy subjects	n/r (20–59)	n/r	n/r	Gastric precancerous lesions	>4
You et al. [41]	China	community	Healthy subjects	46.8 (35–64)	50 %	64 %	Gastric precancerous lesions	14.7
Wong et al. [47]	China	community	Healthy subjects	53 (35–64)	46.4 %	100 %	Gastric precancerous lesions	5
Fukase et al. [52]	Japan	Hospital	Post ESD/EMR EGC patients	68.5 (20–79)	76.5 %	n/r	Metachronous gastric cancer incidence	3

ESD endoscopic submucosal dissection, *EMR* endoscopic mucosal resection, *EGC* early gastric cancer, *n/r* not reported

Study name	Treated	Control	Odds ratio	Lower limit	Upper limit	Z-Value	p-Value	Relative weight
Correa 2000 and 2001	3 / 437	2 / 415	1.427	0.237	8.586	0.389	0.697	3.34
Leung 2004	4 / 295	6 / 292	0.655	0.183	2.346	-0.650	0.516	6.60
Wong 2004	7 / 817	11 / 813	0.630	0.243	1.634	-0.950	0.342	11.84
Saito 2005	2 / 379	3 / 313	0.548	0.091	3.301	-0.656	0.512	3.33
You 2004 and Ma 2012	34 / 1130	52 / 1128	0.642	0.413	0.997	-1.973	0.048	55.39
Wong 2012	3 / 255	1 / 258	3.060	0.316	29.609	0.966	0.334	2.09
Fukase 2008	9 / 272	24 / 272	0.354	0.161	0.776	-2.594	0.009	17.41
			0.610	0.440	0.847	-2.954	0.003	

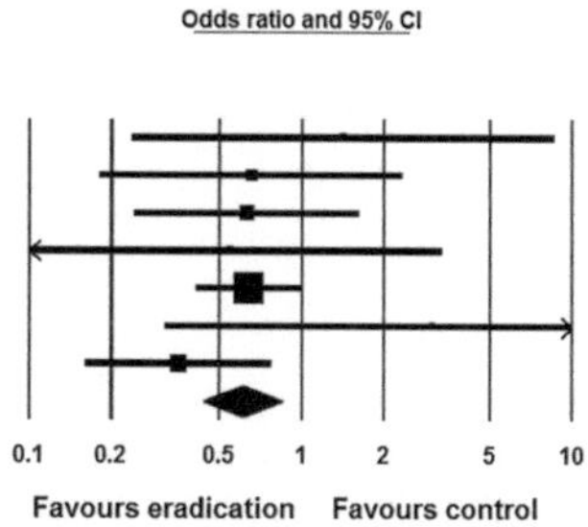

Fig. 12.3 Updated meta-analysis of randomized interventional trials of gastric cancer prevention by *H. pylori* eradication [39–42, 44–49, 52]

In an open-label randomized trial, Fukase et al. assessed whether eradication of *H. pylori* may reduce the risk of subsequent metachronous gastric cancer in patients who receive endoscopic resection compared to untreated control group [52]. They showed that 7 of the 272 patients in the eradication group developed gastric cancer after a 3-year followed-up, compared to 24 of the 272 patients in the control group (Odds ratio 0.35, 95 % CI 0.16–0.78) [52]. An updated meta-analysis which included the result of this trial was shown in Fig. 12.3 [39–42, 44–49, 52]. The results provide evidence that eradication of *H. pylori* in asymptomatic infected subjects or those with early gastric cancer may reduce the risk of gastric cancer.

12.5.2.3 Ongoing Randomized Trials

There are four ongoing large randomized trials being conducted in China, Korea, and Latvia aiming to compare the efficacy of *H. pylori* eradication versus placebo or non-antibiotic treatment in the primary prevention of gastric cancer (Table 12.3). Another placebo controlled randomized trial in UK aimed to assess the efficacy of *H. pylori* eradication in the primary prevention of peptic ulcer bleeding in chronic aspirin users. All of these trials used the triple therapy or the bismuth quadruple therapy for *H. pylori* eradication. Another randomized trial from Taiwan compare the screening and eradication strategy for *H. pylori* versus no screening in the reduction of gastric cancer risk. It is expected that the results from these trials will provide more information on the primary prevention of gastric cancer through *H. pylori* eradication.

Table 12.3 Ongoing trials comparing the efficacy of *H. pylori* eradication on the risk of gastric cancer

Clinical trial registration number	Country	Subjects	Age (years)	Design	Experiment group	Control group	Estimated sample size	Primary outcome
NCT02047994	Latvia	Healthy *H. pylori* infected subjects	40–64	Open label	Triple therapy	No treatment	30,000	Gastric cancer mortality
NCT02112214	Korea	Healthy *H. pylori* infected subjects	40–60	Double blind	10-day Bismuth quadruple therapy	Placebo	11,000	Gastric cancer incidence
NCT01678027	Korea	Sibling or offspring of patients with gastric adenocarcinoma	40–65	Double blind	7-day triple therapy	Placebo	1810	Gastric cancer incidence
ChiCTR-TRC-10000979	China	Healthy residents in Linqu County	25–54	Double blind	10-day bismuth quadruple therapy	10-day bismuth + omeprazole + placebo	184,786	Gastric cancer incidence
NCT01506986	UK	*H. pylori* infected aspirin user	≥60	Double blind	7-day triple therapy	Placebo	33,000	Peptic ulcer bleeding
NCT01741363	Taiwan	Healthy subjects	50–75	Open label	*H. pylori* screening and FIT	FIT alone	40,000	Gastric cancer incidence

FIT fecal immunochemical test

12.6 Screening and Eradication of *H. pylori* for the Primary Prevention of Gastric Cancer in the Community

12.6.1 Three Different Study Designs of Trials Addressing on Different Issues

There are several important issues to be addressed regarding the primary prevention of gastric cancer (Fig. 12.4). The first issue is whether the risk of gastric cancer can be remarkably reduced in *H. pylori* infected subjects after *H. pylori* eradication therapy. Several randomized controlled trials have shown a reduction in the risk of gastric precancerous lesion and gastric cancer in *H. pylori* infected subject after eradication of *H. pylori* (Table 12.2). The second issue is whether the strategy of screening and eradication of *H. pylori* is feasible to reduce the risk of gastric cancer. However, none of the previous randomized trials addressed on this issue. The third issue is whether the incidence of gastric cancer in a population can be reduced after the implementation of mass screening and eradication for *H. pylori*. Randomized control trial is lacking on this issue. Yet, a prospective cohort study in Matsu Island in Taiwan has been reported [38].

12.6.2 Population-Based Mass Screening and Eradication Programs

Lee et al. conducted a mass screening and eradication program in the Matsu Island of Taiwan in 2004. *H. pylori* was positive in 2598 (63 %) of the 4121 participants [38]. A total of 1762 *H. pylori* infected subjects underwent endoscopy and biopsy and eradication therapy. The cumulative eradication rate after first line 7-day clarithromycin triple therapy and second line 10-day levofloxacin triple therapy was 97.7 % [54]. By 2008, the prevalence of *H. pylori* infection has been reduced to 11.2 % (94/841) [38]. The incidence of gastric atrophy was reduced by 61 %. The

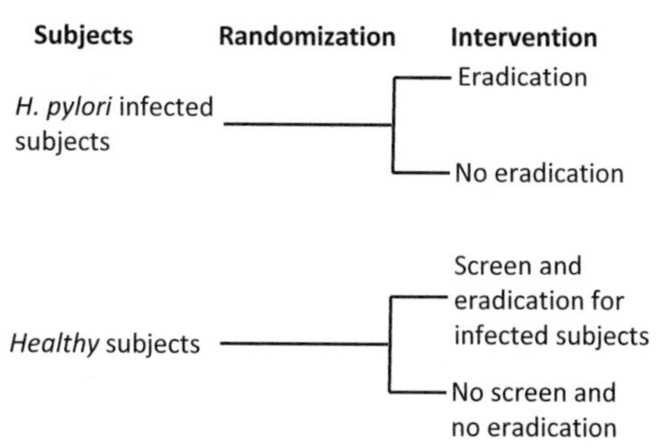

Fig. 12.4 Different designs in the interventional trials

Table 12.4 Factors to be considered for mass screening and eradication of *H. pylori*

Factors	Example 1	Example 2
Participation rate in screening	80 %	50 %
Accuracy of screening test	90 %	80 %
Prevalence of *H. pylori* infection	60 %	20 %
Participation rate in treatment	90 %	70 %
Eradication rate (ITT)	98 %	80 %
Reinfection rate	1 %/year	3 %/year
10 years (cumulative)	10 %	30 %
Prevalence of *H. pylori* 10 years later	25.8 %	17 %
% reduction in the prevalence of *H. pylori*	↓57 %	↓16 %

ITT intention-to-treat

prevalence of peptic ulcer disease was also reduced from 11 % to 3.6 % [38]. However, the prevalence of endoscopic esophagitis was increased from 13.7 % to 27.3 %. There was a trend of reduced incidence of gastric cancer, but the difference was not significant (risk ratio 0.75, 95 % CI 0.37–1.52) [38]. A longer follow-up period would be needed to reach the statistically significance.

12.6.3 *Factors to be Considered in the Mass Screening and Eradication Program*

Whether the strategy of screening and eradication of *H. pylori* infection may reduce the risk of gastric cancer in a region is affected by many factors, including the participation rate in the screening, the accuracy of the screening test, the prevalence of *H. pylori* infection in that region, the participation rate in the eradication therapy, the efficacy of the eradication regimen, and the reinfection (recurrence) rate in that region (Table 12.4). For example, in a region with high prevalence of *H. pylori* infection, the prevalence of *H. pylori* may be reduced by 57 % after the mass screening and eradication if the participation rate in screening and treatment are high, the regimen is effective, and the reinfection rate is low (example 1 in Table 12.4). In contrast, in a region with low prevalence of *H. pylori* infection, the prevalence could be reduced only by 16 % after the mass screening and eradication if the participation rate in screening and treatment are low, the regimen is less effective, and the reinfection rate is higher (example 2 in Table 12.4).

12.7 Regimen to be Used in the Primary Prevention of Gastric Cancer

As discussed in the previous section, regimens with high eradication rate and good compliance is anticipated in the mass eradication program. However, the intention-to-treat eradication rates in the trials conducted in the community were lower than 80 % in trials that used 7-day triple therapy or bismuth quadruple therapy (Table 12.5) [39–42, 46, 47, 54–58]. The efficacy of 14-day triple therapy was about 82 % in Latin America and in Taiwan [56–58]. This indicated that longer treatment duration, less complex and better tolerated regimens might be needed in the asymptomatic *H. pylori* infected subjects in the community. Nevertheless, more randomized trials are warranted to find out the most effective and well tolerated regimen in the community. Development of effective rescue regimens is also needed [58].

12.8 Reinfection or Recrudesces After Successful Eradication of *H. pylori*

Recurrence (reinfection or recrudesces) of *H. pylori* infection after eradication therapy is also an important issue. Reinfection is defined as infection by a new *H. pylori* strain after confirmation of successful eradication by tests that detect active *H. pylori* infection, such as urea breath test. Recrudesces is defined as reactivation of the same strains which became undetected after eradication therapy. Take et al. followed 1609 patients who had received successful eradication therapy for 4.7 years. *H. pylori* became again in 26 patients and 13 (50 %) of them became infected again within the first year [59]. The crude annual reinfection rate was 0.22 % per year. The *H. pylori* strains before eradication therapy and after reinfection or recrudesces were analyzed by random amplification of polymorphic DNA fingerprinting. Of the ten paired strains in patients who became infected again in the first year, 40 % (4/10) of the strains were different from the initial strains (reinfection), whereas another 60 % (6/10) strains were identical with the initial strains (recrudesces) [59]. Of the four paired strains in patients who became infected again after the first year, all of these four strains were different from the initial strains (reinfection). These collectively indicated that half the reinfection or recrudesces occurs in the first year. Of those whose *H. pylori* became positive again within the first year, 40 % was due to reinfection and 60 % was due to recrudesces [59]. Of those whose *H. pylori* became positive after the first year, almost 100 % was due to reinfection of new strains.

The annual reinfection and recrudesces rates vary greatly among different studies and countries [38, 60–62]. Factors that might affect the reinfection rate include the prevalence of *H. pylori* infection in that population, the hygiene status, and the socioeconomic status [38, 60–62]. The probability of recrudesces might be

Table 12.5 The eradication rates of *H. pylori* eradication regimens in the community

Study	Country	Regimen	Duration	Eradication rate
Correa et al. [39]	Columbia	Bismuth/amoxicillin/metronidazole	2 weeks	52.3 % (157/300)
Sung et al. [40]	China	Omeprazole/amoxicillin/clarithromycin	1 week	82.3 % (243/295)
Wong et al. [42]	China	Omeprazole/co-amoxiclav/metronidazole; Retreatment: bismuth/omeprazole /metronidazole/clarithromycin	2 weeks 1 week	1st: 76.4 % (624/817) 2nd: 70.6 % (60/85) Overall: 83.7 % (684/817)
Saito et al. [46]	Japan	Lansoprazole/amoxicillin/clarithromycin	1 week	85.2 % (282/331)
You et al. [41]	China	Omeprazole/amoxicillin; Retreatment: Omeprazole/amoxicillin	2 weeks; 2 weeks	1st: 62 % (703/1130); 2nd: 32.5 % (124/382); Overall: 73.2 % (827/1130)
Wong et al. [47]	China	Omeprazole/amoxicillin/clarithromycin	1 week	63.3 % (323/510)
Lee et al. [54]	Taiwan	Esomeprazole/amoxicillin/clarithromycin; Retreatment: esomeprazole/amoxicillin / levofloxacin	1 week; 10 days	1st: 86.9 % (770/886); 2nd: 91.4 % (96/105) Overall: 97.7 % (866/886)
Greenberg et al. [55]	Latin America	Lansoprazole/amoxicillin/clarithromycin	2 weeks	82.2 % (401/488)
		Lansoprazole/amoxicillin/clarithromycin/ metronidazole (concomitant)	5 days	73.6 % (360/489)
		5 days of lansoprazole and amoxicillin followed by 5 days of lansoprazole, clarithromycin, and metronidazole (sequential therapy)	10 days	76.5 % (372/486)
Pan et al. [56]	China	Bismuth/omeprazole/metronidazole/ tetracycline	10 days	72.9 % (32336/ 44345)
Liou et al. [57]	Taiwan	Lansoprazole/amoxicillin/clarithromycin	2 weeks	82.2 % (213/259)
		5 days of lansoprazole and amoxicillin followed by 5 days of lansoprazole, clarithromycin, and metronidazole (sequential therapy)	10 days	85.3 % (220/258)

attributed to the efficacy of the eradication regimen and the compliance of patients [38, 60–62]. Yet, reinfection or recrudesces could not be differentiated in almost all of the epidemiological studies that reported the reinfection/recrudesces rate due to the lack of strains before and after reinfection or recrudesces. Niv et al. searched the PubMed database up to 2007 and identified 10 and 7 prospective studies addressing on this issue in developed and developing countries, respectively [60]. Meta-analysis of these studies revealed that the annual recurrence rates were 2.7 % and 13 % in developed and developing countries, respectively [60]. In another systemic review including 77 eligible studies and a total of 43,525 follow-up patient-years after successful eradication therapy, recurrence of *H. pylori* infection occurred in 1226 cases [61]. They further showed a correlation of recurrence rate with national Human Development Index (HDI). The annual recurrence rates in countries with very high HDI, high HDI, medium HDI, and low HDI were 1.7 %, 6.1 %, 7.0 %, and 9.6 %, respectively [61].

However, most of the included studies were hospital-based researches and relatively little is known about the recurrence rate in the asymptomatic populations in the community. Recently, a randomized trial comparing the efficacy of 14-day triple therapy, 10-day sequential therapy, and 5-day concomitant therapy in asymptomatic subjects in 7 Latin American communities showed that the recurrence rate 1 year after eradication therapy was as high as 11.5 % of participants who had negative posttreatment UBT result [62]. In a community-based screening and treatment program for gastric cancer prevention in Taiwan, Lee et al. reported that the annual recurrence rate was about 1 % in asymptomatic subjects treated with 7-day triple therapy in Matsu Island [38]. Future studies to identify and block the routes of reinfection, especially in developing countries, are warranted.

12.9 Future Prospects

Elimination of this bacterium from human stomach may prevent the development of gastric cancer. Although improvement in the hygiene may reduce the prevalence of *H. pylori* infection, the implementation of screening and eradication program might hasten the reduction in the prevalence of this infection. The results from the ongoing trials on this issue are highly anticipated. Yet, although mass screening and eradication of *H. pylori* is a promising strategy to prevent gastric cancer, there are some concerns which might limit the application of such program. The most concerned issue is the potential emergence of antibiotic resistance in various bacteria in the community [63, 64]. However, very few studies have addressed on this important issue. In a small cohort study, Sjölund et al. showed a persistence of clarithromycin resistant Enterococcus in three of the five patients 3 years after *H. pylori* eradication [65]. Future randomized trials are needed to clarify the long term impact of the short term *H. pylori* eradication therapy on the antibiotic resistance. The second concern is the development or exacerbation of other diseases, such as gastroesophageal reflux disease, allergic disease, and obesity [64, 66,

67]. However, contradictory results have been reported. Well-designed randomized trials are warranted to clarify these issues. The third concern is the substantial cost of such kind of program, although it has been reported to be cost-effective in areas with high risk of gastric cancer [68–70].

12.10 Conclusion

Ecological studies showed higher cumulative incidence of gastric cancer in countries with higher prevalence of *H. pylori* infection. Meta-analysis of case–control studies nested within prospective cohort showed that *H. pylori* infection was associated with a 5.9-fold increased risk of non-cardia gastric cancer using blood samples collected more than 10 years. *H. pylori* was associated with increased risk of both diffuse type and intestinal type gastric cancer. Cag-A seropositivity was associated with higher risk of gastric cancer. Prospective cohort studies showed that gastric cancer developed in 1–3 % of *H. pylori* infected subjects. Recent meta-analysis of randomized control trials showed a significant reduction in the risk of gastric cancer in *H. pylori* infected subjects who received *H. pylori* eradication therapy. Gastric cancer was successfully induced in Mongolian gerbils and INS-GAS transgenic mice after inoculation of *H. pylori*. The incidence of gastric dysplasia and gastric cancer could be reduced in mice treated with eradication therapy. Based on these evidences, it is well agreed that *H. pylori* is a causal risk factor of gastric cancer. However, large well-designed randomized trials are highly anticipated to assess the effectiveness of the screen and treat strategy as well as the changes in the antibiotic resistance and risks in the development of gastroesophageal reflux disease, obesity, and allergic diseases after *H. pylori* eradication.

References

1. World Health Organization Internal Association of Cancer Registries. The Globocan register. http://globocan.iarc.fr/Pages/age-specific_table_sel.aspx
2. Marshall BJ, Warren JR. Unidentified curved bacilli in the stomach of patients with gastritis and peptic ulceration. Lancet. 1984;1:1311–5.
3. Kuipers EJ, Lundell L, Klinkenberg-Knol EC, Havu N, Festen HP, Liedman B, et al. Atrophic gastritis and *Helicobacter pylori* infection in patients with reflux esophagitis treated with omeprazole or fundoplication. N Engl J Med. 1996;334:1018–22.
4. Parsonnet J, Hansen S, Rodriguez L, Gelb AB, Warnke RA, Jellum E, et al. *Helicobacter pylori* infection and gastric lymphoma. N Engl J Med. 1994;330:1267–71.
5. Kuipers EJ, Uyterlinde AM, Peña AS, Roosendaal R, Pals G, Nelis GF, et al. Long-term sequelae of *Helicobacter pylori* gastritis. Lancet. 1995;345:1525–8.
6. An international association between *Helicobacter pylori* infection and gastric cancer. The EUROGAST Study Group. Lancet. 1993;341:1359–62.
7. Miwa H, Go MF, Sato N. *H. pylori* and gastric cancer: the Asian enigma. Am J Gastroenterol. 2002;97:1106–12.

8. Leung WK, Wu MS, Kakugawa Y, Kim JJ, Yeoh KG, Goh KL, et al. Screening for gastric cancer in Asia: current evidence and practice. Lancet Oncol. 2008;9:279–87.
9. Hansson LE, Engstrand L, Nyrén O, Evans Jr DJ, Lindgren A, Bergström R, et al. *Helicobacter pylori* infection: independent risk indicator of gastric adenocarcinoma. Gastroenterology. 1993;105:1098–103.
10. Huang JQ, Sridhar S, Chen Y, Hunt RH. Meta-analysis of the relationship between *Helicobacter pylori* seropositivity and gastric cancer. Gastroenterology. 1998;114:1169–79.
11. Huang JQ, Zheng GF, Sumanac K, Irvine EJ, Hunt RH. Meta-analysis of the relationship between cagA seropositivity and gastric cancer. Gastroenterology. 2003;125:1636–44.
12. Nomura A, Stemmermann GN, Chyou PH, Kato I, Perez-Perez GI, Blaser MJ. *Helicobacter pylori* infection and gastric carcinoma among Japanese Americans in Hawaii. N Engl J Med. 1991;325:1132–6.
13. Parsonnet J, Friedman GD, Vandersteen DP, Chang Y, Vogelman JH, Orentreich N, et al. *Helicobacter pylori* infection and the risk of gastric carcinoma. N Engl J Med. 1991;325:1127–31.
14. Helicobacter and Cancer Collaborative Group. Gastric cancer and *Helicobacter pylori*: a combined analysis of 12 case control studies nested within prospective cohorts. Gut. 2001;49:347–53.
15. Forman D, Newell DG, Fullerton F, Yarnell JW, Stacey AR, Wald N, et al. Association between infection with *Helicobacter pylori* and risk of gastric cancer: evidence from a prospective investigation. BMJ. 1991;302:1302–5.
16. Wald NJ, Law MR, Morris JK, Bagnall AM. *Helicobacter pylori* infection and mortality from ischaemic heart disease: negative result from a large, prospective study. BMJ. 1997;315:1199–201.
17. Lin JT, Wang LY, Wang JT, Wang TH, Yang CS, Chen CJ. A nested case–control study on the association between *Helicobacter pylori* infection and gastric cancer risk in a cohort of 9775 men in Taiwan. Anticancer Res. 1995;15:603–6.
18. Aromaa A, Kosunen TU, Knekt P, Maatela J, Teppo L, Heinonen OP, et al. Circulating anti-*Helicobacter pylori* immunoglobulin A antibodies and low serum pepsinogen I level are associated with increased risk of gastric cancer. Am J Epidemiol. 1996;144:142–9.
19. Webb PM, Yu MC, Forman D, Henderson BE, Newell DG, Yuan JM, et al. An apparent lack of association between *Helicobacter pylori* infection and risk of gastric cancer in China. Int J Cancer. 1996;67:603–7.
20. Yuan JM, Yu MC, Xu WW, Cockburn M, Gao YT, Ross RK. *Helicobacter pylori* infection and risk of gastric cancer in Shanghai, China: updated results based upon a locally developed and validated assay and further follow-up of the cohort. Cancer Epidemiol Biomarkers Prev. 1999;8:621–4.
21. Simán JH, Forsgren A, Berglund G, Florén CH. Association between *Helicobacter pylori* infection and gastric carcinoma in the city of Malmo, Sweden. A prospective study. Scand J Gastroenterol. 1997;32:1215–21.
22. Watanabe Y, Kurata JH, Mizuno S, Mukai M, Inokuchi H, Miki K, et al. *Helicobacter pylori* infection and gastric cancer. A nested case–control study in a rural area of Japan. Dig Dis Sci. 1997;42:1383–7.
23. Hansen S, Melby KK, Aase S, Jellum E, Vollset SE. *Helicobacter pylori* infection and risk of cardia cancer and non-cardia gastric cancer. A nested case–control study. Scand J Gastroenterol. 1999;34:353–60.
24. Limburg P, Qiao Y, Mark S, Wang G, Perez-Perez G, Blaser M, et al. *Helicobacter pylori* seropositivity and subsite-specific gastric cancer risks in Linxian, China. J Natl Cancer Inst. 2001;93:226–33.
25. Shin A, Shin HR, Kang D, Park SK, Kim CS, Yoo KY. A nested case–control study of the association of *Helicobacter pylori* infection with gastric adenocarcinoma in Korea. Br J Cancer. 2005;92:1273–5.

26. Kamangar F, Dawsey SM, Blaser MJ, Perez-Perez GI, Pietinen P, Newschaffer CJ, et al. Opposing risks of gastric cardia and noncardia gastric adenocarcinomas associated with *Helicobacter pylori* seropositivity. J Natl Cancer Inst. 2006;98:1445–52.
27. Sasazuki S, Inoue M, Iwasaki M, Otani T, Yamamoto S, Ikeda S, et al. Effect of *Helicobacter pylori* infection combined with CagA and pepsinogen status on gastric cancer development among Japanese men and women: a nested case–control study. Cancer Epidemiol Biomarkers Prev. 2006;15:1341–7.
28. Palli D, Masala G, Del Giudice G, Plebani M, Basso D, Berti D, et al. CagA+ *Helicobacter pylori* infection and gastric cancer risk in the EPIC-EURGAST study. Int J Cancer. 2007;120:859–67.
29. Hansson LE, Nyrén O, Hsing AW, Bergström R, Josefsson S, Chow WH, et al. The risk of stomach cancer in patients with gastric or duodenal ulcer disease. N Engl J Med. 1996;335:242–9.
30. Uemura N, Okamoto S, Yamamoto S, Matsumura N, Yamaguchi S, Yamakido M, et al. *Helicobacter pylori* infection and the development of gastric cancer. N Engl J Med. 2001;345:784–9.
31. Hsu P, Lai KH, Hsu PN, Lo GH, Yu HC, Chen WC, et al. *Helicobacter pylori* infection and the risk of gastric malignancy. Am J Gastroenterol. 2007;102(4):725–30.
32. Watanabe T, Tada M, Nagai H, Sasaki S, Nakao M. *Helicobacter pylori* infection induces gastric cancer in mongolian gerbils. Gastroenterology. 1998;115:642–8.
33. Wang TC, Dangler CA, Chen D, Goldenring JR, Koh T, Raychowdhury R, et al. Synergistic interaction between hypergastrinemia and Helicobacter infection in a mouse model of gastric cancer. Gastroenterology. 2000;118:36–47.
34. Lee CW, Rickman B, Rogers AB, Ge Z, Wang TC, Fox JG. *Helicobacter pylori* eradication prevents progression of gastric cancer in hypergastrinemic INS-GAS mice. Cancer Res. 2008;68:3540–8.
35. Romero-Gallo J, Harris EJ, Krishna U, Washington MK, Perez-Perez GI, Peek Jr RM. Effect of *Helicobacter pylori* eradication on gastric carcinogenesis. Lab Investig. 2008;88:328–36.
36. Lee CW, Rickman B, Rogers AB, Muthupalani S, Takaishi S, Yang P, et al. Combination of sulindac and antimicrobial eradication of *Helicobacter pylori* prevents progression of gastric cancer in hypergastrinemic INS-GAS mice. Cancer Res. 2009;69:8166–74.
37. Ohkusa T, Fujiki K, Takashimizu I, Kumagai J, Tanizawa T, Eishi Y, et al. Improvement in atrophic gastritis and intestinal metaplasia in patients in whom *Helicobacter pylori* was eradicated. Ann Intern Med. 2001;134:380–6.
38. Lee YC, Chen TH, Chiu HM, Shun CT, Chiang H, Liu TY, et al. The benefit of mass eradication of *Helicobacter pylori* infection: a community-based study of gastric cancer prevention. Gut. 2013;62:676–82.
39. Correa P, Fontham ETH, Bravo JC, Bravo LE, Ruiz B, Zarama G, et al. Chemoprevention of gastric dysplasia: randomized trial of antioxidant supplements and anti-*Helicobacter pylori* therapy. J Natl Cancer Inst. 2000;92:1881–8.
40. Sung JJY, Lin S-R, Ching JYL, Zhou L-Y, To KF, Wang R-T, et al. Atrophy and intestinal metaplasia one year after cure of *H. pylori* infection: a prospective, randomized study. Gastroenterology. 2000;119:7–14.
41. You W-C, Brown LM, Zhang L, Li J-Y, Jin M-L, Chang Y-S, et al. Randomized double-blind factorial trial of three treatments to reduce the prevalence of precancerous gastric lesions. J Natl Cancer Inst. 2006;98:974–83.
42. Wong BCY, Lam SK, Wong WM, Chen JS, Zheng TT, Feng RE, et al. *Helicobacter pylori* eradication to prevent gastric cancer in a high-risk region of China: a randomized controlled trial. JAMA. 2004;291:187–94.
43. Wu CY, Kuo KN, Wu MS, Chen YJ, Wang CB, Lin JT. Early *Helicobacter pylori* eradication decreases risk of gastric cancer in patients with peptic ulcer disease. Gastroenterology. 2009;137:1641–8.e1-2.

44. Leung WK, Lin S-R, Ching JYL, To K-F, Ng EKW, Chan FKL, et al. Factors predicting progression of gastric intestinal metaplasia: results of a randomised trial on *Helicobacter pylori* eradication. Gut. 2004;53:1244–9.
45. Zhou L. Ten-year follow-up study on the incidence of gastric cancer and the pathological changes of gastric mucosa after *H. pylori* eradication in China. Gastroenterology. 2008;134 suppl 1:A233.
46. Saito D, Boku N, Fujioka T, Fukuda Y, Matsushima Y, Sakaki N, et al. Impact of *H. pylori* eradication on gastric cancer prevention: endoscopic results of the Japanese intervention trial (JITHP-Study): a randomized multi-center trial. Gastroenterology. 2005;128 suppl 2:A4.
47. Wong BCY, Zhang L, Ma J-L, Pan K-F, Li J-Y, Shen L, et al. Effects of selective COX-2 inhibition and *Helicobacter pylori* eradication on precancerous gastric lesions. Gut. 2012;61:812–8.
48. Ma J-L, Zhang L, Brown LM, Li J-Y, Shen L, Pan K-F, et al. Fifteen-year effects of *Helicobacter pylori*, garlic, and vitamin treatments on gastric cancer incidence and mortality. J Natl Cancer Inst. 2012;104:488–92.
49. Fuccio L, Zagari RM, Eusebi LH, Laterza L, Cennamo V, Ceroni L, et al. Meta-analysis: can *Helicobacter pylori* eradication treatment reduce the risk for gastric cancer? Ann Intern Med. 2009;151:121–8.
50. Fischbach LA, Correa P, Ramirez H, Realpe JL, Collazos T, Ruiz B, et al. Anti-inflammatory and tissue-protectant drug effects: results from a randomized placebo-controlled trial of gastritis patients at high risk for gastric cancer. Aliment Pharmacol Ther. 2001;15:831–41.
51. Miehlke S, Kirsch C, Dragosics B, Gschwantler M, Oberhuber G, Antos D, et al. *Helicobacter pylori* and gastric cancer: current status of the Austrian Czech German gastric cancer prevention trial (PRISMA-study). World J Gastroenterol. 2001;7:243–7.
52. Fukase K, Kato M, Kikuchi S, Inoue K, Uemura N, Okamoto S, et al. Effect of eradication of *Helicobacter pylori* on incidence of metachronous gastric carcinoma after endoscopic resection of early gastric cancer: an open-label, randomised controlled trial. Lancet. 2008;372:392–7.
53. Ford AC, Forman D, Hunt RH, Yuan Y, Moayyedi P. *Helicobacter pylori* eradication therapy to prevent gastric cancer in healthy asymptomatic infected individuals: systematic review and meta-analysis of randomised controlled trials. BMJ. 2014;348:g3174.
54. Lee YC, Wu HM, Chen TH, Liu TY, Chiu HM, Chang CC, et al. A community-based study of *Helicobacter pylori* therapy using the strategy of test, treat, retest, and re-treat initial treatment failures. Helicobacter. 2006;11:418–24.
55. Greenberg ER, Anderson GL, Morgan DR, Torres J, Chey WD, Bravo LE, et al. 14-day triple, 5-day concomitant, and 10-day sequential therapies for *Helicobacter pylori* infection in seven Latin American sites: a randomised trial. Lancet. 2011;378:507–14.
56. Liou JM, Chen CC, Chang CY, Chen MJ, Chen CC, Fang YJ, et al. Sequential therapy for 10 days versus triple therapy for 14 days in the eradication of *Helicobacter pylori* in the community and hospital populations: a randomised trial. Gut. 2015. doi:10.1136/gutjnl-2015-310142.
57. Pan KF, Zhang L, Gerhard M, Ma JL, Liu WD, Ulm K, et al. A large randomised controlled intervention trial to prevent gastric cancer by eradication of *Helicobacter pylori* in Linqu County, China: baseline results and factors affecting the eradication. Gut. 2015. doi:10.1136/gutjnl-2015-309197.
58. Liou JM, Chen CC, Chang CY, Chen MJ, Fang YJ, Lee JY, et al. Efficacy of genotypic resistance-guided sequential therapy in the third-line treatment of refractory *Helicobacter pylori* infection: a multicentre clinical trial. J Antimicrob Chemother. 2013;68:450–6.
59. Take S, Mizuno M, Ishiki K, Imada T, Okuno T, Yoshida T, et al. Reinfection rate of *Helicobacter pylori* after eradication treatment: a long-term prospective study in Japan. J Gastroenterol. 2012;47:641–6.
60. Niv Y, Hazazi R. *Helicobacter pylori* recurrence in developed and developing countries: meta-analysis of 13C-urea breath test follow-up after eradication. Helicobacter. 2008;13(1):56–61.

61. Yan TL, Hu QD, Zhang Q, Li YM, Liang TB. National rates of *Helicobacter pylori* recurrence are significantly and inversely correlated with human development index. Aliment Pharmacol Ther. 2013;37:963–8.
62. Morgan DR, Torres J, Sexton R, Herrero R, Salazar-Martínez E, Greenberg ER, et al. Risk of recurrent *Helicobacter pylori* infection 1 year after initial eradication therapy in 7 Latin American communities. JAMA. 2013;309:578–86.
63. Malfertheiner P, Sipponen P, Naumann M, Moayyedi P, Mégraud F, Xiao SD, et al. *Helicobacter pylori* eradication has the potential to prevent gastric cancer: a state-of-the-art critique. Am J Gastroenterol. 2005;100:2100–15.
64. Malfertheiner P, Megraud F, O'Morain CA, Atherton J, Axon AT, Bazzoli F, et al. Management of *Helicobacter pylori* infection – the Maastricht IV/ Florence Consensus Report. Gut. 2012;61:646–64.
65. Sjölund M, Wreiber K, Andersson DI, Blaser MJ, Engstrand L. Long-term persistence of resistant Enterococcus species after antibiotics to eradicate *Helicobacter pylori*. Ann Intern Med. 2003;139:483–7.
66. Cover TL, Blaser MJ. *Helicobacter pylori* in health and disease. Gastroenterology. 2009;136:1863–73.
67. Atherton JC, Blaser MJ. Coadaptation of *Helicobacter pylori* and humans: ancient history, modern implications. J Clin Invest. 2009;119:2475–87.
68. Ford AC, Forman D, Bailey AG, Axon ATR, Moayyedi P. A community screening program for *Helicobacter pylori* saves money: ten-year follow-up of a randomised controlled trial. Gastroenterology. 2005;129:1910–7.
69. Lee YC, Lin JT, Wu HM, Liu TY, Yen MF, Chiu HM, et al. Cost-effectiveness analysis between primary and secondary preventive strategies for gastric cancer. Cancer Epidemiol Biomarkers Prev. 2007;16:875–85.
70. Parsonnet J, Harris RA, Hack HM, Owens DK. Modelling cost-effectiveness of *Helicobacter pylori* screening to prevent gastric cancer: a mandate for clinical trials. Lancet. 1996;348:150–4.

Part IV
Treatment

Chapter 13
Trends in Global Eradication Rates

Makoto Sasaki

Abstract *Helicobacter pylori* (*H. pylori*) eradication eliminates gastritis and prevents gastric cancer and recurrence of peptic ulcers. Therefore, curative treatment of *H. pylori* infection is widely used worldwide. Effective treatment is defined as the achievement of an eradication rate ≥90 % with the first per-protocol treatment. Standard triple therapy with a proton pump inhibitor and amoxicillin, clarithromycin, or metronidazole has been widely used globally and was associated with a cure rate of approximately >90 % when the treatment was first used. However, the eradication rate of *H. pylori* has been declining to unacceptable levels due to antibiotic resistance, and the efficacy of standard triple therapy has been gradually declining in most countries. Although quadruple and sequential therapies are appropriate regimens for initial treatment, these therapies require multiple antibiotic drugs, which might increase multiple antibiotic resistance and reduce the choice of other antibiotic drugs when eradication fails. Eradication success depends on patient factors and the nature of the bacterium itself, such as antibiotic susceptibility, which is the most common factor underlying treatment success. First-line treatment for *H. pylori* infection should be recommended on the basis of an understanding of the local prevalence of antimicrobial resistance and human species.

Keywords Triple therapy • Eradication rate • Antibiotic resistance • *Helicobacter pylori*

13.1 Guidelines for First-Line Treatment of *Helicobacter pylori*

Starting with the Maastricht conference in 1996, discussions regarding the clinical management of *Helicobacter pylori* (*H. pylori*) infection have been conducted, and guidelines or consensus statements have been established or updated globally (Table 13.1). The optimal eradication regimen for *H. pylori* should combine good

M. Sasaki (✉)
Department of Gastroenterology, Aichi Medical University School of Medicine, 1-1 Yazakokarimata, 480-1195 Nagakute, Aichi, Japan
e-mail: msasaki@aichi-med-u.ac.jp

H. Suzuki et al. (eds.), *Helicobacter pylori*, DOI 10.1007/978-4-431-55705-0_13

Table 13.1 Recommended guidelines for *Helicobacter pylori* eradication

Region	First-line therapy	Second-line (salvage) therapy
USA (2007)	Clarithromycin-containing triple therapy for 14 days	Bismuth-containing quadruple therapy for 7–10 days
	Bismuth-containing quadruple therapy for 10–14 days	Levofloxacin-containing triple therapy for 10 days
	Sequential therapy for 10 days	
European Union (2012)	In areas of <20 % clarithromycin resistance:	
	Clarithromycin-containing triple therapy for 10–14 days	Bismuth-containing quadruple therapy for 10–14 days
	Bismuth-containing quadruple therapy for 10–14 days	Levofloxacin-containing triple therapy for 10 days
	In areas of >20 % clarithromycin resistance:	
	Sequential therapy for 10 days	Levofloxacin-containing therapy for 10 days
	Bismuth-containing quadruple therapy for 10–14 days	
	Non-bismuth quadruple therapy for 3–10 days	
Japan (2010)	Clarithromycin-containing triple therapy for 7 days	Metronidazole-containing triple therapy for 7 days
Korea (2014)	Clarithromycin-containing triple therapy for 10–14 days	Bismuth-containing quadruple therapy for 7–14 days
	Bismuth-containing quadruple therapy for 7–14 days	Regimen including ≥2 other antibiotics
China (2013)	Bismuth-containing quadruple therapy for 10–14 days	Bismuth-containing quadruple therapy for 10–14 days

outcomes achieved over a short duration with a favorable side-effect profile, all at low cost. Concerning the first-line treatment of *H. pylori* infection, standard triple therapy regimens consisting of one proton pump inhibitor (PPI) and two antibiotic agents such as amoxicillin, clarithromycin, or metronidazole have been widely used for over a decade. A treatment success rate ≥90 % is generally desirable for bacterial infections, and standard triple therapy started with an acceptable eradication rate. However, similar to other bacterial infections, the rate of eradication tends to gradually decrease with increased chemical tolerance of the bacteria. Because triple therapy is limited by resistance to clarithromycin or metronidazole, other therapeutic regimens have been investigated worldwide. To increase the eradication rate, several strategies have been proposed, including longer treatment durations of 10 or 14 days, increases in the number of drugs (quadruple, sequential, or concomitant treatments), and the use of novel antibiotics, such as levofloxacin. Regimens that achieve higher first-cure rates have recently been devised based on consideration of the geographic regions. However, new therapies are not generally accepted as first-line treatment in several countries because of a lack of national validation studies, lack of studies regarding antibiotic resistance, limited drug use, and limitations of health-care systems.

13.1.1 First-Line and Salvage Therapies for Helicobacter pylori *in the USA*

In the USA, the American College of Gastroenterology guidelines recommend clarithromycin-based triple therapy comprised of a PPI (standard dose twice a day), clarithromycin (500 mg twice a day), and amoxicillin (1 g twice a day) or metronidazole (400 or 500 mg twice a day) for first-line therapy of *H. pylori* infections [1]. Based on a meta-analysis, a better eradication rate was achieved with 14-day therapy compared with 7-day therapy; furthermore, equivalent eradication rates were observed with 7- or 10-day therapy in a large-scale trial. Therefore, treatment lasting 14 days is recommended. Bismuth-containing quadruple therapy with a PPI (standard dose once or twice a day) or H_2-receptor antagonist (H_2RA) (standard dose twice a day), bismuth subsalicylate (525 mg four times a day), metronidazole (250 mg four times a day), and tetracycline (500 mg four times a day) for 10–14 days is also recommended but requires validation within the USA before it can be recommended as first-line therapy. If this regimen is not available, sequential treatment is recommended with a PPI (standard dose twice a day) and amoxicillin (1 g twice a day) for the first 5 days, followed by a PPI (standard dose twice a day), clarithromycin (500 mg twice a day), and tinidazole (500 mg twice a day) for the next 5 days. Recommended salvage therapy includes bismuth-containing quadruple therapy comprised of a PPI, bismuth subsalicylate, metronidazole, and tetracycline for 7–14 days or levofloxacin-containing triple therapy consisting of a PPI, amoxicillin (1 g twice a day), and levofloxacin (500 mg once a day) for 10 days.

13.1.2 Treatment Strategies for Helicobacter pylori *in Europe*

In Europe, the Maastricht IV/Florence Consensus Report recommends that the regimen considers the clarithromycin resistance rate of the area, with regions of high or low resistance identified by a threshold of 15–20 %. In areas with low clarithromycin resistance, clarithromycin-based triple therapy containing a PPI, clarithromycin, and amoxicillin or metronidazole is recommended for first-line empirical treatment [2]. Treatment lasting for 10–14 days is accepted because of the superior rate of *H. pylori* eradication compared with 7 days. Bismuth-containing quadruple therapy consisting of a PPI or H_2RA, bismuth subsalicylate, metronidazole, and tetracycline for 10–14 days is also recommended in both high and low clarithromycin resistance areas. If bismuth-containing quadruple therapy is not available in areas with high clarithromycin resistance, sequential treatment or non-bismuth-containing quadruple treatment is recommended. Bismuth-containing quadruple therapy or levofloxacin-containing triple therapy for 10 days is also recommended as second-line therapy if clarithromycin-containing triple therapy fails. In areas with high clarithromycin resistance, levofloxacin-containing triple therapy is recommended as salvage therapy.

13.1.3 Recommended Therapies for Helicobacter pylori *Infection in Asia*

In Japan, the Japanese Society for *Helicobacter* Research recommends 7-day triple therapy using a PPI (standard dose twice a day; omeprazole, lansoprazole, rabeprazole, or esomeprazole), amoxicillin (750 mg twice a day), and clarithromycin (200 or 400 mg twice a day); this is the only regimen currently covered by the Japanese national health insurance system. Triple therapy comprised of amoxicillin, clarithromycin, and lansoprazole or rabeprazole is available as a one-sheet tablet prepared to improve drug compliance. Metronidazole-containing salvage therapy for failure to eradicate *H. pylori* with clarithromycin-containing triple therapy consists of 7-day triple therapy using a PPI (standard dose twice a day), amoxicillin (750 mg twice a day), and metronidazole (250 mg twice a day) [3].

In Korea, the Korean College of *Helicobacter* and Upper Gastrointestinal Research recommends triple therapy including a standard dose of PPI, 1 g amoxicillin, and 500 mg clarithromycin twice a day for 7–14 days. With suspected resistance to clarithromycin, quadruple therapy including two standard doses of PPI, three doses of 500 mg metronidazole, four doses of 120 mg bismuth, and four doses of 500 mg tetracycline daily for 7–14 days is the recommended alternative primary therapy. When the conventional triple therapy fails, bismuth-containing quadruple therapy is recommended as the secondary regimen. In case of eradication failure with first-line bismuth-containing quadruple therapy, a secondary regimen including two or more antibiotics that were not used in the primary regimen is recommended [4].

In China, the Fourth Chinese National Consensus Report recommends bismuth-containing quadruple therapy for 10 or 14 days with bismuth (standard dose twice a day), a PPI (standard dose twice a day), and two antibiotic drugs (twice a day), chosen from one of the following four combinations: (1) amoxicillin (1 g twice a day) + clarithromycin (500 mg twice a day), (2) amoxicillin (1 g twice a day) + levofloxacin (500 mg once a day or 200 mg twice a day), (3) amoxicillin (1 g twice a day) + furazolidone (100 mg twice a day), or (4) tetracycline (750 mg twice a day) + metronidazole (400 mg twice or three times a day) or furazolidone (100 mg twice a day). In China, a change in the medication choice and regimen to improve the eradication rate is recommended. For example, in a case with amoxicillin hyperesthesia, clarithromycin + levofloxacin/furazolidone/metronidazole or tetracycline + metronidazole/furazolidone is recommended, and the use of standard triple, sequential, or concomitant therapies is approved for bismuth intolerance and in areas with a low rate of antibiotic tolerance. Because of the insufficient data concerning concomitant therapy in China, it is difficult to determine first- and second-line therapies. When the initial treatment fails, the salvage therapy can be chosen from the remaining regimens [5].

13.2 Factors Influencing *Helicobacter pylori* Treatment

There are a number of factors underlying the successful treatment of *H. pylori* infections, including the nature of the bacterium itself, intragastric environment where the bacterium resides, regimens used to eradicate the bacterium, and behavior and reactions of the host. In addition, success depends on patient factors such as age, sex, body mass index (BMI), smoking status, drug compliance, CYP2C19 genotype, intragastric acidity, and bacterial antibiotic susceptibility, which is the most common factor for treatment success [6].

13.2.1 Smoking

Smoking is an important factor for a successful eradication therapy. In a meta-analysis of 22 studies involving 5538 patients, the odds ratio (OR) for eradication failure in smokers was 1.95 (95 % confidence interval [CI], 1.55–2.45; $P < 0.01$), indicating an approximate twofold higher probability of eradication failure [7]. Possible mechanisms include decreased gastric blood flow and mucus secession, which might reduce the delivery of antibiotics to the gastric mucosa. In addition, smoking stimulates acid secretion, which has been associated with treatment failure, and smoking might be associated with reduced compliance [7]. Therefore, smoking cessation during *H. pylori* therapy may improve eradication rates among smokers; treatment efficacy was similar between smokers who stopped smoking during eradication therapy and nonsmokers, whereas those who continued smoking experienced poorer results. An 8.4 % increase in the eradication rate was observed with the cessation of smoking [7, 8].

13.2.2 CYP2C19 Polymorphisms

PPIs result in more stable acid-sensitive antibiotics and increase the concentration of antibiotics in the gastric juice. In a neutral gastric condition, *H. pylori* actively propagate and become more sensitive to antibiotics, especially clarithromycin and amoxicillin [9]. Therefore, PPIs play a crucial role in *H. pylori* eradication therapy. PPIs are substitutes for benzimidazole and are commonly metabolized by hepatic cytochrome P450 enzymes, especially the CYP2C19 genotype. The gene encoding CYP2C19 is polymorphic, and various mutations have been described in different ethnic groups. Based on polymorphism of the CYP2C19 genotype, it is possible to classify individuals into three distinct phenotypes that influence drug metabolism and thereby affect the pharmacodynamics of PPIs: wild-type, homozygous extensive metabolizer (HomEM), which is the most common genotype and consists of two normal alleles with normal enzyme activity; heterozygous EM (HetEM), which consists of one wild-type allele and one mutant allele; and poor metabolizer (PM) genotype, which consists of two mutated alleles. People with the HomEM genotype metabolize PPIs at the fastest rate, followed by the HetEM and PM genotypes. Thus, the CYP2C19 genotype can affect eradication by limiting PPI bioavailability and consequently lowering the antisecretory effect. A significant difference in the *H. pylori* eradication rate has been reported between HetEM and HomEM (OR = 1.90; 95 % CI, 1.38–2.60; $P < 0.0001$) but not between PM and HetEM. However, there are a number of PPIs whose antisecretory efficacy is affected to different degrees by CYP2C19 polymorphisms. Only dual and triple therapies with omeprazole resulted in significantly higher *H. pylori* eradication rates with PM compared with HomEM (OR = 4.03; 95 % CI, 1.97–8.28; $P = 0.0001$) and HetEM (OR = 2.24; 95 % CI, 1.09–4.61; $P = 0.03$). In contrast, the eradication rates were not significantly different between PM and HomEM with rabeprazole and lansoprazole therapy [9]. With the use of pantoprazole or esomeprazole in triple therapy, the eradication rate with PM was significantly higher than with HetEM+HomEM combined (97 % vs. 83 %; $P = 0.016$) [10]. In patients with the HomEM genotype, 7-day triple therapy with clarithromycin, amoxicillin, and 20 mg omeprazole or 40 mg esomeprazole twice daily resulted in significantly higher *H. pylori* eradication rates with esomeprazole than with omeprazole (93 % vs. 76 %; $P < 0.05$) [11]. Therefore, differences in *H. pylori* eradication between omeprazole, pantoprazole, and esomeprazole can depend on the CYP2C19 polymorphism.

13.2.3 Resistance to Antibiotics

A meta-analysis of 93 studies with 10,178 patients clearly indicated the influence of antibiotic resistance on *H. pylori* eradication. Clarithromycin resistance had a greater effect on triple therapy outcomes than metronidazole resistance, with a decrease in treatment efficacy of 66 % or 35 % (25–45 %) with clarithromycin resistance in patients administered triple therapy consisting of a PPI, amoxicillin, and clarithromycin or a PPI, metronidazole, and clarithromycin, respectively. In comparison, metronidazole resistance reduced the rate of eradication by 30 % or 18 % with triple therapy consisting of a PPI, metronidazole, and amoxicillin or clarithromycin, respectively [12].

13.2.4 Other Factors Influencing Eradication Success

Interleukin (IL)-1β polymorphisms affect stomach acidity in the presence of *H. pylori* infection. Regarding background disease, the rate of eradication in patients with peptic ulcers is consistently higher than with functional dyspepsia. Furthermore, several studies have reported fewer strains resistant to clarithromycin in patients with duodenal ulcers than with functional dyspepsia [13]. A higher risk of eradication failure might occur in patients with high BMIs, especially in those who fall in the overweight or obese categories, because the distribution volume of the drugs is larger and the concentration at the gastric mucosal level is lower [14]. Therefore, it could be expected that Asian patients, who usually have lower BMIs, will have better outcomes with *H. pylori* treatment. In relation to bacterial factors, CagA-negative strains, compared with CagA-positive strains, are considered a risk factor for eradication failure (risk ratio of treatment failure 2.0; 95 % CI, 1.6–2.4) [15].

13.3 Trends in Standard Triple Therapy for *Helicobacter pylori* Infection

Although standard triple therapy with PPI, amoxicillin, and clarithromycin is recommended globally and the most widely used first-line treatment for *H. pylori* infection because of the certain cure success and safety, the efficacy of this treatment is declining. The global success of this standard triple therapy has become unacceptably low, with only 18 % exceeding 85 % success and approximately 60 % failing to reach 80 % eradication with intention-to-treat protocols in most studies in the Southern or Central European countries, Mexico, Japan, Iran, and the USA, in which there tends to be a high prevalence (≥18.5 %) of clarithromycin resistance, as indicated by the Maastricht III report [16].

Antibiotic resistance is the most common factor causing treatment failure; although the incidence of antibiotic resistance is increasing worldwide, it is generally higher in developing areas than in developed regions. However, there are also a number of areas in which the details of bacterial resistance remain unknown; therefore, nationwide epidemiological surveillance regarding resistance rates is necessary to determine the optimal treatment strategies in each country.

13.3.1 Antibiotic Resistance and Eradication Rate in Japan

In Japan, eradication with clarithromycin-containing 7-day triple therapy was approved for gastric and duodenal ulcers in 2000 when the initial eradication rate was approximately 90 % [17]. However, the eradication rate significantly decreased to 70 % after 2000 [18]. When rabeprazole, amoxicillin, and clarithromycin were first used as first-line eradication treatment in Japan, a large-scale nationwide multicenter prospective study conducted between 2007 and 2009 recorded a successful eradication rate of 80.7 % [19]. The effectiveness of clarithromycin-containing triple therapy has decreased over time, corresponding with the increase in clarithromycin resistance, as demonstrated by an eradication rate of clarithromycin-resistant bacteria of 0–33 % compared with 90 % for clarithromycin-sensitive bacteria [20, 21]. A working group of the Japanese Society of *Helicobacter* Research undertook a surveillance study to determine the temporal antimicrobial susceptibility profiles of *H. pylori* isolated during 2002–2006 and 2010–2011 in 67 institutions, resulting in the analysis of 7735 isolates. Primary resistance to clarithromycin increased from 19 % to 31 % because of the increase in clarithromycin use [22]. Furthermore, in the 1990s, the primary resistance rate of clarithromycin was 7.3 %, which dramatically increased to 31 % on an annual basis in the 2010s, while the primary eradication rate decreased gradually from 89.0 % (Fig. 13.1). Generally, primary resistance to antibiotics is due to the wider use of these antibiotics in the community as monotherapy for other indications.

In comparison, the rate of primary resistance to amoxicillin/metronidazole increased from 15.2 %/2.8 % to 21.9 %/5.3 % [23, 24]. The low rate of resistance to metronidazole, particularly compared with other countries, likely reflects its restricted use; the Japanese health insurance system has not yet approved metronidazole-containing triple therapy, sequential therapy, or non-bismuth quadruple therapy for first-line eradication treatment of *H. pylori*. Therefore, the

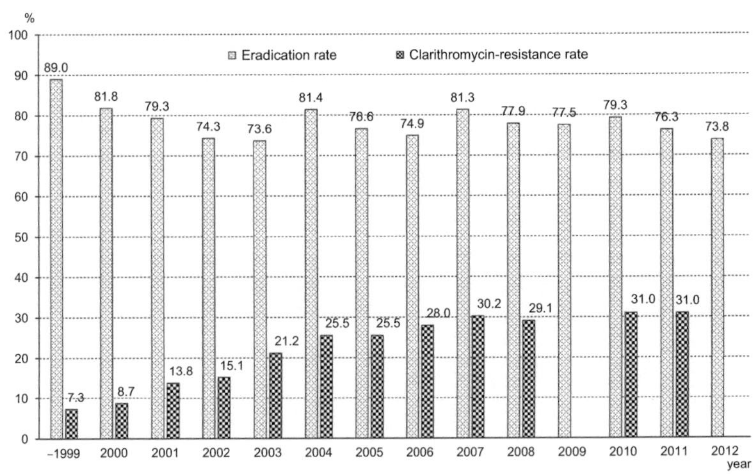

	-1999	2000	2001	2002	2003	2004	2005	2006	2007	2008	2009	2010	2011	2012
Number of eradication	553	154	324	338	326	301	308	478	819	842	925	880	152	130
Number of strains	2335	343	348	1652	1626	1511	1592	1137	281	244	N/A	1953		N/A

Fig. 13.1 Per-protocol treatment success for clarithromycin-containing triple therapy as well as the rates of clarithromycin resistance. Data were compiled from the annual data in published articles [17, 23–31]; the rates of resistance to clarithromycin for 2010 and 2011 were combined

second-line eradication rate using metronidazole-containing triple therapy is approximately 90 %, resulting in failure of first- and second-line therapy in only 2–3 % of cases [25, 32].

Interestingly, the Japanese Ministry of Health, Labour and Welfare approved an eradication therapy for *H. pylori*-related chronic gastritis by the health insurance system in February 2013, which is the first use for gastric cancer treatment and represents a unique strategy for gastric cancer prevention. In Japan, all *H. pylori*-infected persons can currently benefit from the eradication treatment; however, the number of patients who experience eradication failure will also considerably increase. Owing to the decrease in the eradication rate to <80 % with standard clarithromycin-containing triple therapy, this legacy triple therapy should be abandoned as an empiric regimen. Because the Japanese health insurance system does not allow susceptibility testing, the adoption of susceptibility test-guided therapy has been limited, and only standard clarithromycin-containing 7-day triple therapy has been proposed for the last 15 years.

Increasing the duration of therapy could be an option to improve eradication, although the eradication rate remains low; improvements of 77–81 % and 73–75 % are observed by increasing the duration from 7 to 10 days and 7 to 14 days, respectively [33]. Considering that the use of bismuth is unacceptable in Japan, it is necessary to investigate other acceptable first-line therapies, including sequential or non-bismuth quadruple therapy. However, these therapies require three antibiotic drugs, which might not only increase the risk of adverse reactions to the drugs but also reduce the choice of other antibiotic drugs when eradication fails. Based on recent national surveillance by the Japanese Society of *Helicobacter* Research from 2010 to 2011, susceptibility of 2413 *H. pylori* isolates to clarithromycin, metronidazole, and amoxicillin was 34.1 %, 5.0 %, and 17.4 %, respectively. However, antimicrobial resistance was dependent on prior eradication therapies, with significantly more patients who experienced failure with first- ($n = 145$) and second-line ($n = 131$) therapies showing resistance than untreated patients ($n = 1953$) to clarithromycin (86.2 % and 80.2 % vs. 31.0 %, $P < 0.0001$ and $P < 0.0001$, respectively), metronidazole (5.5 % and 68.7 % vs. 3.4 %, $P = 0.165$ and $P < 0.001$, respectively), and amoxicillin (29.7 % and 51.1 % vs. 16.9 %, $P < 0.001$ and $P < 0.0001$, respectively). Owing to the increase in multiple antibiotic resistance, the overuse of antibiotics should be avoided.

13.3.2 Antibiotic Resistance and Eradication Rate in Asia

Primary clarithromycin resistance in Asia is 18.9 %, which is significantly higher than the 11.1 % in Europe and lower than the 29.3 % in North and South America. In Asia, the highest prevalence of clarithromycin resistance is 20.6–40.7 % in Japan, followed by 32 % in China, 9.5–13.5 % in Taiwan, 12.4 % in Korea, and 2.1 % in Malaysia, which is the lowest [34].

In Korea, triple therapy with PPI, clarithromycin, and amoxicillin has been the recommended primary regimen since 1998, when regimens for *H. pylori* eradication were first recommended. Although metronidazole was generally used for *H. pylori* eradication previously, metronidazole-containing triple therapy is not currently recommended as a first-line regimen because of the high rate of antibiotic resistance. An analysis of 652 strains isolated between 1994 and 1999 revealed that resistance to metronidazole and clarithromycin increased from 33.3 % to 47.7 % and 4.8 % to 7.7 %, respectively. However, no strain resistant to amoxicillin was identified during the same period [35]. Similarly, primary resistance to clarithromycin reportedly increased from 2.8 % in 1994 to 14 % in 2003, while that for metronidazole rose from 53 % in 1987 to 66 % in 2003 [36]. In a longitudinal observational study conducted in Korea, the yearly *H. pylori* eradication rates with clarithromycin-containing 7-day triple therapy for the years 1998–2005 were 84 %, 80 %, 81 %, 79 %, 75 %, 78 %, 79 %, and 78 % consecutively [37], demonstrating a nonsignificant, gradual decline of the eradication rate to <80 % and suggesting an immediate need to identify a new first-line regimen that is expecting to result in a higher cure rate.

In Beijing, China, the resistance rates of *H. pylori* to metronidazole and clarithromycin in 1999–2000 were 36 % and 10 %, respectively, and increased to 43 % and 18 %, respectively, in 2001–2002 [38]. More recent rates of resistance to metronidazole and clarithromycin were 60–70 % and 20–38 %, respectively, and the resistance rate to levofloxacin was 30–38 %. However, the rate of resistance to amoxicillin, furazolidone, and tetracycline remains low (1–5 %) [5]. With the increase in *H. pylori* resistance, the eradication rate of standard triple therapy (PPI + clarithromycin + amoxicillin or metronidazole) is now <80 % [5].

In Hong Kong, the prevalence of metronidazole resistance is increasing, with consecutive rates for 1991–1995 of 22.0 %, 52.5 %, 50.5 %, 60.1 %, and 73.2 % [39]. In a more recent study, the proportions of *H. pylori* resistant to metronidazole and clarithromycin were reportedly 39.2–49.4 % and 7.8–10.8 %, respectively [40, 41]. Because of the limited perseverance of resistant strains, the eradication rate with clarithromycin-containing 7-day triple therapy remains >90 % [41, 42].

When considering the increase in primary antibiotic resistance, sampling bias is possible because most studies are from referral centers; therefore, the values might not reflect the true community-based prevalence. Other than Japan, which has more long-term national research center data, there is a need for systematic prospective surveillance of antibiotic resistance rates in Asia.

13.3.3 Antibiotic Resistance and Eradication Rate in Europe

As in other parts of the world, the efficacy of clarithromycin-containing standard triple therapy for *H. pylori* eradication has significantly decreased in many

European countries, with eradication rates <80 % because of the increasing resistance of *H. pylori* to clarithromycin. In some countries with high levels of resistance, the eradication rate with clarithromycin-containing triple therapy has decreased to 50–60 % [43, 44].

The resistance rate to clarithromycin in Europe increased from 9 % in 1998 [45] to 17.5 % in 2008–2009 [46]. In most countries in Central, Western, and Southern Europe, resistance increased, reaching a high prevalence of >20 %, which is considered to be the result of external adaptation of standard clarithromycin-containing triple therapy. However, in Northern European countries, the rate is <10 %, which is considered low and in the application range of standard clarithromycin-containing triple therapy [46]. The highest prevalence of resistance to clarithromycin is 49.2 % in Spain, followed by 16.4–48.2 % in Turkey, 11.0–37.6 % in Italy, 15.4 % in Bulgaria, 11 % in Denmark, 8.8 % in Ireland, 1.5 % in Sweden, and 1 % in the Netherlands, which is the lowest rate [34]. This geographic difference is associated with the prudent use of macrolides in Northern European countries during the previous decades and greater use of clarithromycin in Southern European countries [47–49].

A large German study of over 5000 strains detected a major problem with secondary resistance, especially to fluoroquinolones. The rate of resistance to levofloxacin/ciprofloxacin or triple resistance to metronidazole, clarithromycin, and levofloxacin/ciprofloxacin steadily increased from 2006 (20.9 % or 12.9 %, respectively) and peaked in 2011 (29.1 % or 18.9 %, respectively). Pretreated patients with failed eradication therapy were far more likely to be both quinolone and triple resistant than untreated patients [50].

13.3.4 Antibiotic Resistance and Eradication Rate in the USA

In the USA, the prevalence of clarithromycin resistance was low in the 1990s, with consecutive rates for 1993–1999 of 6.1 %, 8.1 %, 12.1 %, 12.1 %, 14.5 %, 11.1 %, and 9 %. The overall rates of resistance to clarithromycin and metronidazole were 10.6 % and 21.6 %, respectively [51]. In 2006–2009, the primary resistant rate (26.7 %) to metronidazole was relatively stable, while the resistance to clarithromycin increased to 29.3 % (31 % in the North and 20 % in the South), which is higher than both Europe and Asia [34]. A decline in the eradication rate with standard triple therapy has also been reported; however, the cure rate reportedly increases by 8 % with metronidazole instead of clarithromycin treatment. The current eradication rate with triple therapy is 70–85 % in patients who have not previously been administered a macrolide.

13.4 What Can We Do to Improve the Eradication Rate?

Treatment success depends on both bacterial and host factors. Of these, antibiotic resistance is the most important consideration; when resistance is low, previously optimized therapies can be prescribed empirically. However, when resistance begins to undermine success, the regimen should be altered or replaced. Even with the establishment of a best regimen, poor compliance will result in poor outcomes; therefore, the optimal methods to ensure compliance should be considered, such as dosing, duration, and side effects. Moreover, tailored therapies suitable for an individual or community should be discussed.

13.4.1 Antibiotic Susceptibility Test

Using susceptible antibodies, the eradication effect of triple therapy can reach at least 90 %. In Western countries, the approximate eradication rates with susceptible strains are calculated at 94 % and 97 % with 7-day and 14-day clarithromycin-containing triple therapy, respectively. Based on a formula described previously, the success with 7-day or 14-day clarithromycin-containing triple therapy decreased to <90 % when clarithromycin resistance exceeds 5 % or 15 %, respectively [52]. Therefore, 14-day clarithromycin-containing triple therapy is still useful in a few regions in the world where clarithromycin resistance is <15 % (Northern Europe and Thailand). Susceptibility-based eradication therapy is an option in many regions; however, issues with susceptibility-based eradication therapy include the time to initiate treatment and the use of *H. pylori* cultures. Eradication does not have to be initiated quickly because most patients are infected in childhood and have been infected with *H. pylori* for decades before treatment initiation. The agar dilution method, microdilution method, and Epsilometer test (E-test) are used as antibiotic susceptibility tests for *H. pylori*. However, it is difficult to confirm antibiotic resistance owing to bacterial culture conditions, uncertainty about the bacterial minimum inhibitory concentration, and differences in various antibiotic susceptibility tests. For example, when using the E-test, we have to consider the possibility of overestimating metronidazole resistance.

13.4.2 Treatment Duration

A meta-analysis that compared short- and long-term therapy with a PPI, clarithromycin, and amoxicillin for *H. pylori* infection determined that extending the duration from 7 to 10 or 14 days improves the eradication success by 4 % or 5–9 %, respectively [53]. However, this effect is limited, and even 14-day therapy reaches a 50 % eradication rate in patients with clarithromycin resistance

[52]. Clarithromycin has to bind to ribosomes to kill *H pylori*. Because clarithromycin cannot bind to ribosomes in resistant organisms, resistance cannot be overcome by increasing the dose or duration [54]. Similarly, resistance to fluoroquinolones and rifabutin cannot be overcome by increasing the dose or duration. From the viewpoint of the *H. pylori* organism, it is as if antibiotics do not exist when the organism is resistant. Thus, in the presence of clarithromycin resistance, PPI, amoxicillin, and clarithromycin triple therapy becomes PPI and amoxicillin dual therapy. Metronidazole is the only drug in which resistance can be overcome by increasing the dose and duration.

13.4.3 Probiotics

To improve eradication success, the Maastricht IV consensus recommends 10–14 days of administration of antibiotics and high-dose PPI (twice a daily), resulting in an increased incidence of undesirable side effects, such as antibiotic-associated diarrhea, nausea, or vomiting during anti-*H. pylori* treatment, which can lead to reduced compliance. Although probiotics have been used to improve *H. pylori* treatment and reduce the side effects, the Maastricht IV consensus did not emphasize and preclude a positive recommendation because of the poor quality of numerous trials and the limited number of centers. However, a recent meta-analysis of 21 randomized controlled trials involving a total of 3814 participants showed the efficacy and safety of probiotics as adjuvant agents for anti-*H. pylori* standard triple therapy regimens. When *Saccharomyces boulardii* or specific Lactobacillus strains were used, the eradication rate increased significantly by 9–17 %, but when multistrain probiotics were used, the eradication rate was not affected as much [55–57]. It is thought that all of these treatments might indirectly improve the eradication rate by decreasing the adverse events, especially diarrhea. However, *Bifidobacterium infantis*, which has been proposed as having anti-*H. pylori* activity, increased the eradication rate from 68.9 % to 83 %; furthermore, with 2 weeks of *Bifidobacterium infantis* pretreatment, the rate increased to 90.5 % [58].

13.4.4 Intragastric pH Control

PPI effectiveness, which is measured by the degree and duration of acid suppression, is related to *H. pylori* eradication and depends on the PPI type, dose, and frequency of administration as well as the ability of the stomach to produce acid. A meta-analysis showed that PPI administered twice a day is better than a single daily dose of triple therapy [59]. In addition, the eradication rate with standard triple therapy depends on the bioavailability of PPI, which depends on the CYP2C19 and multidrug-resistance (MDR) genetic polymorphisms of the patients. MDR codes P-glycoprotein, which is an ATP-dependent efflux transporter and often influences absorption of orally consumed drugs [60]. In patients with the EM polymorphism or MDR T/T genotype, the eradication rate is lower compared with the HomEM/HetEM or T/C and C/C genotypes [9, 61]. It may be necessary to use a PPI that is not affected by genetic polymorphisms. In Asians, the proportion of the population with PM (15–23 % or more) is higher than in Caucasians (2 %), and the effect of the CYP2C19 gene polymorphism on *H. pylori* eradication therapy might be greater [62]. This raises the question of whether knowledge of the CYP2C19 genotype is required before starting *H. pylori* therapy in Asia. However, because of the cost and limited availability of genotype testing, PPI choice and/or dose could be a more realistic approach than CYP2C19 genotyping to increase the eradication rate of *H. pylori* treatment in a clinical setting.

Increasing the intragastric environment to a pH6 or 7 induces a replicative cycle for bacteria, during which they are susceptible to amoxicillin and clarithromycin. In PPI plus amoxicillin dual treatment, four doses of PPI a day increases the eradication rate approximately 27 % more than two doses of PPI a day [16]. PPI dose escalation or a stronger acid reducer could restore the effect of triple therapy.

Vonoprazan (Takecab®), which is an orally bioavailable potassium-competitive acid blocker (P-CAB), was recently developed by Takeda (Osaka, Japan) and approved in Japan. P-CAB competes with K^+ on the luminal surface and reversibly inhibits gastric acid secretion by binding to the gastric H^+, K^+-ATPase enzyme. Vonoprazan has a slower dissociation rate than other P-CABs and is different from lansoprazole in that ambient pH does not affect the inhibitory effect. Vonoprazan rapidly reaches a high plasma concentration, undergoes comparatively slow metabolism, and strongly inhibits daytime and nighttime acid secretion. Furthermore, the drug might be less affected by CYP2C19 polymorphisms, unlike PPI. The superiority of vonoprazan to lansoprazole as a 7-day first-line triple therapy (200 or 400 mg clarithromycin twice a daily, 750 mg amoxicillin twice a day, and 20 mg vonoprazan twice a day or 30 mg lansoprazole twice a day) was established in a randomized, double-blind phase III study ($n = 650$). The eradication rate of vonoprazan was 92.6 %, which was significantly higher than the rate of 75.9 % with lansoprazole (95 % CI of the difference, 11.2–22.1; $P < 0.0001$). Surprisingly, even in patients with clarithromycin resistance, the eradication rate of vonoprazan was significantly higher than lansoprazole (82.0 % vs. 40.0 %; $P < 0.0001$) [63]. Only the intragastric pH environment affects this superiority of standard triple therapy. With vonoprazan, legacy triple therapy could replicate the high eradication rate first observed in the 1990s.

13.5 Conclusion

Increases in the global *H. pylori* resistance to clarithromycin and metronidazole have resulted in reduced efficacy of PPI-based triple therapy. In particular, clarithromycin-containing triple therapy is no longer adequately effective in most populations. To improve treatment effectiveness, knowledge of the local resistance pattern is important to choose the appropriate therapeutic antibiotic. The eradication rate can be calculated using the *H. pylori* resistance rate to the antibiotics in each regimen; for example, when the clarithromycin resistance rate exceeds 15 %, 14-day clarithromycin-containing triple therapy decreases the success rate to <90 %. Therefore, in regions with high levels of resistance, a change in regimen should be considered. However, quadruple, sequential, and concomitant treatments are all affected by resistance, with clarithromycin-metronidazole resistance rates of 15 %, 5 %, or 9 % decreasing the cure rate to <90 % with 14-day concomitant, sequential, or hybrid therapy, respectively [52]. Individualized treatment based on antibiotic resistance might need to be reviewed, and the efficacy of treatment with relatively unused antibiotics at adequate PPI doses also needs to be evaluated. However, with the use of a new antibiotic, bacteria also develop new tolerances. Smoking cessation and changes in the PPI dose or choice might improve the eradication rates; however, large effects should not be expected. However, sustained suppression of gastric acid secretion with the novel P-CAB, vonoprazan, resulted in significant outcomes with clarithromycin-containing triple therapy even in the presence of resistance. Thus, vonoprazan might be considered the savior of *H. pylori* eradication treatment.

References

1. Chey WD, Wong BC. Practice parameters committee of the American college of G. American college of gastroenterology guideline on the management of *Helicobacter pylori* infection. Am J Gastroenterol. 2007;102(8):1808–25. doi:10.1111/j.1572-0241.2007.01393.x.
2. Malfertheiner P, Mégraud F, O'Morain CA, Atherton J, Axon AT, Bazzoli F, et al. Management of *Helicobacter pylori* infection—the Maastricht IV/ Florence consensus report. Gut. 2012;61(5):646–64. doi:10.1136/gutjnl-2012-302084.
3. Asaka M, Kato M, Takahashi S, Fukuda Y, Sugiyama T, Ota H, et al. Guidelines for the management of *Helicobacter pylori* infection in Japan: 2009 revised edition. Helicobacter. 2010;15(1):1–20. doi:10.1111/j.1523-5378.2009.00738.x.
4. Kim SG, Jung HK, Lee HL, Jang JY, Lee H, Kim CG, et al. Guidelines for the diagnosis and treatment of *Helicobacter pylori* infection in Korea, 2013 revised edition. J Gastroenterol Hepatol. 2014;29(7):1371–86. doi:10.1111/jgh.12607.
5. Chinese Society of Gastroenterology CSGoHp, Liu WZ, Xie Y, Cheng H, Lu NH, Hu FL. Fourth Chinese national consensus report on the management of *Helicobacter pylori* infection. J Dig Dis. 2013;14(5):211–21. doi:10.1111/1751-2980.12034.
6. Lee JY, Kim N, Kim MS, Choi YJ, Lee JW, Yoon H, et al. Factors affecting first-line triple therapy of *Helicobacter pylori* including CYP2C19 genotype and antibiotic resistance. Dig Dis Sci. 2014;59(6):1235–43. doi:10.1007/s10620-014-3093-7.
7. Suzuki T, Matsuo K, Ito H, Sawaki A, Hirose K, Wakai K, et al. Smoking increases the treatment failure for *Helicobacter pylori* eradication. Am J Med. 2006;119(3):217–24. doi:10.1016/j.amjmed.2005.10.003.
8. Matsuo K, Hamajima N, Ikehara Y, Suzuki T, Nakamura T, Matsuura A, et al. Smoking and polymorphisms of fucosyltransferase gene Le affect success of *H. pylori* eradication with lansoprazole, amoxicillin, and clarithromycin. Epidemiol Infect. 2003;130(2):227–33.

9. Padol S, Yuan Y, Thabane M, Padol IT, Hunt RH. The effect of CYP2C19 polymorphisms on *H. pylori* eradication rate in dual and triple first-line PPI therapies: a meta-analysis. Am J Gastroenterol. 2006;101(7):1467–75. doi:10.1111/j.1572-0241.2006.00717.x.
10. Kang JM, Kim N, Lee DH, Park YS, Kim JS, Chang IJ, et al. Effect of the CYP2C19 polymorphism on the eradication rate of *Helicobacter pylori* infection by 7-day triple therapy with regular proton pump inhibitor dosage. J Gastroenterol Hepatol. 2008;23(8 Pt 1):1287–91. doi:10.1111/j.1440-1746.2008.05392.x.
11. Sheu BS, Kao AW, Cheng HC, Hunag SF, Chen TW, Lu CC, et al. Esomeprazole 40 mg twice daily in triple therapy and the efficacy of *Helicobacter pylori* eradication related to CYP2C19 metabolism. Aliment Pharmacol Ther. 2005;21(3):283–8. doi:10.1111/j.1365-2036.2005.02281.x.
12. Fischbach L, Evans EL. Meta-analysis: the effect of antibiotic resistance status on the efficacy of triple and quadruple first-line therapies for *Helicobacter pylori*. Aliment Pharmacol Ther. 2007;26(3):343–57. doi:10.1111/j.1365-2036.2007.03386.x.
13. Broutet N, Tchamgoue S, Pereira E, Lamouliatte H, Salamon R, Megraud F. Risk factors for failure of *Helicobacter pylori* therapy—results of an individual data analysis of 2751 patients. Aliment Pharmacol Ther. 2003;17(1):99–109.
14. Abdullahi M, Annibale B, Capoccia D, Tari R, Lahner E, Osborn J, et al. The eradication of *Helicobacter pylori* is affected by body mass index (BMI). Obes Surg. 2008;18(11):1450–4. doi:10.1007/s11695-008-9477-z.
15. Suzuki T, Matsuo K, Sawaki A, Ito H, Hirose K, Wakai K, et al. Systematic review and meta-analysis: importance of CagA status for successful eradication of *Helicobacter pylori* infection. Aliment Pharmacol Ther. 2006;24(2):273–80. doi:10.1111/j.1365-2036.2006.02994.x.
16. Graham DY, Fischbach L. *Helicobacter pylori* treatment in the era of increasing antibiotic resistance. Gut. 2010;59(8):1143–53. doi:10.1136/gut.2009.192757.
17. Asaka M, Sugiyama T, Kato M, Satoh K, Kuwayama H, Fukuda Y, et al. A multicenter, double-blind study on triple therapy with lansoprazole, amoxicillin and clarithromycin for eradication of *Helicobacter pylori* in Japanese peptic ulcer patients. Helicobacter. 2001;6(3):254–61.
18. Sasaki M, Ogasawara N, Utsumi K, Kawamura N, Kamiya T, Kataoka H, et al. Changes in 12-year first-line eradication rate of *Helicobacter pylori* based on triple therapy with proton pump inhibitor, amoxicillin and clarithromycin. J Clin Biochem Nutr. 2010;47(1):53–8. doi:10.3164/jcbn.10-10.
19. Fujioka T, Aoyama N, Sakai K, Miwa Y, Kudo M, Kawashima J, et al. A large-scale nationwide multicenter prospective observational study of triple therapy using rabeprazole, amoxicillin, and clarithromycin for *Helicobacter pylori* eradication in Japan. J Gastroenterol. 2012;47(3):276–83. doi:10.1007/s00535-011-0487-6.
20. Hoshiya S, Watanabe K, Tokunaga K, Tanaka A, Ninomiya H, Shingaki M, et al. Relationship between eradication therapy and clarithromycin-resistant *Helicobacter pylori* in Japan. J Gastroenterol. 2000;35(1):10–4.
21. Murakami K, Sato R, Okimoto T, Nasu M, Fujioka T, Kodama M, et al. Eradication rates of clarithromycin-resistant *Helicobacter pylori* using either rabeprazole or lansoprazole plus amoxicillin and clarithromycin. Aliment Pharmacol Ther. 2002;16(11):1933–8.
22. Perez Aldana L, Kato M, Nakagawa S, Kawarasaki M, Nagasako T, Mizushima T, et al. The relationship between consumption of antimicrobial agents and the prevalence of primary *Helicobacter pylori* resistance. Helicobacter. 2002;7(5):306–9.
23. Kobayashi I, Murakami K, Kato M, kato S, Azuma T, Takahashi S, et al. The current state of drug resistance of Helicobacter pylori in our country; a report of surveillance of resistant bacteria in 2006 and a total report for 5 years. Jpn J Helicobacter Res. 2009;10(2):98–103.
24. Kobayashi I, Azuma T, Ikeda F, Uemura N, Kato S, Kato M, et al. The current state of drug resistance of Helicobacter pylori in our country: a report of surveillance of resistant bacteria from 2010 to 2011. Jpn J Helicobacter Res. 2013;14(2):102–6.

25. Sasaki M, Kasugai K. The change in the eradication rate of *H. pylori* in Japan. Jpn J Helicobacter Res. 2014;15(2):62–7.
26. Kihira K, Satoh K, Saifuku K, Kawakami S, Fukazawa K, Ishino Y, et al. Rabeprazole, amoxycillin and low- or high-dose clarithromycin for cure of *Helicobacter pylori* infection. Aliment Pharmacol Ther. 2000;14(8):1083–7.
27. Kuwayama H, Luk G, Yoshida S, Nakamura T, Kubo M, Uemura N, et al. Efficacy of a low-dose omeprazole-based triple-therapy regimen for *Helicobacter pylori* eradication independent of cytochrome P450 genotype: the Japanese MACH study. Clin Drug Investig. 2005;25(5):293–305.
28. Masuyama H, Nakamura T, Watanabe H, Sasai T, Saifuku Y, Matsuoka M, et al. Actual condition of *Helicobacter pylori* medical examination; treatment and problem in Tochigi-ken north area. Helicobacter Res. 2012;16(1):67–73.
29. Horiki N, Omata F, Uemura M, Suzuki S, Ishii N, Iizuka Y, et al. Annual change of primary resistance to clarithromycin among *Helicobacter pylori* isolates from 1996 through 2008 in Japan. Helicobacter. 2009;14(5):86–90.
30. Rimbara E, Noguchi N, Tanabe M, Kawai T, Matsumoto Y, Sasatsu M. Susceptibilities to clarithromycin, amoxycillin and metronidazole of *Helicobacter pylori* isolates from the antrum and corpus in Tokyo, Japan, 1995–2001. Clin Microbiol Infect. 2005;11(4):307–11.
31. Murakami K, Fujioka T. Helicobacter pylori cure treatment in our country and drug-resistance. Helicobacter Res. 1998;2(5):21–6.
32. Sasaki H, Nagahara A, Hojo M, Asaoka D, Matsumoto K, Osada T, et al. Ten-year trend of the cumulative *Helicobacter pylori* eradication rate for the 'Japanese eradication strategy'. Digestion. 2013;88(4):272–8. doi:10.1159/000353313.
33. Fuccio L, Minardi ME, Zagari RM, Grilli D, Magrini N, Bazzoli F. Meta-analysis: duration of first-line proton-pump inhibitor based triple therapy for *Helicobacter pylori* eradication. Ann Intern Med. 2007;147(8):553–62.
34. De Francesco V, Giorgio F, Hassan C, Manes G, Vannella L, Panella C, et al. Worldwide *H. pylori* antibiotic resistance: a systematic review. J Gastrointestin Liver Dis: JGLD. 2010;19 (4):409–14.
35. Kim JJ, Reddy R, Lee M, Kim JG, El-Zaatari FA, Osato MS, et al. Analysis of metronidazole, clarithromycin and tetracycline resistance of *Helicobacter pylori* isolates from Korea. J Antimicrob Chemother. 2001;47(4):459–61.
36. Kim JM, Kim JS, Jung HC, Kim N, Kim YJ, Song IS. Distribution of antibiotic MICs for *Helicobacter pylori* strains over a 16-year period in patients from Seoul, South Korea. Antimicrob Agents Chemother. 2004;48(12):4843–7. doi:10.1128/AAC.48.12.4843-4847. 2004.
37. Choi YS, Cheon JH, Lee JY, Kim SG, Kim JS, Kim N, et al. The trend of eradication rates of first-line triple therapy for *Helicobacter pylori* infection: single center experience for recent eight years. Korean J Gastroenterol = Taehan Sohwagi Hakhoe chi. 2006;48(3):156–61.
38. Cheng H, Hu FL. The epidemiology of *Helicobacter pylori* resistance to antibiotics in Beijing. Zhonghua yi xue za zhi. 2005;85(39):2754–7.
39. Ling TK, Cheng AF, Sung JJ, Yiu PY, Chung SS. An increase in *Helicobacter pylori* strains resistant to metronidazole: a five-year study. Helicobacter. 1996;1(1):57–61.
40. Wang WH, Wong BC, Mukhopadhyay AK, Berg DE, Cho CH, Lai KC, et al. High prevalence of *Helicobacter pylori* infection with dual resistance to metronidazole and clarithromycin in Hong Kong. Aliment Pharmacol Ther. 2000;14(7):901–10.
41. Gu Q, Xia HH, Wang JD, Wong WM, Chan AO, Lai KC, et al. Update on clarithromycin resistance in *Helicobacter pylori* in Hong Kong and its effect on clarithromycin-based triple therapy. Digestion. 2006;73(2–3):101–6. doi:10.1159/000094040.
42. Hung IF, Chan P, Leung S, Chan FS, Hsu A, But D, et al. Clarithromycin-amoxycillin-containing triple therapy: a valid empirical first-line treatment for *Helicobacter pylori* eradication in Hong Kong? Helicobacter. 2009;14(6):505–11. doi:10.1111/j.1523-5378.2009. 00722.x.

43. Kadayifci A, Buyukhatipoglu H, Cemil Savas M, Simsek I. Eradication of *Helicobacter pylori* with triple therapy: an epidemiologic analysis of trends in Turkey over 10 years. Clin Ther. 2006;28(11):1960–6. doi:10.1016/j.clinthera.2006.11.011.
44. Tursi A, Elisei W, Giorgetti G, Picchio M, Brandimarte G. Decreasing efficacy of the standard seven-day triple therapy containing amoxycillin and clarithromycin in curing *Helicobacter pylori* infection in clinical setting in Italy: a 10-year follow-up study. Panminerva Med. 2014;56(1):57–61.
45. Glupczynski Y, Megraud F, Lopez-Brea M, Andersen LP. European multicentre survey of in vitro antimicrobial resistance in *Helicobacter pylori*. Eur J Clin Microbiol Infect Dis. 2001;20(11):820–3.
46. Megraud F, Coenen S, Versporten A, Kist M, Lopez-Brea M, Hirschl AM, et al. *Helicobacter pylori* resistance to antibiotics in Europe and its relationship to antibiotic consumption. Gut. 2013;62(1):34–42. doi:10.1136/gutjnl-2012-302254.
47. Debets-Ossenkopp YJ, Herscheid AJ, Pot RG, Kuipers EJ, Kusters JG, Vandenbroucke-Grauls CM. Prevalence of *Helicobacter pylori* resistance to metronidazole, clarithromycin, amoxycillin, tetracycline and trovafloxacin in The Netherlands. J Antimicrob Chemother. 1999;43 (4):511–5.
48. Cabrita J, Oleastro M, Matos R, Manhente A, Cabral J, Barros R, et al. Features and trends in *Helicobacter pylori* antibiotic resistance in Lisbon area, Portugal (1990–1999). J Antimicrob Chemother. 2000;46(6):1029–31.
49. Pilotto A, Rassu M, Leandro G, Franceschi M, Di Mario F. Interdisciplinary group for the study of U. Prevalence of *Helicobacter pylori* resistance to antibiotics in Northeast Italy: a multicentre study. GISU. Interdisciplinary group for the study of ulcer. Dig Liver Dis. 2000;32 (9):763–8.
50. Wueppenhorst N, Stueger HP, Kist M, Glocker EO. High secondary resistance to quinolones in German *Helicobacter pylori* clinical isolates. J Antimicrob Chemother. 2013;68(7):1562–6. doi:10.1093/jac/dkt061.
51. Osato MS, Reddy R, Reddy SG, Penland RL, Malaty HM, Graham DY. Pattern of primary resistance of *Helicobacter pylori* to metronidazole or clarithromycin in the United States. Arch Intern Med. 2001;161(9):1217–20.
52. Graham DY, Lee YC, Wu MS. Rational *Helicobacter pylori* therapy: evidence-based medicine rather than medicine-based evidence. Clin Gastroenterol Hepatol. 2014;12(2):177–86.e3; Discussion e12-3. doi:10.1016/j.cgh.2013.05.028.
53. Yuan Y, Ford AC, Khan KJ, Gisbert JP, Forman D, Leontiadis GI, et al. Optimum duration of regimens for *Helicobacter pylori* eradication. Cochrane Database Syst Rev. 2013;12, CD008337. doi:10.1002/14651858.CD008337.pub2.
54. Megraud F. H pylori antibiotic resistance: prevalence, importance, and advances in testing. Gut. 2004;53(9):1374–84. doi:10.1136/gut.2003.022111.
55. Szajewska H, Horvath A, Piwowarczyk A. Meta-analysis: the effects of Saccharomyces boulardii supplementation on *Helicobacter pylori* eradication rates and side effects during treatment. Aliment Pharmacol Ther. 2010;32(9):1069–79. doi:10.1111/j.1365-2036.2010.04457.x.
56. Shavakhi A, Tabesh E, Yaghoutkar A, Hashemi H, Tabesh F, Khodadoostan M, et al. The effects of multistrain probiotic compound on bismuth-containing quadruple therapy for *Helicobacter pylori* infection: a randomized placebo-controlled triple-blind study. Helicobacter. 2013;18(4):280–4. doi:10.1111/hel.12047.
57. Navarro-Rodriguez T, Silva FM, Barbuti RC, Mattar R, Moraes-Filho JP, de Oliveira MN, et al. Association of a probiotic to a *Helicobacter pylori* eradication regimen does not increase efficacy or decreases the adverse effects of the treatment: a prospective, randomized, double-blind, placebo-controlled study. BMC Gastroenterol. 2013;13:56. doi:10.1186/1471-230X-13-56.

58. Dajani AI, Abu Hammour AM, Yang DH, Chung PC, Nounou MA, Yuan KY, et al. Do probiotics improve eradication response to *Helicobacter pylori* on standard triple or sequential therapy? Saudi J Gastroenterol. 2013;19(3):113–20. doi:10.4103/1319-3767.111953.
59. Vallve M, Vergara M, Gisbert JP, Calvet X. Single vs. double dose of a proton pump inhibitor in triple therapy for *Helicobacter pylori* eradication: a meta-analysis. Aliment Pharmacol Ther. 2002;16(6):1149–56.
60. Pauli-Magnus C, Rekersbrink S, Klotz U, Fromm MF. Interaction of omeprazole, lansoprazole and pantoprazole with P-glycoprotein. Naunyn Schmiedebergs Arch Pharmacol. 2001;364 (6):551–7.
61. Furuta T, Sugimoto M, Shirai N, Matsushita F, Nakajima H, Kumagai J, et al. Effect of MDR1 C3435T polymorphism on cure rates of *Helicobacter pylori* infection by triple therapy with lansoprazole, amoxicillin and clarithromycin in relation to CYP 2C19 genotypes and 23S rRNA genotypes of *H. pylori*. Aliment Pharmacol Ther. 2007;26(5):693–703. doi:10.1111/j.1365-2036.2007.03408.x.
62. Sheu BS, Fock KM. CYP2C19 genotypes and *Helicobacter pylori* eradication. J Gastroenterol Hepatol. 2008;23(8 Pt 1):1163. doi:10.1111/j.1440-1746.2008.05519.x.
63. Garnock-Jones KP. Vonoprazan: first global approval. Drugs. 2015;75(4):439–43. doi:10.1007/s40265-015-0368-z.

Chapter 14
Personalized Therapy in *H. pylori* Eradication

Takahisa Furuta, Mitsushige Sugimoto, Mihoko Yamade, Takahiro Uotani, Shu Sahara, Hitomi Ichikawa, and Takuma Kagami

Abstract Eradication of *H. pylori* is generally achieved by the administration of antibiotics under acid inhibition. The clinical efficacy of proton pump inhibitors, which inhibit gastric acid secretion, differs depending on the activity of metabolizing enzymes. Thus, drugs which are rapidly metabolized will have a decreased therapeutic effect and may require increased doses. Antibiotic resistance of *H. pylori* is also critical, and antibiotic drugs must be selected according to antibiotic resistance. Furthermore, antibiotics have their own characteristics, and some need to be administrated three or four times a day rather than uniformly twice a day. Standard therapy might not be indicated for certain patients, such as those with penicillin allergy and/or deterioration of renal function and receiving dialysis for renal failure. Accordingly, it may be necessary to provide personalized therapy based on the characteristics of each case.

Keywords Personalization • *H. pylori* eradication • CYP2C19 • Susceptibility

14.1 Introduction

Eradication therapy for *H. pylori*, in addition to proton pump inhibitors (PPIs), which inhibit gastric acid secretion, and amoxicillin (AMPC), which is an antibiotic drug, also uses clarithromycin (CAM) and metronidazole (MNZ) as primary and secondary eradication therapies, respectively. Recently, a potassium-competitive acid blocker (P-CAB) has come into use, in addition to PPIs.

Eradication therapy for *H. pylori* requires that gastric acid is sufficiently inhibited to enhance the effectiveness of antibiotics. Furthermore, antibiotics

T. Furuta (✉)
Center for Clinical Research, Hamamatsu University School of Medicine, 1-20-1 Handayama, Higashi-ku, Hamamatsu, Shizuoka 431-3192, Japan
e-mail: furuta@hama-med.ac.jp

M. Sugimoto • M. Yamade • T. Uotani • S. Sahara • H. Ichikawa • T. Kagami
First Department of Medicine, Hamamatsu University School of Medicine, 1-20-1 Handayama, Shizuoka 431-3192, Japan

H. Suzuki et al. (eds.), *Helicobacter pylori*, DOI 10.1007/978-4-431-55705-0_14

need to be selected based on *H. pylori* sensitivity. Moreover, as antibiotics have different characteristics, selection is preferably made according to the individual case. Tolerance to specific drugs also varies among individual cases. The following summarizes the personalization of drug selection for gastric acid inhibitors and antimicrobial agents and the method of administering antibiotics.

14.2 Personalization of Gastric Acid Inhibitors in Eradication Therapy of *H. pylori*

Suppression of gastric acid secretion is an extremely important part of eradication therapy for *H. pylori*. Concurrent administration of a gastric acid secretion suppressant provides an intragastric environment with a neutral pH. Inhibition of gastric acid provides *H. pylori* with a fertile environment, which in turn makes it more sensitive to antibiotics. In other words, with *H. pylori* multiplication, if DNA replication becomes more active, the sensitivity of *H. pylori* to quinolone, a DNA replication-dependent antimicrobial drug, may increase; or if DNA transcription becomes more active, the sensitivity of *H. pylori* to CAM, which targets 23S rRNA, may increase. If cell wall synthesis increases, the sensitivity of *H. pylori* to AMPC, which targets cell wall synthesis enzymes, may increase. In addition, the gastric acid inhibition is effective in preventing the degradation of antibiotics caused by acid in the stomach. Moreover, some reports indicate that gastric acid inhibition by PPI results in increased concentration and delayed excretion of antibiotics. Furthermore, PPIs themselves also provide an anti-*H. pylori* effect, and thus gastric acid secretion suppression by PPI is essential for the eradication of *H. pylori* [1, 2].

One class of gastric acid inhibitors used to eradicate *H. pylori* is PPIs. The metabolism of PPIs primarily involves cytochrome p450 2C19 (CYP2C19), a drug-metabolizing enzyme in the liver. This enzyme shows interindividual, genetically determined variability in activity. CYP2C19 genotypes are classified into rapid metabolizers (RMs), intermediate metabolizers (IMs), and poor metabolizers (PMs) lacking enzyme activity. The existence of CYP2C19 ultrarapid metabolizers in RM has been reported, but is less common in Japanese, and the clinical significance for PPIs has not been determined [3].

With regard to the effect of PPIs, the rapid metabolism in RMs means that high blood PPI levels cannot be maintained (Fig. 14.1A). Gastric acid secretion is insufficiently suppressed (Fig. 14.1B) [4], antibiotics then show insufficient efficacy, and eradication eventually fails [5]. Figure 14.2 shows the relationship between CYP2C19 genetic polymorphism and eradication rates. Rates are significantly lower in RMs than in IMs and PMs [6]. Personalized therapy therefore requires a PPI administration plan which accords with genetic polymorphism in CY2C19.

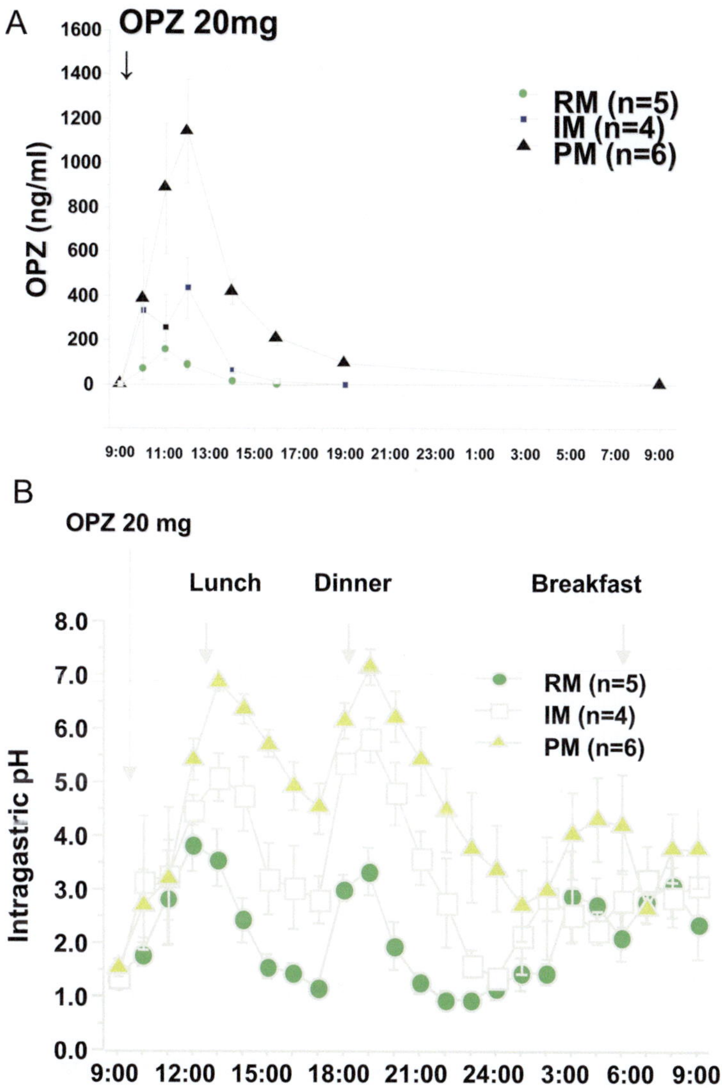

Fig. 14.1 Trends in omeprazole (OPZ) blood concentration (A) and intragastric pH at oral administration of 20 mg omeprazole in rapid metabolizers (RMs), intermediate metabolizers (IMs), and poor metabolizers (PMs) of CYP2C19 (cited from Reference [4])

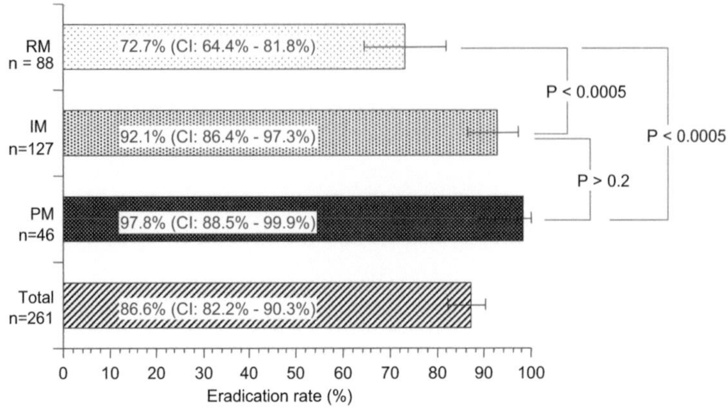

Fig. 14.2 Eradication rate of *H. pylori* by proton pump inhibitor, amoxicillin and clarithromycin, and CYP2C19 genetic polymorphism (cited from Reference [6])

14.3 Dosage Selection of PPIs According to CYP2C19 Genetic Polymorphism

Eradication therapy of *H. pylori* is considered to be optimized at an intragastric pH of 5 or higher. Figure 14.3 shows intragastric pH at 24 h after administration of 15 or 30 mg of lansoprazole (LPZ), a PPI which is given by different administration methods. An intragastric pH 5 or higher can be achieved by administration of LPZ at 30 mg three times a day in RMs, at 15 mg three times a day or 30 mg twice a day in IMs, and at 15 mg twice a day in PMs. Thus, increased PPI doses may be required, particularly in RMs [7].

14.4 Eradication of *H. pylori* and Antibiotic Resistance

Antibiotics used in the eradication of *H. pylori* include AMPC, CAM, and MNZ. CAM and AMPC are generally administered in combination in primary eradication therapy. This treatment can be expected to provide an eradication rate of 90 % or higher with CAM-sensitive strains, but of only around 40 % with CAM-resistant strains. The frequency of CAM-resistant strains exceeds 30 %, resulting in a 70–80 % eradication rate [8], but the formula 90 % $\times$ 0.7 + 40 % $\times$ 0.3 = 75 % may be more appropriate. In contrast, the triple PPI/AMPC/MNZ therapy as secondary eradication after the failure of the primary eradication achieves an eradication rate of 90 % or higher [9]; thus, if MNZ is used in primary eradication with CAM-resistant strains, an eradication rate of 90 % or higher can be expected at the point of primary eradication.

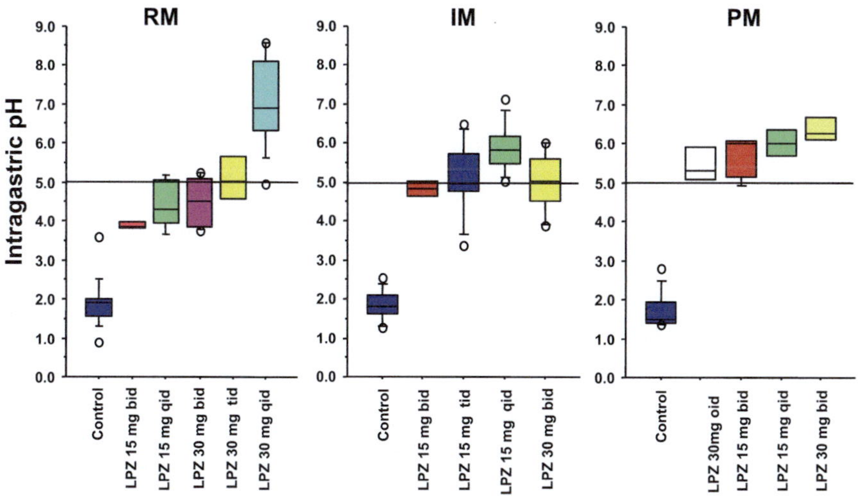

Fig. 14.3 Intragastric pH in the different dosage and administration of lansoprazole in rapid metabolizers (RMs), intermediate metabolizers (IMs), and poor metabolizers (PMs) of CYP2C19. The administration methods of PPI to achieve pH 5 differ by CYP2C19 gene polymorphism (cited from Reference [7])

14.5 Antibiotic Resistance in *H. pylori* Due to Genetic Mutations

The sensitivity of *H. pylori* to antibiotics results from genetic mutation. CAM resistance is due to a single-nucleotide polymorphism (SNP) with adenine (A)-to-guanine (G) transition at position 2142 or 2143 of the 23S rRNA, and the presence or absence of CAM-resistant strains can be detected by testing for this SNP (Fig. 14.4) [10]. Drug resistance against quinolones is detectable by identifying mutations in gyrase A (*gyrA*) [11].

14.6 Concepts of Antibiotic Administration According to Characteristics for *H. pylori*

Antibiotics used in eradication therapy for *H. pylori* are normally administered twice a day. However, the antimicrobial activity of antibiotics with a beta-lactam ring, such as AMPC, is known to be time dependent. In other words, antibiotic activity of these agents depends on the percentage of the dosage interval in which drug concentrations remain above MIC (%time above MIC [%TAM]). The plasma half-life of AMPC in blood is very short, around 1 h, and AMPC does not exert a post-antibiotic effect (PAE) against *H. pylori* [12]. Thus, ensuring sufficient antibiotic activity may require that AMPC is administered frequently. For the dual

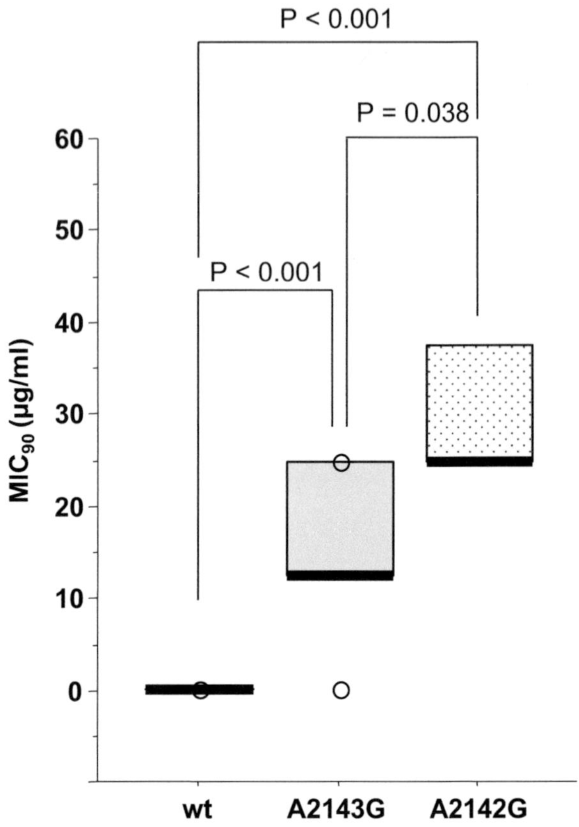

Fig. 14.4 Presence or absence of adenine (A)-to-guanine (G) transition at position 2142 or 2143 of the *H. pylori* 23S rRNA and MIC for clarithromycin (cited from Reference [10])

therapy with AMPC and PPI, eradication rates of 90 % or higher have been largely achieved with AMPC administration four times a day [13, 14]. In contrast, eradication rates with twice daily AMPC are around 50 %. Figure 14.5 shows eradication rates by the frequency of AMPC administration in triple therapy [15]. Rates rise significantly with administration of 500 mg three or four times a day compared with 750 mg twice a day. Thus, AMPC administration at least three times a day may be considered desirable.

14.7 Practical Personalized Therapy

(1) Dealing with primary/secondary eradication

At Hamamatsu University School of Medicine, personalized eradication therapy for *H. pylori* is performed according to CYP2C19 genetic polymorphism and 23S rRNA gene mutation of *H. pylori*. After consent for personalized therapy is obtained, endoscopy is performed. At the same time, gastric

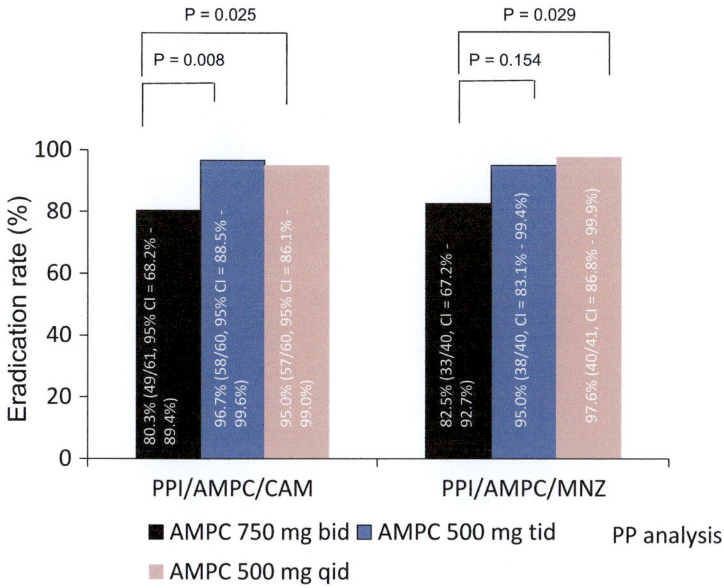

Fig. 14.5 Relationship between the administration method of AMPC in triple-drug therapy and eradication rate. Eradication rate is higher with administration of 500 mg AMPC three or four times a day than with administration of 750 mg AMPC twice a day (cited from Reference [15])

juice sample is collected by aspiration and DNAs are extracted to measure CYP2C19 genetic polymorphism and 23S rRNA gene mutation of *H. pylori* with a rapid SNP analyzer. A drug is then prescribed [16]. Rapid metabolizers of CYP2C19 are generally administered four times the normal PPI dose, while IMs and PMs of CYP2C19 are administered twice the normal amount. When 23S rRNA gene mutation in *H. pylori* is detected, it is considered a CAM-resistant strain and MNZ is prescribed in place of CAM. Detecting the presence or absence of a CAM-resistant strain with a genetic test is similar to that with a sensitivity test. As a rule, a total of 2000 mg AMPC is administered per day by four administrations of 500 mg. Figure 14.6 shows a flowchart of the eradication procedure.

Prescription example of personalized therapy
Dose adjustment of PPI by CYP2C19
CYP2C19 RM: PPI qid
CYP2C19 IM or PM: PPI bid
Choice and administration method of antibiotics
CAM-sensitive strains: CAM 200 mg bid + AMPC 500 mg qid
CAM-resistant strains: MNZ 250 mg bid + AMPC 500 mg qid

(2) Dealing with penicillin allergy

For patients with penicillin allergy, MNZ or CAM can be used in place of AMPC. Similarly, following testing for CYP2C19, PPI dose is increased for RMs, as previously described. If the strain is CAM sensitive, CAM and MNZ

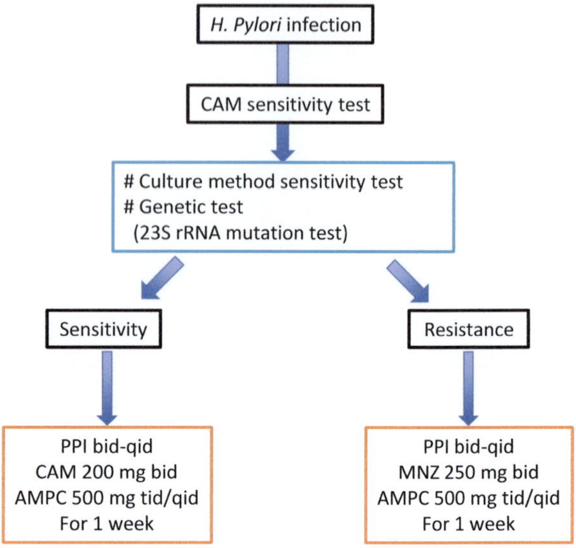

Fig. 14.6 Flowchart of primary eradication therapy in personalized therapy

are used, while if it is CAM resistant, 100 mg sitafloxacin (STFX) in addition to MNZ is administered twice a day [17]. On rare occasions, strains show STFX resistance or certain antibiotics cannot be used because of allergy. For these patients, we concurrently administer 100 mg minocycline (MINO) twice a day (one 100 mg tablet each time).

Eradication example in a patient with penicillin allergy

Dose adjustment of PPI by CYP2C19

CYP2C19 RM: PPI qid

CYP2C19 IM or PM: PPI bid

Antibiotic choice and administration method

CAM-sensitive strains: CAM 200 mg bid + MNZ 250 mg bid for 1 week

CAM-resistant strains: STFX 100 mg bid + MNZ 250 mg bid for 1–2 weeks

Or MINO 100 mg bid + MNZ 250 mg bid for 1week

(3) Dealing with patients with renal dysfunction or receiving dialysis

Several reports have described eradication therapy of *H. pylori* for patients with renal dysfunction. It is important to note that administration of eradication drugs will exacerbate any preexisting renal dysfunction. Sheu et al. [18] compared the PPI/CAM/AMPC regimen with PPI/CAM/MNZ for patients with renal dysfunction and found a mild increase in serum creatinine in the PPI/CAM/AMPC group which did not decrease after the end of administration. The use of AMPC requires particular care, and this agent is in fact better avoided if possible. CAM-sensitive strains can be treated with PPI/CAM/MNZ and CAM-resistant strains with PPI/MNZ/STFX.

Eradication example in patients with renal dysfunction

CAM-sensitive strains: PPI bid + CAM 200 mg bid + MNZ 250 mg bid for 1 week

CAM-resistant strains: PPI bid + STFX 100 mg bid + MNZ 250 mg bid for 1–2 weeks

PPI bid + MINO 100 mg bid + MNZ 250 mg bid for 1 week

In patients receiving dialysis, maximum AMPC concentration in the blood rises to around 3 times that in non-dialytic patients, with a half-life of 8–17 h. Theoretically, therefore, single daily administration of 250 mg AMPC should suffice. In most reports in patients on dialysis, AMPC dose was reduced to one or two 250 mg AMPC capsules given once or twice a day. On dialysis days, AMPC was administered after finishing the dialysis procedure [19–21]. For CAM, single daily administration of 200 mg CAM is sufficient. Also, there are many reports of single daily administration of normal doses of PPI.

Eradication example in cases with dialysis

CAM-sensitive strains: PPI qd + AMPC 500 mg qd + CAM 200 mg qd for 1 week

CAM-resistant strains: PPI qd + AMPC 500 mg qd + MNZ 250 mg bid for 1 week

(4) Personalized therapy in consideration of drug-drug interactions (e.g., warfarin)

PPIs, CAM, and MNZ all influence drug-metabolizing enzymes such as CYPs and transporters. For example, they intensify the effects of warfarin. In addition, the administration of antibiotics impacts intestinal bacteria and thereby additionally intensifies the effects of warfarin. Thus, eradication of *H. pylori* during warfarin administration needs to be carefully controlled. In patients with non-valvular arterial fibrillation, various considerations may be required, such as replacing warfarin with another anticoagulant.

Patients with oral administration of warfarin.

Replacing warfarin with NOAC, such as edoxaban and apixaban.

Prescription example in cases with continuous administration of warfarin.

RPZ 10 mg bid + AMPC 500 mg qid + probiotics qid.

PT-INR should be measured before, during, and after eradication.

(5) Personalization in third-line eradication

In the third-line eradication, antibiotics are selected according to culture results, and PPI dose is adjusted in accordance with CYP2C19 genetic polymorphism.

Basically, most cases have AMPC-sensitive strains until secondary eradication therapy, but some nonsensitive cases are also observed in the third-line eradication therapy. These patients are treated with a combination of STFX and MNZ. If the *H. pylori* strain is AMPC sensitive, a combination of STFX and AMPC is used [22]. Eradication with dual therapy with a PPI and AMPC dosed four times a day is also performed. Personalized therapy is optimized and carried out according to the concurrent drug provided to each patient. The addition of MNZ to the combination of high-dose PPI and AMPC is also effective.

Prescription example in the case with tertiary eradication

Dose adjustment of PPI by CYP2C19

CYP2C19 RM: PPI qid

CYP2C19 IM or PM: PPI bid
Antibiotics
AMPC 500 mg qid + STFX 100 mg bid for 1–2 weeks
AMPC 500 mg qid + MNZ 250 mg bid for 2 weeks
AMPC 500 mg qid + ecabet sodium 1 g qid for 2 weeks
MNZ 250 mg bid + STFX 100 mg bid for 1–2 weeks

(6) Personalized therapy with P-CAB

Vonoprazan (P-CAB) is now available in Japan. In a clinical study, this agent provided eradication rates of 92.6 % and 98 % in primary and secondary eradication therapy, respectively, and thus better values than those with PPI-based eradication therapy. Basic experiments reported that the inhibition of gastric acid secretion by vonoprazan was strong and was not influenced by CYP2C19 genetic polymorphism. In addition, eradication rate was not influenced by CYP2C19 genetic polymorphism. Thus, this drug is considered able to allow the omission of personalization of PPI treatment by CYP2C19 polymorphism and to allow more emphasis on antibiotic sensitivity and antibiotics selection. Further data are expected.

14.8 Conclusion

We have outlined personalized therapy for eradication of *H. pylori* in our department and described a number of specific examples. These personalized approaches described above provide eradication rates of 95 % or higher in both primary and secondary eradication and around 90 % in the third-line eradication therapy. Even when third-line therapy fails, most cases can be successfully treated with other regimen candidates. In addition, we have added P-CAB, a new gastric acid inhibitor, as an adjuvant in eradication treatment. Results obtained to date for P-CAB have been based on one clinical study only, but suggest that personalization based on CYP2C19 genetic polymorphism may become unnecessary; instead, personalization will depend on the greater importance of sensitivity to antibiotics, drug combination, and individual clinical condition.

References

1. Sachs G, et al. Acid, protons and *Helicobacter pylori*. Yale J Biol Med. 1996;69(3):301–16.
2. Furuta T, Graham DY. Pharmacologic aspects of eradication therapy for *Helicobacter pylori* infection. Gastroenterol Clin North Am. 2010;39(3):465–80.
3. Furuta T, Sugimoto M, Shirai N. Individualized therapy for gastroesophageal reflux disease: Potential impact of pharmacogenetic testing based on CYP2C19. Mol Diagn Ther. 2012;16 (4):223–34.
4. Furuta T, et al. CYP2C19 genotype status and effect of omeprazole on intragastric pH in humans. Clin Pharmacol Ther. 1999;65(5):552–61.

5. Furuta T, et al. Effect of genetic differences in omeprazole metabolism on cure rates for *Helicobacter pylori* infection and peptic ulcer. Ann Intern Med. 1998;129(12):1027–30.
6. Furuta T, et al. Effect of genotypic differences in CYP2C19 on cure rates for *Helicobacter pylori* infection by triple therapy with a proton pump inhibitor, amoxicillin, and clarithromycin. Clin Pharmacol Ther. 2001;69(3):158–68.
7. Furuta T, et al. Pharmacogenomics-based tailored versus standard therapeutic regimen for eradication of H. pylori. Clin Pharmacol Ther. 2007;81(4):521–8.
8. Sasaki M, et al, Changes in 12-year first-line eradication rate of *Helicobacter pylori* based on triple therapy with proton pump inhibitor, amoxicillin and clarithromycin. J Clin Biochem Nutr. 47(1):pp. 53–8.
9. Shirai N, et al. Dual therapy with high doses of rabeprazole and amoxicillin versus triple therapy with rabeprazole, amoxicillin, and metronidazole as a rescue regimen for *Helicobacter pylori* infection after the standard triple therapy. Eur J Clin Pharmacol. 2007;63(8):743–9.
10. Furuta T, et al. Influence of CYP2C19 polymorphism and *Helicobacter pylori* genotype determined from gastric tissue samples on response to triple therapy for H. pylori infection. Clin Gastroenterol Hepatol. 2005;3(6):564–73.
11. Nishizawa T, et al. Gatifloxacin resistance and mutations in gyra after unsuccessful *Helicobacter pylori* eradication in Japan. Antimicrob Agents Chemother. 2006;50 (4):1538–40.
12. Hassan IJ, et al. Absence of a post-antibiotic effect (PAE) of beta-lactams against *Helicobacter pylori* NCTC 11637. J Antimicrob Chemother. 1998;42(5):661–3.
13. Furuta T, et al. The dual therapy with 4 times daily dosing of rabeprazole and amoxicillin as the 3rd rescue regimen for eradication of H. pylori. Hepatogastroenterology. 2010;57 (102–103):1314–9.
14. Miehlke S, et al. A prospective, randomized study of quadruple therapy and high-dose dual therapy for treatment of *Helicobacter pylori* resistant to both metronidazole and clarithromycin. Helicobacter. 2003;8(4):310–9.
15. Furuta T, et al. Effect of dosing schemes of amoxicillin on eradication rates of *Helicobacter pylori* with amoxicillin-based triple therapy. J Clin Pharmacol. 2014;54(3):258–66.
16. Furuta T, et al. Personalized therapy for H. pylori infection based on automated SNP analysis. Gastroenterology. 2014;146(5 Suppl 1):104.
17. Furuta T, et al. Eradication of H. pylori infection in patients allergic to penicillin using triple therapy with a PPI, metronidazole and sitafloxacin. Intern Med. 2014;53(6):571–5.
18. Sheu BS, et al. The selection of triple therapy for *Helicobacter pylori* eradication in chronic renal insufficiency. Aliment Pharmacol Ther. 2003;17(10):1283–90.
19. Araki H, et al. Significance of serum pepsinogens and their relationship to *Helicobacter pylori* infection and histological gastritis in dialysis patients. Nephrol Dial Transplant. 1999;14 (11):2669–75.
20. Tokushima H, et al. Eradication of *Helicobacter pylori* in patients with end-stage renal disease undergoing dialysis treatment. Nihon Jinzo Gakkai Shi. 1996;38(8):349–55.
21. Chang WC, et al. *Helicobacter pylori* eradication with a 7-day low-dose triple therapy in hemodialysis patients. Clin Exp Nephrol. 2010;14(5):469–73.
22. Furuta T, et al. Sitafloxacin-based third-line rescue regimens for *Helicobacter pylori* infection in Japan. J Gastroenterol Hepatol. 2014;29(3):487–93.

Chapter 15
Quadruple Regimens for *Helicobacter pylori* Infection

Javier Molina-Infante

Abstract An effective therapy against *Helicobacter pylori* (*H. pylori*) is defined as one achieving at least a 90 % eradication rate. Treatment results are best when reliable schemes are based on susceptibility testing; however, empiric eradication therapies are generally prescribed. Standard triple therapy has been the most recommended regimen, but its efficacy is currently suboptimal worldwide due to rising clarithromycin resistance. Consequently, the scientific community has recently taken on the task of rescuing old empiric quadruple regimens to overcome antibiotic resistance.

The choice of empiric therapy may depend on patient previous antibiotic treatment, local patterns of antibiotic resistance, and drug availability. Currently, the preferred first-line choices are bismuth quadruple therapy and non-bismuth quadruple concomitant therapy. A 14-day bismuth quadruple therapy is expected to have a 95 % efficacy regardless of metronidazole resistance, but its Achilles heel is compliance, besides drug availability. On the contrary, the efficacy of concomitant therapy is challenged by dual clarithromycin and metronidazole resistance. Optimization of all regimens (increased duration, adequate proton pump inhibitor and antibiotic doses, and dosing intervals) is indispensable to maximize their efficacy. Treatment failures should be managed with an alternate regimen using an optimized combination of different antibiotics. Bismuth-containing fluoroquinolone quadruple therapy and bismuth quadruple therapy (if not used previously) have recently shown the best results. Due to regional variation in *H. pylori* resistance patterns, the golden rule for choice of treatment is only to use therapies that work locally (>90–95 % success) and to monitor their effectiveness over time.

Keywords *Helicobacter pylori* • Quadruple • Eradication • Bismuth • Concomitant

---

Disclosure Statement: The author has no conflict of interest to declare.

J. Molina-Infante (✉)
Department of Gastroenterology, Hospital San Pedro de Alcantara, C/ Pablo Naranjo s/n, 10003 Caceres, Spain
e-mail: xavi_molina@hotmail.com

H. Suzuki et al. (eds.), *Helicobacter pylori*, DOI 10.1007/978-4-431-55705-0_15

15.1 Introduction

Thirty years after the transcendental discovery of the originally termed *Campylobacter pyloridis* as a causative agent for gastritis and peptic ulceration in 1984 [1], *Helicobacter pylori* (*H. pylori*) remains the most common bacterial infection in humans, affecting 50 % of the world's population. Eradication of *H. pylori* infection has dramatically changed the natural history of peptic ulcer disease, may influence symptoms in patients with uninvestigated dyspepsia, and even cure patients with gastric mucosa associated lymphoid tissue lymphoma [2]. In addition, *H. pylori* is a well-recognized carcinogen, and *H. pylori* eradication is a major target for primary and secondary gastric cancer prevention programs in high-risk geographical areas. In fact, Japan has recently embarked on population-wide *H. pylori* eradication coupled with surveillance targeted to those with significant remaining risk [3].

Despite the number of studies evaluating *Helicobacter pylori* therapy, the optimal therapeutic regimen has not yet been defined. The standard triple therapy, consisting of a proton pump inhibitor (PPI) plus clarithromycin and either amoxicillin or metronidazole, has been the gold standard therapy for *H. pylori* infection over the last two decades. However, the efficacy of triple therapy at the present time is seriously challenged in many parts of the world, where eradication rates have declined to unacceptably low levels (<80 %), largely related to the development of resistance to clarithromycin [4]. Moreover, rising resistance of *H. pylori* to other antimicrobial drugs, such as metronidazole (e.g., South America, Turkey, Iran, China) and especially fluoroquinolones, has limited the efficacy of alternate regimes without clarithromycin [5]. Complicating it even more, some of the drugs included in first-line or rescue eradication regimens (e.g., bismuth, tetracycline, furazolidone) may be unavailable in some geographical areas. As such, the scientific community has just recently taken on the task of exploring or, better said, rescuing old empiric quadruple therapeutic schemes to overcome antibiotic resistance. New guidelines on *H. pylori* therapy published after 2012, starting with the European Maastricht IV/Florence Consensus Report, have highlighted this evolving trend over the past 3 years. Table 15.1 summarizes these changes, comparing recommended first- and second-line therapy in the most important international guidelines from 2007 to 2014 [6–13].

15.2 Choice of Therapy

The strongest predictor of *H pylori* treatment failure using a regimen proven to be effective elsewhere is antimicrobial resistance [5]. From a microbiological standpoint, treatment results are best when regimens are used to treat patients with organisms susceptible to the antimicrobials chosen. Pretreatment susceptibility testing, either by direct culture of the organism from gastric biopsies or indirectly by molecular testing in gastric biopsies/stools, can be used for this purpose.

Table 15.1 Changing trends in recommendations for *H. pylori* eradication therapy in international consensus guidelines over the past 7 years

	First-line therapy		Second-line therapy
American College of Gastroenterology, 2007 [6]	Triple therapy 7–14 days		Bismuth quadruple therapy
Maastricht III European Consensus Report, 2007 [7]	Triple therapy 7–14 days *Alternate regimen* Bismuth quadruple therapy 10–14 days		Bismuth quadruple therapy *Alternate regimen* PPI and amoxicillin plus metronidazole or tetracycline
Second Asia-Pacific Consensus, 2009* [8]	Triple therapy 7 days *Alternate regimen* Bismuth quadruple therapy 10–14 days		Bismuth quadruple therapy *Alternate regimens* Levofloxacin-containing triple therapy Rifabutin-containing triple therapy
Revised edition guidelines in Japan, 2010* [9]	Triple therapy 7 days		PPI, amoxicillin, and metronidazole for 5–10 days
Maastricht IV/Florence European Consensus Report, 2012 [10]	*CLA-R < 15–20 %* Optimized** triple therapy *Alternate regimen* Bismuth quadruple therapy	*CLA-R > 15–20 %* Bismuth quadruple therapy *Alternate regimen* Sequential or concomitant therapy, if bismuth quadruple not available	Bismuth quadruple therapy, if not used previously Levofloxacin-containing triple therapy
III Spanish Consensus Conference, 2013*** [11]	*Triple therapy still effective* Optimized** triple therapy	*Poor efficacy of triple therapy* Optimized** non-bismuth concomitant therapy	Levofloxacin-based triple therapy
IV Chinese National Consensus Report, 2013 [12]	*Bismuth-based quadruple therapies* Triple therapy and non-bismuth quadruple regimens (e.g., sequential, concomitant) not recommended due to high rates of dual clarithromycin and metronidazole resistance		*Bismuth-containing quadruple therapy*, combining two or more different antibiotics from the first regimen
Revised edition guidelines in Korea, 2014 [13]	*CLA-R < 15–20 %* Triple therapy 7–14 days	*CLA-R > 15–20 %* Bismuth quadruple therapy	*Bismuth quadruple therapy*, if not used previously, or a bismuth-containing regimen combining two or more antibiotics not previously used

**Increasing antibiotic resistance or decreasing efficacy of triple therapy was considered to likely alter the recommendation for first-line therapy in the next version of these guidelines.*
***Optimization stands for extended duration (10–14 days) and high-dose PPI therapy*
****Tetracycline and Pylera® are mostly unavailable in Spain*

Nonetheless, this strategy is hampered by the wide unavailability of these techniques, besides the need for an invasive procedure (endoscopy), cost, or time consumption [14]. As such, one often must choose antibiotic therapy empirically and the best approach is using regimens that have proven to be reliably excellent locally. Considering *H. pylori* is an infectious disease and 100 % success might be obtainable, the efficacy of an antimicrobial therapy should be scored as excellent (>95 % success), good (>90 % success), borderline acceptable (85–89 % success), poor (80–84 % success), or unacceptably low (<80 %) [15].

The choice of therapy should take advantage of knowledge of local resistance patterns, clinical experience, and patient history. The history of the patient's prior antibiotic use and any prior therapies will help identify which antibiotics are likely to be successful and those where resistance is probable [5]. In fact, treatment outcome in a population or a patient can be calculated based on the effectiveness of a regimen for infections with susceptible and with resistant strains coupled with knowledge of the prevalence of resistance (i.e., based on formal measurement, clinical experience, or both). The formula and calculations for predicting the outcome of any antimicrobial therapy have been recently reported [5]. The therapeutic expectations for an individual patient with all available first-line regimens (triple, sequential, concomitant, and bismuth quadruple therapy), depending on the duration of therapy and the rate of clarithromycin and metronidazole resistance, are shown in Fig. 15.1 [5].

Another choice would be empirically using bismuth quadruple therapy, since tetracycline resistance is negligible and metronidazole resistance can be partially overcome by increasing doses and duration. On account of it, bismuth quadruple therapy (the first real effective therapy against *H. pylori* [16]) has resurfaced, and it is gradually becoming the regimen of choice in the era of increasing antibiotic resistance [12, 17].

At the present time, all eradication therapies should be optimized in order to maximize their efficacy [18]. High-dose proton pump inhibitor (PPI) therapy (i.e., 40 mg of omeprazole or equivalent b.i.d.) is recommended in order to guarantee effective and prolonged gastric acid suppression, which may increase the effectiveness of amoxicillin and clarithromycin and avoid the selection of non-replicative *H. pylori* strains. In addition, the prevalence of CYP2C19 rapid metabolizer genotype has been shown to be highest in Europe and in North America (56–81 %), while the proportion is lower (27–38 %) in the Asian population. All PPI molecules are metabolized by cytochrome P450 (CYP) 2C19. Plasma PPI levels and intragastric pHs during PPI treatment are inversely related to CYP2C19 genotype. Accordingly, several meta-analyses have shown that the ultrarapid and rapid metabolizer groups have lower eradication rate compared to other groups [19, 20]. As such, it is conceivable that all patients, especially in Europe and North America, should receive high-dose PPI therapy with a proper dosing interval (twice or above a day) to achieve similar effects in either rapid or poor CYP2C19 metabolizers. Moreover, a 14-day duration is recommended, given the fact it has proven a therapeutic benefit for triple, bismuth quadruple and non-bismuth quadruple, sequential, and concomitant therapy [10, 18, 21].

Antimicrobial prediction	7-day triple therapy	14-day triple therapy	10-day sequential therapy	14-day sequential therapy	14-day concomitant therapy	14-day bismuth quadruple therapy
Clarithromycin and metronidazole susceptible	94%	97%	95%	98%	97%	99%
Clarithromycin resistant-metronidazole susceptible	<20%	50%	80%	88%	97%	99%
Clarithromycin susceptible-metronidazole resistant	94%	97%	75%	75%	97%	95%
Clarithromycin and metronidazole resistant	<20%	50%	<20%	<20%	<50%	95%

Fig. 15.1 Efficacy of clarithromycin-containing regimens for an individual patient, based on predicted resistance to clarithromycin and metronidazole [5]

15.3 First-Line Regimens

Available first-line regimens, with preferred drug doses and dosing intervals, along with caveats for each treatment, are summarized in Table 15.2. The preferred empirical choices are currently 14-day bismuth quadruple therapy or 14-day non-bismuth quadruple concomitant or hybrid (sequential-concomitant) therapy, depending on the local resistance pattern, clinical experience, and patient history of antibiotic use. The selected regimen must be effective, but considerations such as cost, side effects, availability, and ease of administration should also be taken into account.

The Achilles heel of triple therapy is clarithromycin resistance, the Achilles heel of sequential therapy is metronidazole resistance, and the Achilles heel of concomitant therapy is dual *H. pylori* resistance to both clarithromycin and metronidazole. In areas of low clarithromycin resistance (<5 %), 14-day triple therapy, sequential therapy, and concomitant therapy would all be equivalent. Increasing clarithromycin resistance would undermine triple therapy, whereas increasing metronidazole resistance would undermine sequential therapy [5]. It is important to emphasize that cure rates with sequential, hybrid, and concomitant therapy will always be < 90 % when the rate of dual resistant strains is >5 %, >9 %, or

Table 15.2 Current first-line therapeutic recommendations in the era of increasing clarithromycin and metronidazole resistance

	Preferred doses and dosing intervals	Caveat
14-day bismuth-containing classical quadruple therapy	Bismuth salts q.i.d or b.i.d. PPI (double doses) b.i.d. Tetracycline 500 mg q.i.d. Nitroimidazole 500 mg t.i.d.	Availability Complexity Side effects Compliance
14-day bismuth-containing quadruple therapy using Pylera®	PPI (double doses) b.i.d. Pylera 3 pills q.i.d.	Availability Cost Relatively low tetracycline doses
14-day non-bismuth quadruple concomitant therapy	PPI (double doses) b.i.d. Amoxicillin 1 g b.i.d. Clarithromycin 500 mg b.i.d. Nitroimidazole 500 mg b.i.d.	Cure rates ≤90 % if dual resistance rate ≥15 %
14-day non-bismuth quadruple hybrid therapy	*7 days* PPI (double doses) b.i.d. Amoxicillin 1 g b.i.d. *7 days* PPI (double doses) b.i.d. Amoxicillin 1 g b.i.d. Clarithromycin 500 mg b.i.d. Nitroimidazole 500 mg b.i.d.	Cure rates < 90 % if dual resistance rate > 9 %
14-day non-bismuth quadruple sequential therapy	*7 days* PPI (double doses) b.i.d. Amoxicillin 1 g b.i.d. *7 days* PPI (double doses) b.i.d. Clarithromycin 500 mg b.i.d. Nitroimidazole 500 mg b.i.d.	Cure rates < 90 % if dual resistance rate > 5 %
14-day triple therapy	PPI (double doses) b.i.d Amoxicillin 1 g b.i.d Clarithromycin 500 mg b.i.d	Cure rates < 90 % if clarithromycin resistance > 15 %

***Dual resistant H. pylori**: microorganism resistant to both clarithromycin and metronidazole; **b.i.d.**: bis in die, twice a day; **t.i.d.**: ter in die, three times a day; **q.i.d.**: quater in die, four times a day*

>15 %, respectively [5]. As such, any empirical non-bismuth quadruple therapy would likely be a poor choice in settings with documented high clarithromycin and metronidazole resistance (e.g., China, Turkey) or in some specific high-risk patients for acquired antibiotic resistance (e.g., patients who previously took clarithromycin- and/or metronidazole-containing eradication regimens, women where metronidazole has been used for *Trichomonas* infections, immigrants from developing countries) [5]. A decision-making algorithm taking all these consideration into account for first-line eradication regimens is displayed in Fig. 15.2.

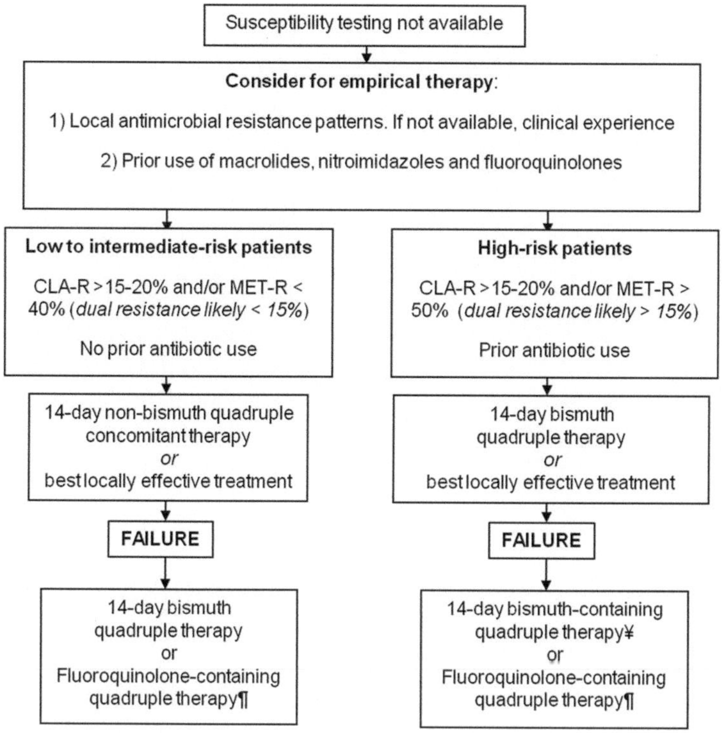

Fig. 15.2 Recommended individualized approach for *Helicobacter pylori* first- and second-line quadruple eradication regimen. ¥ *A quadruple regimen combining bismuth and two other antibiotics not previously used.* ¶ *Only if not used previously and bismuth is not available, taking into consideration local fluoroquinolone resistance rates*

15.3.1 Standard Triple Therapy

The Achilles heel of standard triple therapy is clarithromycin resistance. Due to worldwide increasing clarithromycin resistance, triple therapy should be no longer prescribed on an empirical basis, and its use should be strictly restricted to a minority of settings where clarithromycin resistance is known to be low or where cure rates >90–95 % have been documented in clinical practice (e.g., Northern Europe [22] or Thailand [23]) When used, it should be given for 14 days with double-dose PPI twice a day [10, 18].

15.3.2 Non-Bismuth Quadruple Sequential Therapy

Sequential therapy was developed in Italy in 2000 as a replacement for triple therapy, in order to overcome the problem of clarithromycin resistance. It initially consisted of 5 days of PPI therapy plus amoxicillin, followed by a further 5 days of PPI with two other antibiotics, usually clarithromycin and a nitroimidazole [24]. Several trials, meta-analyses, and pooled data analysis, mostly from Italy, confirmed between 2007 and 2009 the advantage of 10-day sequential over 7- or 10-day triple therapy, with mean cure rates > 90 % [25–27]. In 2012 and 2013, two updated meta-analysis [28, 29] and a systematic review [30], including studies from Asia, Latin America, and European regions different from Italy, exhibited mean eradication rates notably lower (79–84 %) than those reported in early Italian trials. These poor results were further confirmed in a global meta-analysis in 2013 and a further systematic review in 2015 with overall cure rates of 84 % (95 %CI 82.1–86.4 %), besides confirming the lack of advantage over 14-day triple therapy [31, 32]. Further analyses have shown that the efficacy of sequential therapy was undermined by metronidazole resistance and dual clarithromycin-metronidazole resistance [5, 33]. As such, sequential therapy is not a good empiric choice and should be limited to geographical areas where clarithromycin resistance is high, but metronidazole resistance low [5].

15.3.3 Non-Bismuth Quadruple Concomitant Therapy

The concept of a "non-bismuth quadruple regimen" or "concomitant" regimen consists of converting standard triple therapy into a quadruple therapy by the addition of 500 mg of metronidazole or tinidazole twice daily. This therapeutic regimen was first described in 1998 [34, 35], but resurfaced in 2010 as an alternative therapy to triple and sequential therapy. Recent meta-analyses have consistently shown its advantage over triple therapy [36, 37]. Unlike sequential therapy, several recent studies over the last 5 years have proven the efficacy of concomitant therapy (cure rates ≥ 90 %) in some specific Asian (Thailand [38], Japan [39], Taiwan [40]) and European (Spain [41, 42], Greece [43]) geographical areas. Nonetheless, it is important to acknowledge it has failed to achieve good results in regions where dual resistance is known to be high (Latin America [44], Korea [45], Turkey [46], and likely Italy [47, 48]).

Currently, 14-day concomitant therapy should be the preferred non-bismuth quadruple therapy because it is less complicated than sequential therapy since the drugs are not changed halfway through the treatment course, besides it is the most effective to overcome dual resistance. When facing *H. pylori* dual resistant strains, cure rates for concomitant therapy have been notably higher (18/23 (78 %)) when compared to those of sequential therapy (9/27 (33 %)) [49]. Therefore, concomitant therapy would always be equal or superior to sequential therapy [5]. However, it is important to stress that concomitant therapy will not achieve successful cure rates in settings with high rates of metronidazole resistance (>50–60 %) along with high clarithromycin resistance.

15.3.4 Non-Bismuth Quadruple Hybrid (Sequential-Concomitant) Therapy

The hybrid sequential-concomitant regimen is a therapeutic innovation which includes a proton pump inhibitor (PPI) plus amoxicillin for 14 days, adding clarithromycin and a nitroimidazole for the final 7 days [50]. Several studies have been conducted with good results in Spain, Iran, and Taiwan [41, 50–52], but poor results in Italy and Korea [47, 48, 53]. It could be considered in the same populations where concomitant therapy is recommended; however, 14-day hybrid therapy is expected to fall below 90 % when clarithromycin-metronidazole resistance exceeds 9 % [5]. Hybrid therapy provides the advantage of improved safety, convenience, and better compliance. A recent first meta-analyses on hybrid therapy has not show relevant differences in terms of efficacy with the sequential and concomitant therapy [54]. Further studies, however, are required to validate this therapy in settings with different patterns of resistance.

15.3.5 Bismuth Quadruple Therapy

This is the oldest effective therapy [16], mostly underappreciated and a resurfacing one on the account of increasing failure of clarithromycin-containing therapies. Doses and duration are critical to outcome, especially in the presence of metronidazole resistance. Using this regimen at full doses and for 14 days, one can expect 95 % or greater treatment success, irrespective of the level of metronidazole resistance [5]. In fact, recent studies in China identified bismuth-based regimens that were highly effective despite the high prevalence of resistance to metronidazole, fluoroquinolones, and macrolides [55]. Accordingly, updated 2013 Chinese guidelines have proposed this regimen as first-line therapy in China [12].

Bismuth salts or tetracycline is not widely available at the present time, and it is important to acknowledge that doxycycline cannot successfully substitute for tetracycline [5]. The main disadvantages of bismuth quadruple therapy are its complexity and frequent side effects, both of which may hamper compliance with therapy. As such, bismuth quadruple therapy is the regimen most in need for optimization and simplification to improve convenience and minimize side effects. Recent studies conducted in Italy and China have shown similar success rates using twice a day bismuth and full q.i.d. antibiotic doses [56, 57]. In this context, a combination capsule containing bismuth subcitrate 140 mg, metronidazole 125 mg, and tetracycline 125 mg, sold under license as Pylera®, has been recently approved. This new capsule has proven effective [58] and could improve compliance since the new regimen will use three capsules four times daily plus PPI twice daily, instead of four to eight pills four times daily and a PPI twice daily in the standard bismuth quadruple therapy. However, its use is currently limited by its high cost, the fact that only a 10-day regimen is available in a prepackaged form and the dose of tetracycline is only 1500 mg. Moreover, it remains unknown if these more convenient formulations produce improved compliance, especially when considering that side effects are often the reason for poor compliance with bismuth therapy. Head-to-head comparisons with standard bismuth triple therapy and with b.i.d. dosing for bismuth are needed especially in population with high rates of metronidazole resistance.

15.4 Second-Line Regimens

An initial attempt at eradicating *H. pylori* fails in approximately 20 % of patients [5]. Despite the number of studies, the optimal retreatment regimen has not yet been defined. Our therapeutic target, similarly to first-line regimens, should be at least 90 % cure rates. The empiric choice of a second-line therapy primarily depends on which treatment was used initially, the local rate of fluoroquinolone resistance, and availability of either bismuth or tetracycline. For patients failing an initial course of therapy, an alternate regimen, preferably quadruple, using a different combination of two antibiotics for 14 days should be prescribed. After failure of a clarithromycin- or metronidazole-containing treatment, clinicians should assume that *H. pylori* are likely resistance to the antibiotic used, so it is not appropriate to repeat the same antibiotics. Exceptions to this rule are amoxicillin and bismuth. Most patients will be cured by using bismuth quadruple therapy after a failure of clarithromycin-containing regimens [59]. In geographical areas where bismuth quadruple therapy is unavailable, a fluoroquinolone-containing therapy is the most commonly used regimen for a second line. This should only be considered if no fluoroquinolone, including ciprofloxacin, was used before [5]. The available regimens for second-line therapy, with preferred duration, doses, and interval dosing, besides potential caveats, are summarized in Table 15.3. A recommended therapeutic algorithm for *H. pylori* second-line therapy is displayed in Fig. 15.2.

15.4.1 Fluoroquinolone-Containing Therapies

Levofloxacin is a fluoroquinolone with a broad spectrum of activity against *H. pylori*. Fluoroquinolone-containing therapy has been proposed in several international guidelines as a second-line therapy [8, 10, 11], and it remains the most commonly used regimen in geographical areas where bismuth quadruple therapy is unavailable. This should only be considered if no fluoroquinolone, including ciprofloxacin, was used before [5]. Of note, there is also concern about rapidly increasing prevalence of fluoroquinolone resistance *H. pylori* strains in recent years, mostly due to widespread use of levofloxacin for the ear, nose, and throat, respiratory tract, and urinary infections. Recent studies have reported levofloxacin resistance rates ranging from 63 % in China [12], 31 % in the USA [60], and 14 % in

Table 15.3 Current recommendations for second-line eradication therapy in the era of increasing fluoroquinolone resistance, in patients with high risk of acquired clarithromycin and/or metronidazole resistance after failure of a first eradication regimen

	Preferred doses and dosing intervals	Caveat
14-day fluoroquinolone triple therapy	PPI (double doses) b.i.d. Amoxicillin 1 g b.i.d. Levofloxacin 500 mg	Cure rates ≤ 90 % if levofloxacin resistance ≥ 12 %
14-day bismuth-containing fluoroquinolone quadruple therapy	Bismuth salts 240 mg b.i.d. PPI (double doses) b.i.d. Amoxicillin 1 g b.i.d. Levofloxacin 500 mg	Cure rates ≤ 90 % if levofloxacin resistance ≥ 25 %
14-day bismuth-containing quadruple therapy	PPI (double doses) b.i.d. Bismuth salts 240 mg b.i.d. *Plus a combination of 2 antibiotics among* Amoxicillin 1 g t.i.d. or b.i.d. Nitroimidazole 500 mg t.i.d. Tetracycline 500 mg q.i.d. Furazolidone 100 mg t.i.d.	Availability Side effects Complexity Compliance Potential genotoxic and carcinogenetic effects of furazolidone

b.i.d.*: bis in die, twice a day;* ***t.i.d.****: ter in die, three times a day;* ***q.i.d.****: quater in die, four times a day*

Europe [22]. In line with this trend, a recent review revealed a weighted suboptimal efficacy of 76 % for levofloxacin-containing rescue regimens [61]. The most prescribed regimen, a triple therapy including PPI, amoxicillin, and levofloxacin, provides cure rates of typically <80 % with a 7-day duration, and extending the duration to 10 days improves outcome, but typically below 90 % [62, 63]. Treatment success > 90 % with 14-day fluoroquinolone triple therapy can be expected when fluoroquinolone resistance rates are < 12 % [64], whereas addition of bismuth leading to 14-day bismuth-containing fluoroquinolone quadruple therapy can guarantee cure rates > 90 % in areas with a fluoroquinolone resistance of up to approximately 25 % [64, 65]. Therefore, awareness of local resistance rates or close monitoring of cure rates is mandatory in order to promptly detect inefficacy of these therapies. The role of newer fluoroquinolones, such as sitafloxacin and gemifloxacin, to overcome fluoroquinolone resistance needs to be validated in further studies.

15.4.2 Bismuth Quadruple Therapy, Including Furazolidone-Containing Regimens

Classical bismuth quadruple regimen has been also proposed in international guidelines as a rescue therapy [6–8, 10, 12, 13]. In a recent meta-analysis with a systematic review, its weighted efficacy was 77 %, once again suboptimal, but including a wide range of different durations, drug doses, and interval dosing [61]. This study also showed no significant differences regarding efficacy between rescue fluoroquinolone-containing and bismuth quadruple therapies, albeit fluoroquinolone therapy was significantly better tolerated and induced significantly fewer side effects [61].

Concerns with this therapy are complexity, patient compliance, and availability. Bismuth quadruple therapy, however, is the regimen with the most unanswered questions regarding what are the optimal doses and frequencies of drug administration. A recent study from China has evaluated the efficacy of 2-week bismuth quadruple therapies (PPI, bismuth salts, and two antibiotics) with different antibiotic combinations in patients with previous eradication treatment failure [57]. They used for all four evaluated regimens twice a day bismuth (220 mg b.i.d.) and full doses and adequate dosing intervals for the antibiotics (amoxicillin 1000 t.i.d., tetracycline 500 mg q.i.d., metronidazole 400 mg t.i.d., furazolidone 100 mg t.i.d.). Cure rates were excellent (>90 %), regardless of the presence of clarithromycin, levofloxacin, or metronidazole resistance, with the best results for the furazolidone-containing regimens. Of note, these bismuth quadruple regimens were equally effective for patients allergic to penicillin, combining tetracycline with either metronidazole or furazolidone. However, furazolidone is seldom available in developed countries, and concern about possible genotoxic and carcinogenetic effects may prevent its implementation. This study, however, importantly highlights that bismuth quadruple therapy compliance and efficacy can be improved through optimization of therapy.

References

1. Marshall BJ, Warren JR. Unidentified curved bacilli in the stomach of patients with gastritis and peptic ulceration. Lancet. 1984;1:1311–5.
2. Graham DY. History of *Helicobacter pylori*, duodenal ulcer, gastric ulcer and gastric cancer. World J Gastroenterol. 2014;20:5191–204. doi:10.3748/wjg.v20.i18.5191.
3. Shiotani A, Cen P, Graham DY. Eradication of gastric cancer is now both possible and practical. Semin Cancer Biol. 2013;23:492–501. doi:10.1016/j.semcancer.2013.07.004.
4. Graham DY, Fischbach L. *Helicobacter pylori* treatment in the era of increasing antibiotic resistance. Gut. 2010;59:1143–53. doi:10.1136/gut.2009.192757.
5. Graham DY, Lee YC, Wu MS. Rational *Helicobacter pylori* therapy: Evidence based medicine rather than medicine based evidence. Clin Gastroenterol Hepatol. 2014;12:177–86. doi:10.1016/j.cgh.2013.05.028.
6. Chey WD, Wong BC, Practice Parameters Committee of the American College of Gastroenterology. American college of gastroenterology guideline on the management of *Helicobacter pylori* infection. Am J Gastroenterol. 2007;102:1808–25.
7. Malfertheiner P, Megraud F, O'Morain C, Bazzoli F, El-Omar E, Graham D, et al. Current concepts in the management of *Helicobacter pylori* infection: The Maastricht III consensus report. Gut. 2007;56:772–81.
8. Fock KM, Katelaris P, Sugano K, Ang TL, Hunt R, Talley NJ, et al. Second Asia-Pacific conference. Second Asia-Pacific consensus guidelines for *Helicobacter pylori* infection. J Gastroenterol Hepatol. 2009;24:1587–600. doi:10.1111/j.1440-1746.2009.05982.x.

9. Asaka M, Kato M, Takahashi S, Fukuda Y, Sugiyama T, Ota H, et al. Guidelines for the management of *Helicobacter pylori* infection in Japan: 2009 revised edition. Helicobacter. 2010;15:1–20. doi:10.1111/j.1523-5378.2009.00738.x.
10. Malfertheiner P, Megraud F, O'Morain CA, Atherton J, Axon AT, Bazzoli F, Gensini GF, Gisbert JP, Graham DY, Rokkas T, El-Omar EM, Kuipers EJ, European Helicobacter Study Group. Management of *Helicobacter pylori* infection--the Maastricht IV/ Florence consensus report. Gut. 2012;61:646–64. doi:10.1136/gutjnl-2012-302084.
11. Gisbert JP, Calvet X, Bermejo F, Boixeda D, Bory F, Bujanda L, et al. III Spanish consensus conference on *Helicobacter pylori* infection. Gastroenterol Hepatol. 2013;36:340–74. doi:10.1016/j.gastrohep.2013.01.011.
12. Chinese Society of Gastroenterology, Chinese Study Group on Helicobacter pylori, Liu WZ, Xie Y, Cheng H, Lu NH, Hu FL, Zhang WD, et al. Fourth Chinese national consensus report on the management of *Helicobacter pylori* infection. J Dig Dis. 2013;14:211–21. doi:10.1111/1751-2980.12034.
13. Kim SG, Jung HK, Lee HL, Jang JY, Lee H, Kim CG, Korean College of Helicobacter and Upper Gastrointestinal Research, et al. J Gastroenterol Hepatol. 2014;29:1371–86. doi:10.1111/jgh.12607.
14. Gisbert JP. Is culture necessary before first-line treatment for *Helicobacter pylori* infection? Intern Med. 2011;50:2717.
15. Graham DY, Lu H, Yamaoka Y. A report card to grade *Helicobacter pylori* therapy. Helicobacter. 2007;12:275–8.
16. Borody TJ, Cole P, Noonan S, Morgan A, Lenne J, Hyland L, et al. Recurrence of duodenal ulcer and *Campylobacter pylori* infection after eradication. Med J Aust. 1989;151:431–5.
17. Megraud F. The challenge of *Helicobacter pylori* resistance to antibiotics: The comeback of bismuth-based quadruple therapy. Ther Adv Gastroenterol. 2012;5:103–9. doi:10.1177/1756283X11432492.
18. Molina-Infante J, Gisbert JP. Optimizing clarithromycin-containing therapy for *Helicobacter pylori* in the era of antibiotic resistance. World J Gastroenterol. 2014;20:10338–47. doi:10.3748/wjg.v20.i30.10338.
19. Zhao F, Wang J, Yang Y, Wang X, Shi R, Xu Z, et al. Effect of CYP2C19 genetic polymorphisms on the efficacy of proton pump inhibitor based triple therapy for *Helicobacter pylori* eradication: A meta-analysis. Helicobacter. 2008;13:532–41. doi:10.1111/j.1523-5378.2008.00643.x.
20. Tang HL, Li Y, Hu YF, Xie HG, Zhai SD. Effects of CYP2C19 loss-of function variants on the eradication of *H. pylori* infection in patients treated with proton pump inhibitor-based triple therapy regimens: A meta analysis of randomized clinical trials. PLoS One. 2013;8:e62162. doi:10.1371/journal.pone.0062162.
21. Yuan Y, Ford AC, Khan KJ, Gisbert JP, Forman D, Leontiadis GI, Tse F, Calvet X, Fallone C, Fischbach L, Oderda G, Bazzoli F, Moayyedi P. Optimum duration of regimens for *Helicobacter pylori* eradication. Cochrane Database Syst Rev. 2013;12, CD008337. doi:10.1002/14651858.CD008337.pub2.
22. Megraud F, Coenen S, Versporten A, Kist M, Lopez-Brea M, Hirschl AM, Andersen LP, Goossens H, Glupczynski Y, Study Group participants. *Helicobacter pylori* resistance to antibiotics in Europe and its relationship to antibiotic consumption. Gut. 2013;62:34–42. doi:10.1136/gutjnl-2012-302254.
23. Prasertpetmanee S, Mahachai V, Vilaichone RK. Improved efficacy of proton pump inhibitor – Amoxicillin – Clarithromycin triple therapy for *Helicobacter pylori* eradication in low Clarithromycin resistance areas or for tailored therapy. Helicobacter. 2013;18:270–3. doi:10.1111/hel.12041.
24. Zullo A, Rinaldi V, Winn S, Meddi P, Lionetti R, Hassan C, Ripani C, Tomaselli G, Attili AF. A new highly effective short-term therapy schedule for *Helicobacter pylori* eradication. Aliment Pharmacol Ther. 2000;14:715–8.

25. Zullo A, De Francesco V, Hassan C, Morini S, Vaira D. The sequential therapy regimen for *Helicobacter pylori* eradication: A pooled-data analysis. Gut. 2007;56:1353–7.
26. Jafri NS, Hornung CA, Howden CW. Meta-analysis: Sequential therapy appears superior to standard therapy for *Helicobacter pylori* infection in patients naive to treatment. Ann Intern Med. 2008;148:923–31.
27. Gatta L, Vakil N, Leandro G, Di Mario F, Vaira D. Sequential therapy or triple therapy for *Helicobacter pylori* infection: Systematic review and meta-analysis of randomized controlled trials in adults and children. Am J Gastroenterol. 2009;104:3069–79. doi:10.1038/ajg.2009.555.
28. Horvath A, Dziechciarz P, Szajewska H. Meta-analysis: Sequential therapy for *Helicobacter pylori* eradication in children. Aliment Pharmacol Ther. 2012;36:534–41. doi:10.1111/j.1365-2036.2012.05229.x.
29. Yoon H, Lee DH, Kim N, Park YS, Shin CM, Kang KK, Oh DH, Jang DK, Chung JW. Meta-analysis: Is sequential therapy superior to standard triple therapy for *Helicobacter pylori* infection in Asian adults? J Gastroenterol Hepatol. 2013;28:1801–9. doi:10.1111/jgh.12397.
30. Zullo A, Hassan C, Ridola L, De Francesco V, Vaira D. Standard triple and sequential therapies for *Helicobacter pylori* eradication: An update. Eur J Intern Med. 2013;24:16–9. doi:10.1016/j.ejim.2012.07.006.
31. Gatta L, Vakil N, Vaira D, Scarpignato C. Global eradication rates for *Helicobacter pylori* infection: Systematic review and meta-analysis of sequential therapy. BMJ. 2013;28:1801–9. doi:10.1136/bmj.f4587.
32. Feng L, Wen MY, Zhu YJ, Men RT, Yang L. Sequential therapy or standard triple therapy for *Helicobacter pylori* infection: An updated systematic review. Am J Ther. 2015 Jan 7. [Epub ahead of print].
33. Liou JM, Chen CC, Chen MJ, Chen CC, Chang CY, Fang YJ, Taiwan Helicobacter Consortium, et al. Sequential versus triple therapy for the first-line treatment of *Helicobacter pylori*: A multicentre, open-label, randomised trial. Lancet. 2013;381:205–13. doi:10.1016/S0140-6736(12)61579-7.
34. Treiber G, Ammon S, Schneider E, Klotz U. Amoxicillin/metronidazole/omeprazole/clarithromycin: A new, short quadruple therapy for *Helicobacter pylori* eradication. Helicobacter. 1998;3:54–8.
35. Okada M, Oki K, Shirotani T, Seo M, Okabe N, Maeda K, et al. A new quadruple therapy for the eradication of *Helicobacter pylori*. Effect of pretreatment with omeprazole on the cure rate. J Gastroenterol. 1998;3:640–5.
36. Essa AS, Kramer JR, Graham DY, Treiber G. Meta-analysis: Four-drug, three-antibiotic, non-bismuth-containing "concomitant therapy" versus triple therapy for *Helicobacter pylori* eradication. Helicobacter. 2009;14:109–18. doi:10.1111/j.1523-5378.2009.00671.x.
37. Gisbert JP, Calvet X. Update on non-bismuth quadruple (concomitant) therapy for eradication of *Helicobacter pylori*. Clin Exp Gastroenterol. 2012;5:23–34. doi:10.2147/CEG.S25419.
38. Kongchayanun C, Vilaichone RK, Pornthisarn B, Amornsawadwattana S, Mahachai V. Pilot studies to identify the optimum duration of concomitant *Helicobacter pylori* eradication therapy in Thailand. Helicobacter. 2012;17:282–5. doi:10.1111/j.1523-5378.2012.00953.x.
39. Yanai A, Sakamoto K, Akanuma M, Ogura K, Maeda S. Non-bismuth quadruple therapy for first-line *Helicobacter pylori* eradication: A randomized study in Japan. World J Gastrointest Pharmacol Ther. 2012;3:1–6. doi:10.4292/wjgpt.v3.i1.1.
40. Hsu PI, Wu DC, Chen WC, Tseng HH, Yu HC, Wang HM, et al. Randomized controlled trial comparing 7-day triple, 10-day sequential, and 7-day concomitant therapies for *Helicobacter pylori* infection. Antimicrob Agents Chemother. 2014;58:5936–42. doi:10.1128/AAC.02922-14.
41. Molina-Infante J, Romano M, Fernandez-Bermejo M, Federico A, Gravina AG, Pozzati L, et al. Optimized nonbismuth quadruple therapies cure most patients with *Helicobacter pylori* infection in populations with high rates of antibiotic resistance. Gastroenterology. 2013;145:121–8. doi:10.1053/j.gastro.2013.03.050.

42. Molina-Infante J, Lucendo AJ, Angueira T, Rodriguez-Tellez M, Perez-Aisa A, Balboa A, et al. Optimized triple and concomitant therapy for *Helicobacter pylori* eradication: The OPTRICON study. Aliment Pharmacol Ther. 2015;41:581–9. doi:10.1111/apt.13069.
43. Georgopoulos SD, Xirouchakis E, Martinez-Gonzalez B, Sgouras DN, Spiliadi C, Mentis AF, Laoudi F. Clinical evaluation of a ten-day regimen with Esomeprazole, Metronidazole, Amoxicillin, and Clarithromycin for the eradication of *Helicobacter pylori* in a high Clarithromycin resistance area. Helicobacter. 2013;18:459–67. doi:10.1111/hel.12062.
44. Greenberg ER, Anderson GL, Morgan DR, et al. 14-day triple, 5-day concomitant, and 10-day sequential therapies for *Helicobacter pylori* infection in seven Latin American sites: A randomised trial. Lancet. 2011;378:507–14. doi:10.1016/S0140-6736(11)60825-8.
45. Lim JH, Lee DH, Choi C, et al. Clinical outcomes of two-week sequential and concomitant therapies for *Helicobacter pylori* eradication: A randomized pilot study. Helicobacter. 2013;18:180–6. doi:10.1111/hel.12034.
46. Sharara AI, Sarkis FS, El-Halabi MM, et al. Challenging the dogma: A randomized trial of standard vs. half-dose concomitant nonbismuth quadruple therapy for *Helicobacter pylori* infection. United Eur Gastroenterol J. 2014;2:179–88. doi:10.1177/2050640614530919.
47. Zullo A, Scaccianoce G, De Francesco V, et al. Concomitant, sequential, and hybrid therapy for *H. pylori* eradication: A pilot study. Clin Res Hepato Gastroenterol. 2013;37:647–50. doi:10.1016/j.clinre.2013.04.003.
48. De Francesco V, Hassan C, Ridola L, Giorgio F, Ierardi E, Zullo A. Sequential, concomitant and hybrid first-line therapies for *H. pylori* eradication: A prospective, randomized study. J Med Microbiol. 2014;63:748–52. doi:10.1099/jmm.0.072322-0.
49. Georgopoulos SD, Xirouchakis E, Mentris A. Is there a nonbismuth quadruple therapy that can reliably overcome bacterial resistance? Gastroenterology. 2013;145:1496–7.
50. Hsu PI, Wu DC, Wu YC, et al. Modified sequential *Helicobacter pylori* therapy: Proton pump inhibitor and amoxicillin for 14 days with clarithromycin and metronidazole added as a quadruple (hybrid) therapy for the final 7 days. Helicobacter. 2011;16:139–45. doi:10.1111/j.1523-5378.2011.00828.x.
51. Sardarian H, Fakheri H, Hosseini V, et al. Comparison of hybrid and sequential therapies for *Helicobacter pylori* eradication in Iran: A prospective randomized trial. Helicobacter. 2013;18:129–34. doi:10.1111/hel.12017.
52. Wu JY, Hsu PI, Wu DC, et al. Feasibility of shortening 14-day hybrid therapy while maintaining an excellent *Helicobacter pylori* eradication rate. Helicobacter. 2014;19:207–13. doi:10.1111/hel.12113.
53. Oh DH, Lee DH, Kang KK, et al. The efficacy of hybrid therapy as first-line regimen for *Helicobacter pylori* infection compared with sequential therapy. J Gastroenterol Hepatol. 2014;29:1171–6.
54. Wang B, Wang YH, Lv ZF, et al. Efficacy and safety of hybrid therapy for *Helicobacter pylori* infection: A systematic review and meta-analysis. Helicobacter. 2014 Nov 8. doi:10.1111/hel.12180. [Epub ahead of print].
55. Lu H, Zhang W, Graham DY. Bismuth-containing quadruple therapy for *Helicobacter pylori*: Lessons from China. Eur J Gastroenterol Hepatol. 2013;25:1134–40. doi:10.1097/MEG.0b013e3283633b57.
56. Dore MP, Farina V, Cuccu M, Mameli L, Massarelli G, Graham DY. Twice-a-day bismuth-containing quadruple therapy for *Helicobacter pylori* eradication: A randomized trial of 10 and 14 days. Helicobacter. 2011;16:295–300. doi:10.1111/j.1523-5378.2011.00857.x.
57. Liang X, Xu X, Zheng Q, Zhang W, Sun Q, Liu W, Xiao S, Lu H. Efficacy of bismuth containing quadruple therapies for clarithromycin-, metronidazole-, and fluoroquinolone resistant *Helicobacter pylori* infections in a prospective study. Clin Gastroenterol Hepatol. 2013;11:802–7. doi:10.1016/j.cgh.2013.01.008.
58. Malfertheiner P, Bazzoli F, Delchier JC, Celiñski K, Giguère M, Rivière M, Mégraud F, Pylera Study Group. Helicobacter pylori eradication with a capsule containing bismuth subcitrate

potassium, metronidazole, and tetracycline given with omeprazole versus clarithromycin-based triple therapy: A randomised, open-label, non-inferiority, phase 3 trial. Lancet. 2011;377:905–13. doi:10.1016/S0140-6736(11)60020-2.
59. Delchier JC, Malfertheiner P, Thieroff-Ekerdt R. Use of a combination formulation of bismuth, metronidazole and tetracycline with omeprazole as a rescue therapy for eradication of *Helicobacter pylori*. Aliment Pharmacol Ther. 2014;40:171–7. doi:10.1111/apt.12808.
60. Shiota S, Reddy R, Alsarraj A, El-Serag HB, Graham DY. Antibiotic resistance of *Helicobacter pylori* among male US veterans. Clin Gastroenterol Hepatol. 2015 Feb 11. pii: S1542–3565(15)00122-6. doi:10.1016/j.cgh.2015.02.005.
61. Marín AC, McNicholl AG, Gisbert JP. A review of rescue regimens after clarithromycin-containing triple therapy failure (for *Helicobacter pylori* eradication). Expert Opin Pharmacother. 2013;14:843–61. doi:10.1517/14656566.2013.782286.
62. Gisbert JP, Pérez-Aisa A, Bermejo F, Castro-Fernández M, Almela P, Barrio J, H. pylori Study Group of the Asociación Española de Gastroenterología (Spanish Gastroenterology Association). Second-line therapy with levofloxacin after failure of treatment to eradicate helicobacter pylori infection: Time trends in a Spanish multicenter study of 1000 patients. J Clin Gastroenterol. 2013;47:130–5. doi:10.1097/MCG.0b013e318254ebdd.
63. Chuah SK, Tai WC, Hsu PI, Wu DC, Wu KL, Kuo CM, Chiu YC, Hu ML, Chou YP, Kuo YH, Liang CM, Chiu KW, Hu TH. The efficacy of second-line anti-Helicobacter pylori therapy using an extended 14-day levofloxacin/amoxicillin/proton-pump inhibitor treatment--a pilot study. Helicobacter. 2012;17:374–81. doi:10.1111/j.1523-5378.2012.00960.x.
64. Liao J, Zheng Q, Liang X, Zhang W, Sun Q, Liu W, Xiao S, Graham DY, Lu H. Effect of fluoroquinolone resistance on 14-day levofloxacin triple and triple plus bismuth quadruple therapy. Helicobacter. 2013;18:373–7. doi:10.1111/hel.12052.
65. Gisbert JP, Romano M, Gravina AG, Solís-Muñoz P, Bermejo F, Molina-Infante J, et al. *Helicobacter pylori* second-line rescue therapy with levofloxacin- and bismuth-containing quadruple therapy, after failure of standard triple or non-bismuth quadruple treatments. Aliment Pharmacol Ther. 2015 Feb 23. doi:10.1111/apt.13128.

Chapter 16
Third-Line Eradication Therapy

Toshihiro Nishizawa and Hidekazu Suzuki

Abstract The success rates of *Helicobacter pylori* eradication treatment may be decreasing in clinical practice, mainly because of the widespread use of antibiotics. Fortunately, the resistance to amoxicillin, tetracycline, and rifabutin has remained low. After the failure of second-line treatment, subsequent treatment should be guided by antimicrobial susceptibility testing whenever possible. With the addition of newer rapid molecular tests to detect *H. pylori* and the determination of the presence of point mutations, resistance-guided therapy may play a more important role in the future. During empirical third-line therapy, antibiotics used previously should be avoided. Third-line treatment options include fluoroquinolone, rifabutin, tetracycline, furazolidone, and high-dose proton pump inhibitor/amoxicillin therapy. Sitafloxacin shows activities 8–16-fold or greater than those of levofloxacin and could overcome the resistance of *H. pylori* with *gyrA* mutations. However, sitafloxacin is not widely available in many countries and has predominantly been employed in Japan. The use of rifabutin should be reserved for the treatment of multiresistant *Mycobacterium tuberculosis* strains, so that rifabutin is preferably used only as a last resort after amoxicillin, clarithromycin, metronidazole, tetracycline, and fluoroquinolone have failed to eradicate *H. pylori*.

Keywords *Helicobacter pylori* • Third-line therapy • Fluoroquinolone • Rifabutin

16.1 Introduction

Triple treatment, including the proton pump inhibitor (PPI)/amoxicillin and clarithromycin or metronidazole, as proposed at the first Maastricht conference, was globally accepted as first-line *Helicobacter pylori* eradication therapy. The

T. Nishizawa (✉)
Division of Research and Development for Minimally Invasive Treatment, Cancer Center, Keio University School of Medicine, 35 Shinanomachi, Shinjuku-ku, Tokyo 160-8582, Tokyo, Japan
e-mail: nisizawa@kf7.so-net.nejp

H. Suzuki
Medical Education Center, Keio University School of Medicine, 35 Shinanomachi, Shinjuku-ku, Tokyo, 160-8582 Tokyo, Japan

H. Suzuki et al. (eds.), *Helicobacter pylori*, DOI 10.1007/978-4-431-55705-0_16

updated Asia-Pacific Consensus published in 2009 acknowledged that there was an increasing rate of resistance to clarithromycin and metronidazole and that this had led to a lower efficacy of clarithromycin-based triple therapy [1]. The Maastricht IV consensus recommended that if the treatment population was known to have a clarithromycin resistance rate above 20 %, clarithromycin-containing triple therapy should be avoided. Instead, bismuth-containing quadruple therapy or non-bismuth quadruple therapy (either sequential or concomitant) could be used as first-line empirical treatment. After failure of the quadruple therapy, levofloxacin-containing triple therapy is recommended [2]. After failure of the second-line treatment, and whenever possible, subsequent treatment should be guided by testing for antimicrobial susceptibility. The promising candidates for third-line therapy are tetracycline, rifabutin, sitafloxacin, furazolidone, and high-dose PPI/amoxicillin therapy [3].

This chapter reviews the evidence accumulated thus far on the efficacy and safety of the third-line therapy for *H. pylori* infection.

16.2 Tetracycline

16.2.1 Tetracycline Resistance

Fortunately, the prevalence of resistance to tetracycline has also remained low; it was reported to be less than 2 % in most studies [4, 5], and Heep et al. have found no secondary resistance to tetracycline in *H. pylori* isolates from patients who failed one or more eradication therapies [6]. Tetracyclines are bacteriostatic drugs that exert their antimicrobial effect on the 30S subunit of the ribosome and block the binding of aminoacyl-tRNA, resulting in impaired protein biosynthesis. The resistance of *H. pylori* to tetracycline is reported to be caused by mutations in the *16S rRNA* gene. Levels of resistance are proportional to the number of changes at the AGA codon at nucleotide positions 965–967. Single- and double-point mutations are associated with low and intermediate minimum inhibitory concentration (MIC) values, respectively. Simultaneous triple point mutations at positions 965–967 are recognized as the major genomic alterations responsible for tetracycline resistance [7].

16.2.2 Tetracycline As Third-Line Eradication Therapy

Investigators at Korea University studied the efficacy of retrial with 14-day quadruple therapy comprising omeprazole or esomeprazole (20 mg b.i.d.), tripotassium dicitrate bismuthate (600 mg b.i.d.), metronidazole (500 mg b.i.d.), and tetracycline (1 g o.d.) as a third-line eradication therapy. Patients with failure of a first trial of

PPI-clarithromycin-amoxicillin and a second trial of PPI-bismuth-metronidazole-tetracycline for 14 days were enrolled in the study. Intention-to-treat and per-protocol analyses of eradication rates were 66.7 % (30/45) and 75.0 % (30/40), respectively [8]. These results suggest that retreatment with the same combination is a feasible choice. However, re-administration of previously used antibiotics is generally not recommended.

A prospective multicenter study in Spain assessed the efficacy of a 7–14-day quadruple therapy comprising PPI (standard dose b.i.d.), bismuth subcitrate (120 mg q.i.d. or 240 mg b.i.d.), tetracycline (250 mg q.i.d., 500 mg t.i.d., or 500 mg q.i.d.), and metronidazole (250 mg t.i.d., 250 mg q.i.d., 500 mg t.i.d., or 500 mg q.i.d.). Two hundred patients with failure of a first trial of PPI-clarithromycin-amoxicillin and a second trial of PPI-amoxicillin-levofloxacin were enrolled. Intention-to-treat and per-protocol analyses of eradication rates were 65 % (131/200) and 67 % (128/192), respectively. Cure rates were not statistically significant with respect to therapy duration (69 % for 7 days, 55 % for 10 days, and 76 % for 14 days) or dose of bismuth, tetracycline, and metronidazole. Adverse effects were reported in 22 % of patients, the most common being nausea (12 %), abdominal pain (11 %), metallic taste (8.5 %), and diarrhea (8 %), and none of the effects were severe [9].

16.3 High-Dose PPI/Amoxicillin Therapy

16.3.1 Amoxicillin Resistance

The prevalence of resistance to amoxicillin has fortunately remained low (<2 %) in most countries, except in Bangladesh (6.6 %) [10]. Although *H. pylori* has been considered to seldom become resistant to amoxicillin, Nishizawa et al. reported that the MIC_{90} of amoxicillin showed a twofold increase with the failure of each eradication treatment [11]. When the interpretive standard (susceptibility, 0.03 μg/ml) established by the Japanese Society of Chemotherapy was used, the resistance rates to amoxicillin in the groups with no history of eradication treatment, one treatment failure, two treatment failures, and three treatment failures were 13.6 %, 26.5 %, 49.5 %, and 72.7 %, respectively [11].

Amoxicillin acts by interfering with peptidoglycan synthesis, especially by blocking transporters, namely, penicillin-binding proteins (PBP) [11]. Multiple mutations in the *pbp1* gene are the major reason for amoxicillin resistance. Although production of β-lactamase is rare and almost inactive in *H. pylori*, Tseng et al. reported that high-level amoxicillin resistance is associated with β-lactamase production in *H. pylori* [12].

16.3.2 High-Dose PPI/Amoxicillin Therapy As Third-Line Eradication Therapy

A multicenter randomized controlled study in Germany investigated the efficacy of high-dose dual therapy and quadruple therapy as third-line eradication therapies (Table 16.1). Eighty-four patients with at least one treatment failure who were infected with an *H. pylori* strain resistant to both metronidazole and clarithromycin were randomized to receive either omeprazole at 40 mg q.i.d; amoxicillin at 750 mg q.i.d for 14 days; or omeprazole at 20 mg b.i.d, bismuth citrate at 107 mg q.i.d,

Table 16.1 Randomized controlled trial as third-line therapy

Author, year, country	Main inclusion criteria	Third-line therapy	Duration (days)	Eradication rate (%)	
				ITT	PP
Miehlke [13] 2003 Germany	At least one failure Infection with *H. pylori* resistant to CLR and MNZ	OPZ (40mgx4) AMX (750mgx4)	14	75.6 (31/41)	83.8 (31/38)
		OPZ (20mgx2), Bismuth (107mgx4) MNZ (500mgx4), TET (500mgx4)	14	81.4 (35/43)	92.1 (35/38)
Miehlke [20] 2006 Germany	At least one failure Infection with *H. pylori* resistant to CLR and MNZ	OPZ (40mgx3) AMX (1000mgx3)	14	70 (50/72)	74.6 (50/67)
		OPZ (20mgx2), AMX (1000mgx2) RFB (150mgx2)	7	74 (54/73)	78.3 (54/69)
Gisbert [31] 2006 Spain	Failures of OPZ-AMX-CLR and OPZ-bismuth-TC-MNZ	OPZ (20mgx2), AMX (1000mgx2) RFB (150mgx2)	10	45 (9/20)	45 (9/20)
		OPZ (20mgx2), AMX (1000mgx2) LVX (500mgx2)	10	85## (17/20)	81.3# (13/16)
Murakami [23] 2013 Japan	Failures of PPI-AMX-CLR and PPI-AMX-MNZ	LPZ (30mgx4), AMX (500mgx4)	14	54.3 (38/70)	56.7 (38/67)
		LPZ (30mgx2), AMX (750mgx2) LVX (300mgx2)	7	43.1 (28/65)	43.7 (28/64)
		LPZ (30mgx2), AMX (750mgx2) STX (100mgx2)	7	70***,+ (49/70)	72.1*** (49/68)

ITT intention-to-treat, *PP* per-protocol, *CLR* clarithromycin, *MNZ* metronidazole, *AMX* amoxicillin, *TET* tetracycline, *RFB* rifabutin, *LVX* levofloxacin, *STX* sitafloxacin, *OPZ* omeprazole, *LPZ* lansoprazole, *PPI* proton pump inhibitor

#OPZ-AMX-RFB vs. OPZ-AMX-LVX; $p < 0.05$

##OPZ-AMX-RFB vs. OPZ-AMX-LVX; $p < 0.01$

+LPZ-AMX vs. LPZ-AMX-STX; $p < 0.05$

***LPZ-AMX-LVX vs. LPZ-AMX-STX; $p < 0.001$

metronidazole at 500 mg q.i.d, and tetracycline at 500 mg q.i.d. for 14 days. Intention-to-treat and per-protocol eradication rates were 75.6 % (31/41) and 83.8 % (31/38), respectively, with high-dose PPI-amoxicillin, and 81.4 % (35/43) and 92.1 % (35/38), respectively, with PPI-bismuth-citrate-metronidazole-tetracycline (per-protocol analysis $p=0.71$, intention-to-treat analysis $p=0.60$) [13]. High-dose dual therapy and quadruple therapy were comparable as third-line eradication therapies.

A prospective study in Japan assessed the efficacy of high-dose PPI-amoxicillin comprising rabeprazole (10 mg q.i.d.) and amoxicillin (500 mg q.d.s.) for 14 days as a third-line eradication therapy. Forty-six patients with failure of a first trial of PPI-clarithromycin-amoxicillin and a second trial of PPI-metronidazole-amoxicillin were enrolled. Intention-to-treat and per-protocol eradication rates were 63.0 % (29/46) and 65.9 % (29/44) [14].

In Korea, Park et al. also investigated the efficacy of esomeprazole (40 mg b.i.d) and amoxicillin (1 g b.i.d) for 14 days as a third-line eradication therapy. Intention-to-treat and per-protocol analyses of eradication rates were both 66.7 % (14/21). Minor side effects were reported in three of the 21 patients (14.3 %). These side effects consisted mainly of nausea and epigastric discomfort [15].

16.4 Rifabutin

16.4.1 Rifabutin Resistance

A recent systematic review demonstrated that the mean *H. pylori* rifabutin resistance rate (calculated from 11 studies including 2982 patients) was 1.3 %. When only studies including pretreatment patients were considered, the primary resistance rate of rifabutin was 0.6 % [16]. In a German study, resistance to rifabutin increased to 6.2 % in patients who received more than two previous eradication therapies [17]. Higher rates of rifabutin resistance were reported in patients who had undergone previous rifampicin treatment for pulmonary tuberculosis [18]. It has been suggested that the use of rifabutin should be reserved for the treatment of multiresistant *Mycobacterium tuberculosis* strains. Rifabutin is preferably used only as a rescue after amoxicillin, clarithromycin, metronidazole, tetracycline, and fluoroquinolone have failed to eradicate *H. pylori*.

Rifabutin is an antituberculosis agent, which is structurally similar to rifampicin and is derived from rifamycin S. Rifabutin inhibits the expression of the β-subunit of DNA-dependent RNA polymerase, which is encoded by the *rpoB* gene. Rifabutin-resistant isolates of *H. pylori* showed mutations in codons 524–545 or codon 585 [19].

16.4.2 Rifabutin As Third-Line Eradication Therapy

A multicenter randomized controlled study in Germany investigated the efficacy of rifabutin-based triple therapy and high-dose dual therapy for rescue treatment (Table 16.1). One hundred and forty-five patients infected with *H. pylori* resistant to both metronidazole and clarithromycin were randomized to either esomeprazole 20 mg b.i.d, amoxicillin 1 g b.i.d, and rifabutin 150 mg b.i.d for 7 days or to omeprazole 40 mg t.i.d and amoxicillin 1000 mg t.i.d for 14 days. Intention-to-treat and per-protocol eradication rates were 74 % (54/73) and 78 % (54/69), respectively, with PPI-amoxicillin-rifabutin, and were 70 % (50/72) and 75 % (50/67), respectively, with high-dose PPI-amoxicillin. PPI-amoxicillin-rifabutin and high-dose PPI-amoxicillin showed comparable effectiveness [20].

A randomized controlled study compared two rifabutin doses as third-line eradication therapies. When the rifabutin dose was reduced from 300 mg/day to 150 mg/day, it resulted in a significant drop in eradication rate, from 86.6 % to 66.6 % [21]. A prospective study in Keio University assessed the efficacy of 10-day and 14-day rifabutin-based triple therapy as a third- or fourth-line rescue therapy. Intention-to-treat and per-protocol analyses of eradication rates were 83.3% and 81.8% for the 10-day group and 94.1% and 91.7% for the 14-day group, respectively. Therapy was stopped due to adverse events in 8.3% and 29.3% of patients in the 10-day and 14-day groups, respectively. The 10-day treatment may be enough to obtain a successful eradication rate, considering the tolerability of therapy [22].

In 2014, two rifabutin-based eradication regimens were compared as a third-line eradication therapy. Patients were randomly assigned to either the triple therapy with lansoprazole 30 mg b.i.d, amoxicillin 1000 mg t.i.d, and rifabutin 150 mg b.i.d for 1 week or lansoprazole 60 mg b.i.d, amoxicillin 1000 mg t.i.d, and rifabutin 150 mg b.i.d for 1 week. Intention-to-treat and per-protocol analyses of eradication rates were 78.1 % (25/32) and 80.6 % (25/31), respectively, with lansoprazole 60 mg-amoxicillin-rifabutin, and were 96.3 % (26/27) and 100 % (26/26), respectively, for lansoprazole 120 mg-amoxicillin-rifabutin (per-protocol analysis $p = 0.047$, intention-to-treat analysis $p = 0.051$) [23]. The key to successful rescue therapy with a PPI-amoxicillin-rifabutin regimen may be to increase the doses of PPI.

According to a recent meta-analysis, the mean eradication rate of rifabutin-containing regimens as third-line therapy (in 342 patients with two eradication failures) was 66 % (95 % confidence interval [CI], 55–77 %) with significant heterogeneity ($I^2 = 87$ %). Exclusion of one study prescribing moxifloxacin instead of amoxicillin decreased the heterogeneity ($I^2 = 63$ %), and the mean eradication rate was slightly lower (63 %). The mean eradication rate, as fourth-line or fifth-line therapy (in 95 patients with three or four eradication failures), was 70 % (95 % CI, 60–79 %) without heterogeneity ($I^2 = 2$ %). Contrary to expectations, rifabutin efficacy did not decrease with increasing number of failed previous therapies [16]. The mean rate of adverse effects of rifabutin-containing regimens was 22 % (95 % CI, 19–25 %). Although myelotoxicity is rare, it is more likely when a high-dose (600 mg/day) and prolonged duration therapy is used.

16.5 Fluoroquinolone

16.5.1 Fluoroquinolone Resistance

Primary resistance of *H. pylori* to fluoroquinolones has been reported in the range of 2–22 % in different countries or regions. The prevalence of fluoroquinolone resistance is higher in Italy, Japan, and Korea (15–22 %), and the prevalence is very low in Egypt and China (about 2 %) [3]. The resistance rate is relatively high in countries with a high consumption rate of these drugs. Primary, secondary, and tertiary resistances to levofloxacin were 11.7 %, 17.6 %, and 36.4 % in Germany [17]. The levofloxacin resistance rate prior to third-line eradication therapy was 57.0 % (61/107) in Japan [24].

Fluoroquinolones exert their antimicrobial activity by inhibiting the enzyme DNA gyrase. The bacterial gyrase is essential for maintaining the DNA helicoidal structure in addition to being involved in DNA replication, recombination, and transcription [25]. The bacterial enzyme gyrase is a tetramer consisting of two A and two B subunits encoded by the *gyrA* and *gyrB* genes, respectively. Fluoroquinolones exert their antimicrobial activity at the level of the A subunit of the DNA gyrase, which is responsible for DNA cleavage and rejoining [26].

Point mutations in the quinolone resistance-determining region (QRDR) of *gyrA* prevent binding between the antibiotic and the enzyme, conferring antibiotic bacterial resistance. *H. pylori* does not possess a gene encoding topoisomerase IV, which is an important fluoroquinolone target in other bacteria. There is a significant association between fluoroquinolone resistance in *H. pylori* and mutations of the QRDR of the *gyrA* gene [27]. The MICs of sitafloxacin in *gyrA* mutation-positive strains were found to differ depending on the position of the *gyrA* mutation [28]. The MICs were higher in N87-mutated strains (0.21 ± 0.16 μg/ml) than in D91-mutated strains (0.12 ± 0.11 μg/ml, $p = 0.03$). The mutation at position 463 in *gyrB* might be a novel mechanism of fluoroquinolone resistance in *H. pylori* [29].

The traditional culture test for bacterial susceptibility to antibiotics is expensive and requires 10–14 days and is not routinely performed in clinical practice, and MIC-based individualized treatment for *H. pylori* infection is not prevalent among general practitioners. Molecular techniques for antibiotic susceptibility testing can determine bacterial susceptibility to some antibiotics within 6 h. HelicoDR® (Hain Life Science, Germany) is a DNA strip genotyping test combining PCR and hybridization that allows for the molecular detection of mutations in the *gyrA* gene and *23S rRNA* [30]. The detectable mutations are at residues 87 (Asn to Lys) and 91 (Asp to Gly, Asn, or Tyr) in *gyrA* and A2142G, A2143G, and A2142C in *23S rRNA*. The sensitivity and specificity of the HelicoDR® test for *gyrA* mutations were 98.2 % and 80.0 %, compared to direct sequencing, and those for the 23S *rRNA* mutation were 94.9 % and 87.1 %, respectively. Compared to the MIC test, the sensitivity and specificity of the HelicoDR® test were 87–74.4 % and 98.5–70 % for levofloxacin and 94–55 % and 99–80 % for clarithromycin, respectively [30, 31]; HelicoDR® is now commercially available.

16.5.2 Levofloxacin As a Third-Line Eradication Therapy

In Spain, levofloxacin-based and rifabutin-based 10-day regimens were compared as third-line eradication therapies (Table 16.1). Forty patients with failure of a first trial of omeprazole-clarithromycin-amoxicillin and a second trial of omeprazole-bismuth-tetracycline-metronidazole (or ranitidine bismuth citrate with these antibiotics) were enrolled. The first 20 patients were treated with rifabutin (150 mg b.i.d), amoxicillin (1 g b.i.d), and omeprazole (20 mg b.i.d), while the next successive 20 patients were treated with levofloxacin (500 mg b.i.d), amoxicillin (1 g b.i.d), and omeprazole (20 mg b.i.d) for 10 days. Intention-to-treat and per-protocol analyses of eradication rates were 45 % (9/20) and 45 % (9/20) with PPI-amoxicillin-rifabutin and were 85 % (17/20) and 81 % (13/16) with PPI-amoxicillin-levofloxacin ($p < 0.05$). The incidence of side effects in the rifabutin-treated group was 60 %, whereas the levofloxacin-treated group showed side effects in 50 % of the cases. Five patients (25 %) treated with rifabutin presented with leucopenia, and six (30 %) treated with levofloxacin presented with myalgias. A 10-day triple levofloxacin-based regimen was superior to the same regimen with rifabutin as third-line treatment [32].

A prospective study in Taiwan assessed the efficacy of a 10-day quadruple therapy comprising rabeprazole (20 mg b.i.d.), bismuth subcitrate (300 mg q.d.s.), amoxicillin (500 mg q.d.s.), and levofloxacin (500 mg o.d.) as a third-line eradication therapy. *H. pylori* was successfully eradicated in 84 % (31/37) of the patients, as determined by both intention-to-treat analysis and per-protocol analysis. Mild-to-moderate adverse events were reported in 7 of the 37 patients (19 %) [33]. This study suggested that a 10-day regimen of levofloxacin-based quadruple therapy is well tolerated and yields as high an eradication rate as a third-line empirical eradication therapy.

16.5.3 Novel Fluoroquinolones As Third-Line Eradication Therapy

Recently, novel fluoroquinolones were developed, and a superior in vitro activity of gatifloxacin over that of levofloxacin against *H. pylori* was reported. Furthermore, sitafloxacin showed activities 8- to 16-fold or greater than those of levofloxacin [34].

Nishizawa et al. investigated the efficacy of a 7-day regimen of gatifloxacin, amoxicillin, and rabeprazole as a third-line eradication therapy [35]. Successful eradication was achieved in 63.6 % (7/11) of cases. The eradication rates were 100 % in the patients infected with *gyrA* mutation-negative *H. pylori*, but were only 33.3 % in those infected with *gyrA* mutation-positive *H. pylori* ($p < 0.05$).

Another prospective study in Keio University assessed the efficacy of a 7-day regimen of sitafloxacin (100 mg b.i.d.), amoxicillin (500 mg q.d.s.), and rabeprazole (10 mg q.d.s.) as a third-line eradication therapy [28]. Patients with failure of a first trial of PPI-clarithromycin-amoxicillin and a second trial of PPI-metronidazole-amoxicillin were enrolled. Intention-to-treat and per-protocol analyses of eradication rates were 78.2 % (61/78) and 83.6 % (61/73), respectively. Among the patients infected with *gyrA* mutation-negative *H. pylori*, intention-to-treat and per-protocol analyses of eradication rates were 93.5 % (29/31) and 96.7 % (29/30), respectively. Even among the patients infected with *gyrA* mutation-positive *H. pylori*, intention-to-treat and per-protocol analyses of eradication rates were 68.1 % (32/47) and 74.4 % (32/43), respectively. Mild and transient adverse effects such as diarrhea (33.3 %), abdominal pain (6.9 %), and taste disturbance (6.9 %) were reported.

In 2014, two sitafloxacin-based eradication regimens were compared as third-line eradication therapies [36]. One hundred and eighty patients with failure of a first trial of PPI-clarithromycin-amoxicillin and a second trial of PPI-metronidazole-amoxicillin were enrolled. Patients were assigned to either the triple therapy with rabeprazole 10 mg b.i.d./q.i.d., amoxicillin 500 mg q.i.d, and sitafloxacin 100 mg b.i.d. for 1 or 2 weeks or the triple therapy with rabeprazole 10 mg b.i.d./q.i.d., metronidazole 250 mg b.i.d., and sitafloxacin 100 mg b.i.d. for 1 or 2 weeks. Intention-to-treat and per-protocol analyses of eradication rates were 84.1 % (37/44) and 86.4 % (37/43) with PPI-amoxicillin-sitafloxacin for 1 week, were 88.9 % (40/45) and 90.9 % (40/44) with PPI-amoxicillin-sitafloxacin for 2 weeks, were 90.9 % (40/44) and 90.9 % (40/44) PPI-metronidazole-sitafloxacin for 1 week, and were 87.2 % (41/47) and 91.1 % (41/45) with PPI-metronidazole-sitafloxacin for 2 weeks. Although there were no statistically significant differences in eradication rates among the four regimens, per-protocol analysis showed that PPI-amoxicillin-sitafloxacin for 2 weeks and PPI-metronidazole-sitafloxacin for 1 or 2 weeks could attain rescue eradication rates higher than 90 %. Mori et al. also reported no significant difference between esomeprazole-amoxicillin-sitafloxacin and esomeprazole-metronidazole-sitafloxacin for 10 days. *GyrA* mutation status was an important factor in predicting successful eradication with sitafloxacin-containing rescue therapies [37].

Murakami et al. conducted a multicenter randomized control study to establish the standard third-line regimen in Japan (Table 16.1) [23]. Two hundred and four patients with failure of a first trial of PPI-clarithromycin-amoxicillin and a second trial of PPI-metronidazole-amoxicillin were enrolled. Patients were randomly assigned to one of the following third-line eradication therapy groups: [1] lansoprazole 30 mg q.i.d and amoxicillin 500 mg q.i.d for 2 weeks; [2] lansoprazole 30 mg b.i.d, amoxicillin 750 mg b.i.d, and levofloxacin 300 mg b.i.d for 1 week; and [3] lansoprazole 30 mg b.i.d, amoxicillin 750 mg b.i.d, and sitafloxacin 100 mg b.i.d for 1 week. Intention-to-treat analysis of third-line eradication therapy eradication rates showed a significantly higher rate in the PPI-amoxicillin-sitafloxacin condition (70.0 %) compared with PPI-amoxicillin treatment for 2 weeks (54.3 %; $p < 0.05$) and PPI-amoxicillin-levofloxacin (43.1 %; $p < 0.001$). Patients for whom these therapies failed underwent a crossover fourth-line eradication regimen.

The fourth-line eradication rate was significantly higher ($p < 0.01$) among PPI-amoxicillin-sitafloxacin-treated patients (60 %; 25 patients from the 2-week PPI-amoxicillin group for whom third-line therapy failed), than for 2-week PPI-amoxicillin-treated patients (27.5 %; 40 patients from the PPI-amoxicillin-sitafloxacin and PPI-amoxicillin-levofloxacin groups for whom third-line therapy failed). These data suggest that triple therapy with PPI, amoxicillin, and sitafloxacin for 1 week would be an effective standard third-line eradication regimen in Japan. However, sitafloxacin is not widely available in many countries, and its use has mainly been restricted to Japan.

16.6 Furazolidone

16.6.1 The Prevalence and Mechanism of Fluoroquinolone Resistance

Resistance of furazolidone was detected in 1.8–8 % of *H. pylori* isolates [38]. Furazolidone is derived from nitrofurans and acts by inhibiting bacterial enzymes, mainly those involved in the tricarboxylic acid cycle. The resistance to furazolidone is mediated by point mutations of the bacteria *porD* and *oorD* genes, resulting in amino acid changes [39].

16.6.2 Furazolidone As Third-Line Eradication Therapy

According to a recent meta-analysis, the mean eradication rate of furazolidone-containing regimens as third-line therapy (in 148 patients with two eradication failures) was 65.5 % (95 % CI, 56.3–75.5 %) [36]. Mean eradication rates as first-line and second-line therapy were 80.3 % (95 % CI, 69.6–88.7 %) and 76.1 % (95 % CI, 66.4–85.3 %). Furazolidone efficacy decreased with increasing number of failed previous therapies. Adverse effects occurred in 34.7 % (95 % CI, 27.6–43.1 %) of patients, such as taste disturbance (17.1 %), nausea (16.9), and diarrhea (7.6 %).

16.7 Resistance-Guided Therapy as Third-Line Eradication

Although susceptibility testing is recommended before third-line treatment, the efficacies of susceptibility-guided therapy were reported only in some studies, with reported efficacies in the range of 36–90 % [40–43] (Table 16.2). Gomollon et al. reported that the eradication rates of susceptibility-guided third-line quadruple

Table 16.2 Resistance-guided third-line therapy

Author, year, country	Determination of resistance	Third-line therapy	Eradication rate (%)			
			ITT		PP	
Gomollon [40] 2000 Spain	E-test of CLR and MNZ	OPZ (20mgx2), Bismuth (120mgx4) TET (500mgx4), CLR (500mgx2) for 14 days	48 (14/29)		35.7 (5/14)	50 (14/28)
		OPZ (20mgx2), Bismuth (120mgx4) TET (500mgx4), MNZ for 14 days			0 (0/1)	
		OPZ (20mgx2), Bismuth (120mgx4) TET (500mgx4), AMX (1000mgx2) for 14 days			66.7 (8/12)	
		OPZ (20mgx2), Bismuth (120mgx4) TET (500mgx4), CPX for 14 days			100 (1/1)	
Vicente [41] 2002 Spain	E-test of CLR, MNZ, AMX, and TET	OPZ (20mgx2), Bismuth (120mgx4) TET (500mgx4), CLR (500mgx2) for 14 days	47.4 (9/19)	60 (24/40)	63.2 (24/38)	
		OPZ (20mgx2), Bismuth (120mgx4) TET (500mgx4), AMX (1000mgx2) for 14 days	70 (14/20)			
		OPZ (20mgx2), Bismuth (120mgx4) TET (500mgx4), CPX (500mgx2) for 14 days	100 (1/1)			
Cammarota [42] 2004 Italy	E-test of CLR, MNZ, AMX, TET, and LVX	OPZ (20mgx2), Bismuth (120mgx4) AMX (1000mgx2), DOX (100mgx2) for 7 days	91 (81/89)	90.4 (85/94)	92 (81/88)	91.4 (85/93)
		OPZ (20mgx2), AMX (1000mgx2) LVX (500mgx2) for 7 days	75 (3/4)		75 (3/4)	
		OPZ (20mgx2), AMX (1000mgx2) CLR (500mgx2) for 7 days	100 (1/1)		100 (1/1)	
Liou [43] 2013 Taiwan	Sequencing of *23S rRNA* and *gyrA*	EPZ (20mgx2), AMX (500mgx2) for 7 days, followed by EPZ (20mgx2), MNZ (250mgx2), CLR (250mgx2) for 7 days	78.9 (15/19)	80.7 (109/135)	83.3 (15/18)	82.6 (109/132)
		EPZ (20mgx2), AMX (500mgx2) for 7 days, followed by EPZ (20mgx2), MNZ (250mgx2), LVX (125mgx2) for 7 days	92.2 (47/51)		94 (47/50)	
		EPZ (20mgx2), AMX (500mgx2) for 7 days, followed by EPZ (20mgx2), MNZ (250mgx2), TET (250mgx2) for 7 days	72.3 (47/65)		73.4 (47/64)	

ITT intention-to-treat, *PP* per-protocol, *CLR* clarithromycin, *MNZ* metronidazole, *AMX* amoxicillin, *TET* tetracycline, *LVX* levofloxacin, *OPZ* omeprazole, *EPZ* esomeprazole

therapies containing omeprazole, tetracycline, and bismuth with either clarithromycin or amoxicillin for 14 days were only 50 % (13/26), showing that in vitro susceptibility did not predict eradication success [40]. Antibiotic susceptibility and *H. pylori* eradication may be discrepant due to the possibility of coinfection with different strains. On the other hand, *H. pylori* eradication may be achieved in the presence of metronidazole- or clarithromycin-resistant strains, even with a drug combination including these antibiotics. Therefore, in vitro resistance to metronidazole or clarithromycin could be overcome in vivo in a significant proportion of patients by prescribing the same antibiotics.

A prospective multicenter study in 2013 investigated the efficacy of genotypic resistance-guided third-line sequential therapy [43]. Genotypic resistance was assessed by PCR with direct sequencing (*23S rRNA* and *gyrA*), while phenotypic resistance was determined by the agar dilution test. One hundred and thirty-five patients were treated with sequential therapy including esomeprazole 20 mg b.i.d. and amoxicillin 500 mg b.i.d. for the first 7 days, followed by esomeprazole 20 mg b.i.d. and metronidazole 250 mg b.i.d. plus clarithromycin 250 mg b.i.d., levofloxacin 125 mg b.i.d., or tetracycline 250 mg b.i.d. for another 7 days, depending on the characterized genotypic resistance. Intention-to-treat and per-protocol analyses of eradication rates were 80.7 % (109/135) and 82.6 % (109/132), respectively. A simple molecular method for guiding sequential therapy could achieve a high eradication rate as a third-line therapy [43]. Genotypic resistance testing is more convenient and rapid than standard culture susceptibility testing, offering the possibility to determine resistance even from stool samples [44].

16.8 Conclusion

The resistance of *H. pylori* to amoxicillin, tetracycline, and rifabutin has remained low. Third-line treatment options include fluoroquinolones (especially sitafloxacin), rifabutin, tetracycline, furazolidone, and high-dose PPI/amoxicillin therapy. It is desirable to avoid previously used antibiotics, for empiric third-line therapy. In randomized control trials, fluoroquinolone-based regimens were more effective than a rifabutin-based triple regimen, which should be reserved for the treatment of multiresistant *Mycobacterium tuberculosis* strains.

Although culture and drug susceptibility tests are time-consuming and expensive, they appear to be useful in selection of third-line therapy. With the addition of newer rapid molecular tests to detect *H. pylori* and to determine the presence of point mutations, resistance-guided therapy may play a bigger role in the future.

References

1. Fock KM, Katelaris P, Sugano K, Ang TL, Hunt R, Talley NJ, et al. Second Asia-Pacific consensus guidelines for *Helicobacter pylori* infection. J Gastroenterol Hepatol. 2009;24 (10):1587–600.
2. Malfertheiner P, Megraud F, O'Morain CA, Atherton J, Axon AT, Bazzoli F, et al. Management of *Helicobacter pylori* infection--the Maastricht IV/ Florence consensus report. Gut. 2012;61(5):646–64.
3. Nishizawa T, Suzuki H, Hibi T. Quinolone-based third-line therapy for *Helicobacter pylori* eradication. J Clin Biochem Nutr. 2009;44(2):119–24.
4. Sharara AI, Chedid M, Araj GF, Barada KA, Mourad FH. Prevalence of *Helicobacter pylori* resistance to metronidazole, clarithromycin, amoxycillin and tetracycline in Lebanon. Int J Antimicrob Agents. 2002;19(2):155–8.
5. Almeida N, Romaozinho JM, Donato MM, Luxo C, Cardoso O, Cipriano MA, et al. *Helicobacter pylori* antimicrobial resistance rates in the central region of Portugal. Clin Microbiol Infect. 2014;20(11):1127–33.
6. Heep M, Kist M, Strobel S, Beck D, Lehn N. Secondary resistance among 554 isolates of *Helicobacter pylori* after failure of therapy. Eur J Clin Microbiol Infect Dis. 2000;19 (7):538–41.
7. Gerrits MM, Berning M, Van Vliet AH, Kuipers EJ, Kusters JG. Effects of 16S rRNA gene mutations on tetracycline resistance in *Helicobacter pylori*. Antimicrob Agents Chemother. 2003;47(9):2984–6.
8. Lee SK, Lee SW, Park JY, Kwon BS, Kim SY, Hyun JJ, et al. Effectiveness and safety of repeated quadruple therapy in *Helicobacter pylori* infection after failure of second-line quadruple therapy: Repeated quadruple therapy as a third-line therapy. Helicobacter. 2011;16(5):410–4.
9. Gisbert JP, Perez-Aisa A, Rodrigo L, Molina-Infante J, Modolell I, Bermejo F, et al. Third-line rescue therapy with bismuth-containing quadruple regimen after failure of two treatments (with clarithromycin and levofloxacin) for *H. pylori* infection. Dig Dis Sci. 2014;59(2):383–9.
10. Nahar S, Mukhopadhyay AK, Khan R, Ahmad MM, Datta S, Chattopadhyay S, et al. Antimicrobial susceptibility of *Helicobacter pylori* strains isolated in Bangladesh. J Clin Microbiol. 2004;42(10):4856–8.
11. Nishizawa T, Suzuki H, Tsugawa H, Muraoka H, Matsuzaki J, Hirata K, et al. Enhancement of amoxicillin resistance after unsuccessful *Helicobacter pylori* eradication. Antimicrob Agents Chemother. 2011;55(6):3012–4.
12. Tseng YS, Wu DC, Chang CY, Kuo CH, Yang YC, Jan CM, et al. Amoxicillin resistance with beta-lactamase production in *Helicobacter pylori*. Eur J Clin Invest. 2009;39(9):807–12.
13. Miehlke S, Kirsch C, Schneider-Brachert W, Haferland C, Neumeyer M, Bastlein E, et al. A prospective, randomized study of quadruple therapy and high-dose dual therapy for treatment of *Helicobacter pylori* resistant to both metronidazole and clarithromycin. Helicobacter. 2003;8(4):310–9.
14. Nishizawa T, Suzuki H, Maekawa T, Harada N, Toyokawa T, Kuwai T, et al. Dual therapy for third-line *Helicobacter pylori* eradication and urea breath test prediction. World J Gastroenterol. 2012;18(21):2735–8.
15. Park HK, Lee DH, Suh S, Seo PJ, Kim N, Jeong SH, et al. Dual therapy trial using esomeprazole and amoxicillin as third-line rescue therapy for *Helicobacter pylori* infection. Clin Endoscopy. 2011;44(1):33–7.
16. Gisbert JP, Calvet X. Review article: Rifabutin in the treatment of refractory *Helicobacter pylori* infection. Aliment Pharmacol Ther. 2012;35(2):209–21.
17. Selgrad M, Meissle J, Bornschein J, Kandulski A, Langner C, Varbanova M, et al. Antibiotic susceptibility of *Helicobacter pylori* in central Germany and its relationship with the number of eradication therapies. Eur J Gastroenterol Hepatol. 2013;25(11):1257–60.

18. Suzuki S, Suzuki H, Nishizawa T, Kaneko F, Ootani S, Muraoka H, et al. Past rifampicin dosing determines rifabutin resistance of *Helicobacter pylori*. Digestion. 2009;79(1):1–4.
19. Nishizawa T, Suzuki H, Matsuzaki J, Muraoka H, Tsugawa H, Hirata K, et al. *Helicobacter pylori* resistance to rifabutin in the last 7 years. Antimicrob Agents Chemother. 2011;55 (11):5374–5.
20. Miehlke S, Hansky K, Schneider-Brachert W, Kirsch C, Morgner A, Madisch A, et al. Randomized trial of rifabutin-based triple therapy and high-dose dual therapy for rescue treatment of *Helicobacter pylori* resistant to both metronidazole and clarithromycin. Aliment Pharmacol Ther. 2006;24(2):395–403.
21. Perri F, Festa V, Clemente R, Villani MR, Quitadamo M, Caruso N, et al. Randomized study of two "rescue" therapies for *Helicobacter pylori*-infected patients after failure of standard triple therapies. Am J Gastroenterol. 2001;96(1):58–62.
22. Mori H, Suzuki H, Matsuzaki J, Tsugawa H, Fukuhara S, Miyoshi S, et al. Rifabutin-based 10-day and 14-day triple therapy as a third-line and fourth-line regimen for *Helicobacter pylori* eradication: a pilot study. United Eur Gastroenterol J. 2015a. doi:10.1177/2050640615618043.
23. Lim HC, Lee YJ, An B, Lee SW, Lee YC, Moon BS. Rifabutin-based high-dose proton-pump inhibitor and amoxicillin triple regimen as the rescue treatment for *Helicobacter pylori*. Helicobacter. 2014;19(6):455–61.
24. Murakami K, Furuta T, Ando T, Nakajima T, Inui Y, Oshima T, et al. Multi-center randomized controlled study to establish the standard third-line regimen for *Helicobacter pylori* eradication in Japan. J Gastroenterol. 2013;19(6):455–61.
25. Moore RA, Beckthold B, Wong S, Kureishi A, Bryan LE. Nucleotide sequence of the gyrA gene and characterization of ciprofloxacin-resistant mutants of *Helicobacter pylori*. Antimicrob Agents Chemother. 1995;39(1):107–11.
26. Matsuzaki J, Suzuki H, Tsugawa H, Nishizawa T, Hibi T. Homology model of the DNA gyrase enzyme of *Helicobacter pylori*, a target of quinolone-based eradication therapy. J Gastroenterol Hepatol. 2010;25 Suppl 1:S7–10.
27. Nishizawa T, Suzuki H, Kurabayashi K, Masaoka T, Muraoka H, Mori M, et al. Gatifloxacin resistance and mutations in gyra after unsuccessful *Helicobacter pylori* eradication in Japan. Antimicrob Agents Chemother. 2006;50(4):1538–40.
28. Matsuzaki J, Suzuki H, Nishizawa T, Hirata K, Tsugawa H, Saito Y, et al. Efficacy of sitafloxacin-based rescue therapy for *Helicobacter pylori* after failures of first- and second-line therapies. Antimicrob Agents Chemother. 2012;56(3):1643–5.
29. Rimbara E, Noguchi N, Kawai T, Sasatsu M. Fluoroquinolone resistance in *Helicobacter pylori*: Role of mutations at position 87 and 91 of GyrA on the level of resistance and identification of a resistance conferring mutation in GyrB. Helicobacter. 2012;17(1):36–42.
30. Cambau E, Allerheiligen V, Coulon C, Corbel C, Lascols C, Deforges L, et al. Evaluation of a new test, genotype HelicoDR, for molecular detection of antibiotic resistance in *Helicobacter pylori*. J Clin Microbiol. 2009;47(11):3600–7.
31. Lee JW, Kim N, Nam RH, Park JH, Choi YJ, Kim JM, et al. Genotype HelicoDR test in the determination of antimicrobial resistance of *Helicobacter pylori* in Korea. Scand J Gastroenterol. 2014;49(9):1058–67.
32. Gisbert JP, Gisbert JL, Marcos S, Moreno-Otero R, Pajares JM. Third-line rescue therapy with levofloxacin is more effective than rifabutin rescue regimen after two *Helicobacter pylori* treatment failures. Aliment Pharmacol Ther. 2006;24(10):1469–74.
33. Hsu PI, Wu DC, Chen A, Peng NJ, Tseng HH, Tsay FW, et al. Quadruple rescue therapy for *Helicobacter pylori* infection after two treatment failures. Eur J Clin Invest. 2008;38(6):404–9.
34. Suzuki H, Nishizawa T, Muraoka H, Hibi T. Sitafloxacin and garenoxacin may overcome the antibiotic resistance of *Helicobacter pylori* with gyrA mutation. Antimicrob Agents Chemother. 2009;53(4):1720–1.
35. Nishizawa T, Suzuki H, Nakagawa I, Iwasaki E, Masaoka T, Hibi T. Gatifloxacin-based triple therapy as a third-line regimen for *Helicobacter pylori* eradication. J Gastroenterol Hepatol. 2008;23 Suppl 2:S167–70.

36. Furuta T, Sugimoto M, Kodaira C, Nishino M, Yamade M, Uotani T, et al. Sitafloxacin-based third-line rescue regimens for *Helicobacter pylori* infection in Japan. J Gastroenterol Hepatol. 2014;29(3):487–93.
37. Mori H, Suzuki H, Matsuzaki J, Tsugawa H, Fukuhara S, Miyoshi S, et al. Helicobacter. 2015b. doi:10.1111/hel.12286.
38. Buzas GM, Jozan J. Nitrofuran-based regimens for the eradication of *Helicobacter pylori* infection. J Gastroenterol Hepatol. 2007;22(10):1571–81.
39. Su Z, Xu H, Zhang C, Shao S, Li L, Wang H, et al. Mutations in *Helicobacter pylori* porD and oorD genes may contribute to furazolidone resistance. Croat Med J. 2006;47(3):410–5.
40. Gomollon F, Sicilia B, Ducons JA, Sierra E, Revillo MJ, Ferrero M. Third line treatment for *Helicobacter pylori*: A prospective, culture-guided study in peptic ulcer patients. Aliment Pharmacol Ther. 2000;14(10):1335–8.
41. Vicente R, Sicilia B, Gallego S, Revillo MJ, Ducons J, Gomollon F. *Helicobacter pylori* eradication in patients with peptic ulcer after two treatment failures: A prospective culture-guided study. Gastroenterol Hepatol. 2002;25(7):438–42.
42. Cammarota G, Martino A, Pirozzi G, Cianci R, Branca G, Nista EC, et al. High efficacy of 1-week doxycycline- and amoxicillin-based quadruple regimen in a culture-guided, third-line treatment approach for *Helicobacter pylori* infection. Aliment Pharmacol Ther. 2004;19 (7):789–95.
43. Liou JM, Chen CC, Chang CY, Chen MJ, Fang YJ, Lee JY, et al. Efficacy of genotypic resistance-guided sequential therapy in the third-line treatment of refractory *Helicobacter pylori* infection: A multicentre clinical trial. J Antimicrob Chemother. 2013;68(2):450–6.
44. Rimbara E, Sasatsu M, Graham DY. PCR detection of *Helicobacter pylori* in clinical samples. Methods Mol Biol. 2013;943:279–87.

MIX
Papier aus verantwortungsvollen Quellen
Paper from responsible sources
FSC® C105338

If you have any concerns about our products,
you can contact us on
ProductSafety@springernature.com

In case Publisher is established outside the EU,
the EU authorized representative is:
Springer Nature Customer Service Center GmbH
Europaplatz 3, 69115 Heidelberg, Germany

Printed by Libri Plureos GmbH
in Hamburg, Germany